Walter Klöpffer and Burkhard O. Wagner

Atmospheric Degradation of Organic Substances

Data for Persistence and
Long-range Transport Potential

WILEY-VCH Verlag GmbH & Co. KGaA

The Authors

Prof. Dr. Walter Klöpffer
Am Dachsberg 56 E
60435 Frankfurt am Main
Germany

Dr. Burkhard O. Wagner
Am Hirschsprung 37 a
14195 Berlin
Germany

All books published by Wiley-VCH are carefully produced. Nevertheless, authors, editors, and publisher do not warrant the information contained in these books, including this book, to be free of errors. Readers are advised to keep in mind that statements, data, illustrations, procedural details or other items may inadvertently be inaccurate.

Library of Congress Card No.: applied for

British Library Cataloguing-in-Publication Data:
A catalogue record for this book is available from the British Library.

Bibliographic information published by the Deutsche Nationalbibliothek
The Deutsche Nationalbibliothek lists this publication in the Deutsche Nationalbibliografie; detailed bibliographic data are available in the Internet at http://dnb.d-nb.de.

© 2007 WILEY-VCH Verlag GmbH & Co. KGaA, Weinheim

All rights reserved (including those of translation into other languages). No part of this book may be reproduced in any form – by photoprinting, microfilm, or any other means – nor transmitted or translated into a machine language without written permission from the publishers. Registered names, trademarks, etc. used in this book, even when not specifically marked as such, are not to be considered unprotected by law.

Typesetting   Manuela Treindl, Laaber
Printing   Strauss GmbH, Mörlenbach
Binding   Litges & Dopf Buchbinderei GmbH, Heppenheim
Wiley Bicentennial Logo: Richard J. Pacifico

Printed in the Federal Republic of Germany
Printed on acid-free paper

ISBN: 978-3-527-31606-9

Walter Klöpffer
Burkhard O. Wagner
**Atmospheric Degradation
of Organic Substances**

1807–2007 Knowledge for Generations

Each generation has its unique needs and aspirations. When Charles Wiley first opened his small printing shop in lower Manhattan in 1807, it was a generation of boundless potential searching for an identity. And we were there, helping to define a new American literary tradition. Over half a century later, in the midst of the Second Industrial Revolution, it was a generation focused on building the future. Once again, we were there, supplying the critical scientific, technical, and engineering knowledge that helped frame the world. Throughout the 20th Century, and into the new millennium, nations began to reach out beyond their own borders and a new international community was born. Wiley was there, expanding its operations around the world to enable a global exchange of ideas, opinions, and know-how.

For 200 years, Wiley has been an integral part of each generation's journey, enabling the flow of information and understanding necessary to meet their needs and fulfill their aspirations. Today, bold new technologies are changing the way we live and learn. Wiley will be there, providing you the must-have knowledge you need to imagine new worlds, new possibilities, and new opportunities.

Generations come and go, but you can always count on Wiley to provide you the knowledge you need, when and where you need it!

William J. Pesce
President and Chief Executive Officer

Peter Booth Wiley
Chairman of the Board

Foreword

Who should read a book on photo-degradation? Only the few scientists experienced in this very specific area? No, this book is a storehouse for students, risk assessors in administration and industry and other interested people who want to learn why we can still breathe fairly clean air in most parts of the world. The book contains a concise and understandable scientific presentation of the various mechanisms that contribute to the photochemical degradation of organic molecules in the troposphere, their kinetics and the variables that influence the efficacy of the degradation processes. Experimental data on 1081 substances make this book valuable for readers who want to assess chemicals, to develop simulation models and to predict the environmental fate and distribution of pollutants.

Photochemical oxidation in the atmosphere by hydroxyl and nitrate radicals and by ozone is obviously the most effective mechanism to decompose xenobiotics (and also several natural substances) which partition to air to a certain extent. This is why in the 1970s and 1980s numerous scientists began to measure rate constants for photochemical degradation. Their experiments were aimed at obtaining a first estimation as to whether atmospheric pollutants would accumulate and subsequently pose a risk. The results were encouraging and relieving. The great majority of chemicals are attacked by the cleansing agents of the atmosphere, oxidized and finally eliminated. However, some molecules, such as the chlorofluorocarbons (CFCs) are recalcitrant and are only slightly or non-reactive.

By 1991 the German Federal Environment Agency (UBA) had compiled the photo-degradation data of 544 substances, in a study that was co-financed by the German Chemical Industry Association (VCI), demonstrating a significant scientific interest in such experimental studies at that time (UBA-Texte 51/91). In the intervening years, AOP (Atmospheric Oxidation Potential) and other quantitative structure activity relationship programmes (QSAR) have been developed, which allow plausible predictions of the photo reactivity of chemical substances. This led to a significant decline in experimental interest. Calculations were easier, cheaper and more rapid than experimental determinations, but sometimes the limitations of the models were neglected. Since then the number of substances that have been examined experimentally for their photo-degradation kinetics has only doubled. However, this broadening of the data basis, which is documented in this book, is the reason for compiling the data again and may

Atmospheric Degradation of Organic Substances. W. Klöpffer and B. O. Wagner
Copyright © 2007 WILEY-VCH Verlag GmbH & Co. KGaA, Weinheim
ISBN: 978-3-527-31606-9

give rise to refinements and modifications to the existing models for calculating the rate constants.

This book does not simply list and explain various rate constants but discusses the importance of these data generated by environmental scientists for environmental policy and regulation. This makes it fascinating reading for risk assessors in administration and industry. For many years discussion on environmental risks of chemicals was focused on their toxicological and ecotoxicological properties. Only in the last ten years has persistence been given increased attention and become important in environmental risk assessment. Substances that are not degradable may accumulate in environmental media and biota irreversibly, thus posing a long-term risk that is not predictable. The cornerstones of this discussion are the POP Protocol (1998) and the Stockholm Convention (2001) (POP = Persistent Organic Pollutant), which prohibit certain extremely persistent and hazardous chemicals worldwide, or at least minimize their use and emissions. Another branch of regulatory discussions of persistence were the marine conventions OSPAR and HELCOM, because the marine environment represents a sink for chemicals discharged into the sea by rivers, atmospheric deposition or human activities offshore. OSPAR was the first international entity which proposed an assessment scheme for PBT (persistent, bioaccumulating and toxic) substances in order to identify and prohibit those substances, followed by inclusion of this PBT assessment in the technical guidance document (TGD) for assessing the risks of new and existing chemicals at a European level. This was taken up by REACH, the new European Chemicals Legislation, which will come into force in 2007. Most experts today accord that chemicals which are very persistent and have a great potential to accumulate in biota pose a risk, and their consumer use and release into the environment should be restricted or even prohibited even in those cases where toxic or ecotoxic effects are (still) not known. Chemicals exhibiting these characteristics are not inherently sustainable.

Most assessment and regulatory schemes for persistent substances focus on (biotic) degradation in water, sediment and soil, however, abiotic degradation in the atmosphere has attracted little attention. Looking at REACH, atmospheric lifetimes are missing in the basic data set. This is clearly a deficiency because atmospheric degradation is an important mechanism not only for gases and volatile substances but also for all chemicals with a vapour pressure $> 10^{-6}$ Pa, which could vaporize and spread via the atmosphere to a significant extent.

The atmosphere is not only a wash-house for photo-degradable substances but also the most important medium for long-range transport of persistent substances. Most POPs are predominantly transported via air in remote regions. They are semi-volatile, i.e., in cold climate zones they may condense and be deposited on soil or water where organisms can take them up. When they volatise again this effect is known as the grasshopper effect, which explains why concentrations in polar regions are often higher than in the temperate and tropical latitudes where they are emitted. Therefore, Annex D of the Stockholm Convention lists environmental long-range transport as one of four criteria for identifying POPs.

The semi-volatility makes experimental determination of the atmospheric lifetime difficult, because a significant proportion of the POPs in the atmosphere are not free gaseous molecules but adsorbed onto particles. In this state the attack of oxidising reactive species is hindered and this is why many calculations of atmospheric stability of semi-volatile compounds underestimate the lifetime. This is a topic where the initial experimental findings resulted in interesting insights. However, more research is necessary.

The Stockholm Convention is discussing persistence in the various environmental compartments individually, but has also given a strong incentive for the development of multimedia models as a new approach to the determination of persistence and the potential for long-range transport. These models take into account the degradation rates in the various compartments as well as the partitioning between these compartments for calculation of overall lifetimes and characteristic travel distances. They demonstrate that persistence and long-range transport are not independent of each other. With their rapid development over recent years, multimedia models represent a very valuable screening tool for identification of POP candidates.

However, the overall lifetime is neither an intrinsic property of a substance nor is it more precise than compartment-specific lifetimes. The preciseness depends upon how exactly the lifetimes in air, soil, sediment or soil were determined experimentally or by calculation. As atmospheric lifetime is, in many cases, a very important mechanism its precise determination is extremely significant to the final result. Furthermore, the result may vary depending on which assumptions about the mode of emission are made. When a substance is emitted to soil, e.g., together with sewage sludge, it may take a long time before it is evaporated and the overall lifetime may be longer or shorter than for emissions directly into air, depending upon which is the more effective degradation mechanism – biological degradation in soil or abiotic degradation in air.

Therefore, the multimedia models are an additional tool for assessment which do not replace compartment-specific risk estimation but supplement them. If an evaluation is carried out when a pesticide is exposed to groundwater, comprehensive information on sorption to soil and degradability in soil is needed, but no overall lifetime. The power of the multimedia models is clearly the identification of persistent chemicals and the determination of their potential to be transported over long distances.

When persistent substances are distributed and transported in various environmental media, not only environmental organisms, populations or biocenosis will be exposed and adversely affected but also humans. Looking at the modes of action for chronic toxicity, it is apparent that animals and other organisms do not differ significantly from humans. Prominent examples are endocrine disruptors, which act on humans and other organisms. Traditionally, human health risk assessment is carried out separately from environmental risk assessment. There is not much dialogue between toxicologists and ecotoxicologists, but integrated views are necessary. In 2001 the WHO called for an integrated risk assessment. A common view on effects and risks will result in synergies when exposure assessors flag a substance for its persistency. This may make determination of degradability a trigger for the necessity for an integrated evaluation of chemicals. This book is an important step on the way.

Klaus Günter Steinhäuser
Head of Division IV "Chemical and Biological Safety"
Federal Environment Agency (Umweltbundesamt), Dessau, Germany
August 2006

Preface

The roots of this book extend back to the late 1970s, when the European Community (EC, now the European Union) discussed the common legislation on new industrial chemicals. At the same time (1977), the OECD Chemicals Group began to harmonize test methods. What seemed straightforward from experience with the (older) pesticide and drug regulations, e.g., the determination of toxicity, turned out to be very difficult in the new field of ecotoxicity. How should this new field be defined and what should be measured? OECD nevertheless successfully solved this problem [1] within a short period, with one notable exception: abiotic degradation in the atmosphere. Although reaction rates in the gas phase had already been measured in the 1970s in the United States and in several European countries, the OECD working group on degradation processes was unable to come up with a scientifically defendable draft test guideline. Degradation meant, at that time, solely biodegradation (see Chapter 1) and abiotic degradation (with the exception of hydrolysis) was *terra incognita*.

Because of this situation, Germany pledged for more research and promised a science-based draft OECD testing guideline, which was presented at an international workshop in Berlin [2]. We remember that Roger Atkinson from the Statewide Air Pollution Research Center in Riverside, California, even then a leading member of the "rate constants community", was present at this meeting. The research programme, on which the draft guideline was based, united virtually all members of this community in Germany, and gave one of us (W.K.) the opportunity to build a special smog chamber designed for future use by industrial clients of the Battelle Institute in Frankfurt/Main, a daughter company of Battelle Memorial Institute, Columbus, Ohio. The final scientific results were published in a report by Kernforschungsanlage (KFA) Jülich, the institution co-ordinating the project [3]. This is the place to thank all colleagues who worked on this project, especially Karl-Heinz Becker, Reinhard Zellner and Cornelius Zetzsch. A parallel project aimed at the development of a large aerosol chamber for measuring semi-volatile compounds, which, after several modifications and turbulent times, is still running at the University of Bayreuth under Cornelius Zetzsch [4]. After several years of international review, the guideline was finally published as an OECD monography [5].

The next step in the prehistory of this book was the databank "ABIOTIKx", initiated by the German Association of the Chemical Industry (VCI), and co-financed, with equal contributions, by Umweltbundesamt (UBA). Here, our thanks are due to Ian Meerkamp van Embden (VCI) and Petra Greiner (UBA). This databank has been published in the series UBA Texte [6] and was also made available as the dBase III Plus (Ashton-Tate) file for UBA and the member companies of VCI. It contained ca. 550 substances showing at least one of the 2nd order rate constants $k_{OH}, k_{O_3}, k_{NO_3}$ or the quantum efficiency Φ of direct photolysis (alternatively the 1st order rate constant k_{hv}). The basic philosophy of presenting only one selected value for each entry in a standard format, expressed by a quality index and full references, is still the same in the present book. The idea was, and is here, to give rapid information to the user and to provide references and additional information to the interested reader or researcher. At this point, we would like to thank Roger Atkinson for his invaluable help in providing reviewed data, especially in the form of manuscripts, which were still in print from 1989 and 1990. Also, other colleagues in North America and Europe who provided data in addition to the data search carried out at Battelle should be thanked, so that the collection of data was almost complete.

In 1992, the next step was the transfer of the ABIOTIKx data into UNEP's IRPTC databank (International Register of Potentially Toxic Chemicals) by one of us (B.O.W.), then at UNEP in Geneva. This was done by C.A.U. GmbH (W.K.) with the help of Ute Schock-Schmelzer from the Fachinformationszentrum Berlin.

There was a marked decline in interest in the 1990s and then a revival at the end of the decade. One reason for this decline, which was especially dramatic in Europe, was the misuse of Quantitative Structure-Reactivity Relationship (QSRR) calculations beyond their limits of applicability. Given the large number of industrial chemicals, QSRRs cannot be avoided, but there are strict rules, which have to be observed in establishing and using these estimation tools: only structurally related compounds should be estimated and a robust learning set of data is needed to establish the correlations. Obviously, the rate constants for a completely new chemical structure cannot be estimated with any degree of certainty, they have to be measured. For this reason, a continuous effort in measuring the degradation rate constants is still of paramount importance.

The development leading to a renewed interest in abiotic degradation in recent years as described in Chapter 1 of this book resulted from the policy discussion on Persistent Organic Pollutants (POPs). The photochemically induced abiotic degradation is one important indicator for persistence and the long-range transport potential, two criteria of the Stockholm Convention on POPs. Another reason is the introduction of multimedia fate models, which need as input the abiotic degradation in air. The scientific development went on, driven partly by the three big atmospheric challenges of our time: stratospheric ozone depletion, climate change and the phase out of Persistent Organic Pollutants (POPs). B.O.W. wishes to thank Martin Scheringer (ETH Zürich), who critically read Chapter 1 of the book.

Given these challenges, it is not surprising that the need for actualizing the data of "ABIOTIKx" came again from Umweltbundesamt. The update was performed in 2003–2005. The old data were critically revised and many new ones added so that the total number of entries nearly doubled. For help in the successful update we thank Isa Renner (C.A.U., Dreieich, Germany), Hans-Peter Schenck and Marcus Oenicke (Chemie Daten, Strachau, Germany) and the colleagues Cornelia Leuschner, Axel Finck and Wolfgang Koch (Umweltbundesamt, Dessau) to name only a few.

Mid-2005, most of the data presented here were incorporated into the Umweltbundesamt chemicals databank "Informationssystem Chemikaliensicherheit (ICS)". In order to make them available to interested parties, the idea was born to produce a book containing all data on photo-degradation in the ICS together with a regulatory and scientific introduction to photo-degradation. Some 60 more substances from the most recent papers published up to the summer of 2006 were added to the book. Umweltbundesamt granted permission to publish these data. The publisher Wiley-VCH, Weinheim, accepted the book immediately. Our thanks are due to Frank O. Weinreich and Rainer Münz (Wiley-VCH, Weinheim) for their advice, help and patience.

The book consists of three parts: Chapter 1 describes the political and regulatory development, as mentioned above. The science behind the photochemical degradation in the atmosphere is presented in Chapter 2. This short treatise is distinguished from other, more voluminous monographies by focussing on the fate of organic molecules, including semi-volatile organic compounds (SOC). The degradation in the droplet phase is discussed, although hardly any data are available for industrial chemicals that may enter into this much neglected phase. The situation is slightly better for compounds adsorbed onto artificial aerosol particles, but virtually no information at all exists about the degradation of SOCs adsorbed on or absorbed in natural aerosol particles. Chapter 3 contains the data for 1081 substances in the form of a Table, with ca. 300 footnotes and 553 references.

A few words should be said about the selection of the substances and our target readership. Completeness was sought with regard to organic chemicals. The Table in Chapter 3 also contains a selection of inorganic molecules of atmospheric significance, e.g., SF_6 and volatile acids. As the emphasis of this book is on "real" chemicals, it does not contain instable radicals and deuterated substances investigated for mechanistic studies. Not all substances listed in the Table in Chapter 3 are produced by the chemical industry; there are also some research products, e.g., potential Freon substitution products (hydrofluorocarbons and -ethers). It also contains a number of transformation products of photochemical degradation processes, studied in order to elucidate processes following the primary attack by radicals, ozone or photons.

The presentation of the data in the Table in Chapter 3 aims at the rapid finding of one preferred rate constant or quantum efficiency for each substance and reaction. Additional information is contained in the footnotes. Each entry is followed by at least one, and often several, literature references.

Who should have this book on the shelf? The main target group consists in environmental experts and administrators in industry and government looking for existing data. They also have access to a survey of the research field from the regulator's perspective (Chapter 1) and from the researcher's perspective (Chapter 2). For researchers at universities and specialized laboratories, who have access to most of the original and reviewed literature, this book may help in the rapid access to data and references. It is therefore complimentary to the great books and reviews (e.g., the monographs of Roger Atkinson) and databanks (NIST) cited in Chapter 3 and is not in the slightest meant to replace them.

Frankfurt/Main and Berlin *Walter Klöpffer*
December 2006 *Burkhard O. Wagner*

References

1 Organization for Economic Co-operation and Development (Ed.): OECD Guidelines for Testing of Chemicals. Paris (1981) and updates. http://www.oecd.org/document/40/0,2340,en_2649_34377_37051368_1_1_1_1,00.html
2 Umweltbundesamt (Ed.): Draft OECD-Test Guideline on Photochemical Oxidative Degradation in the Atmosphere, 2nd Revision. Berlin, January 1987.
3 Becker, K. H.; Biehl, H. M.; Bruckmann, P.; Fink, E. H.; Führ, F.; Klöpffer, W.; Zellner, R.; Zetzsch, C. (Eds.): Methods of the Ecotoxicological Evaluation of Chemicals. Photochemical Degradation in the Gas Phase. Vol. 6: OH Reaction Rate Constants and Tropospheric Lifetimes of Selected Environmental Chemicals. Report 1980-1983. Kernforschungsanlage Jülich GmbH, Projektträgerschaft Umweltchemikalien. Jül-Spez 279, 1984, ISSN 0343 7639.
4 Behnke, W.; Holländer, W.; Koch, W.; Nolting, F.; Zetzsch, C.: A Smog Chamber for Studies of the Photochemical Degradation of Chemicals in the Presence of Aerosols. Atmos. Environ. 22 (1988) 1113–1120.
5 Organization for Economic Co-operation and Development (Ed.): The Rate of Photochemical Transformation of Gaseous Organic Compounds in Air under Tropospheric Conditions. OECD Environment Monographs No. 61 OCDE/GD (92)172. Paris 1992. http://www.oecd.org/ehs.
6 Klöpffer, W.; Daniel, B.: Reaktionskonstanten zum abiotischen Abbau von organischen Chemikalien. Umweltbundesamt (Datenbank ABIOTIKx), UBA Texte 51/91. Berlin 1991.

Contents

Foreword V

Preface IX

Chapter 1 Significance of Photo-degradation in Environmental Risk Assessment 1
1 Introduction 1
2 Persistence and Long-range Transport Potential in Chemicals Regulation 2
3 Multimedia Models as Tools to Estimate Persistence and Long-range Transport Potential 8
4 Data Requirements for Multimedia Models 10
5 Estimation of the Rate Constant of Organic Substances with Hydroxyl Radicals 11
6 Research Requirements for Photo-degradation of Semi-volatile Substances 13
7 Conclusions 15
References 15

Chapter 2 Abiotic Degradation in the Atmosphere 21
1 Introduction 21
2 Photo-degradation in the Homogenous Gas Phase of the Troposphere 22
2.1 Indirect Photochemical Reactions 22
2.1.1 The Reaction with OH-Radicals 22
2.1.1.1 Sources and Sinks of the OH-Radical 22
2.1.1.2 Reactions of OH with Organic Compounds 24
2.1.2 The Reaction with NO_3-Radicals 28
2.1.2.1 Sources and Sinks of the NO_3-Radical 28
2.1.2.2 Reactions of NO_3 with Organic Compounds 31
2.1.3 The Reaction with Ozone 32
2.1.3.1 Sources and Sinks of O_3 in the Troposphere 32
2.1.3.2 Reactions of O_3 with Organic Compounds 34

Atmospheric Degradation of Organic Substances. W. Klöpffer and B. O. Wagner
Copyright © 2007 WILEY-VCH Verlag GmbH & Co. KGaA, Weinheim
ISBN: 978-3-527-31606-9

2.2	Direct Photochemical Reactions	35
2.2.1	Quantum Efficiency	35
2.2.2	Examples of Photochemical Reactions in the Gas Phase	39
3	Heterogeneous Degradation	43
3.1	Degradation on Solid Surfaces	43
3.1.1	Introduction	43
3.1.2	Degradation on Fly Ash and Soot	45
3.1.3	Degradation on Artificial Aerosols	45
3.2	Degradation in Droplets	50
3.2.1	Direct Photochemical Transformation	50
3.2.2	Reactive Trace Compounds in Cloud, Fog and Rainwater	51
3.2.3	Reactions of Organic Molecules	57
3.2.4	Summary	59
4	Experimental	60
4.1	Indirect Photochemical Degradation	60
4.1.1	Bimolecular Reaction with OH	60
4.1.1.1	Direct Methods for Measuring k_{OH}	60
4.1.1.2	Indirect Methods for the Measurement of k_{OH}	62
4.1.2	Bimolecular Reaction with NO_3	68
4.1.2.1	Introduction	68
4.1.2.2	Absolute Measurement	69
4.1.2.3	Relative Measurements	69
4.1.3	Bimolecular Reaction with Ozone	70
4.2	Direct Photo-transformation	71
4.2.1	Determination of the Quantum Efficiency in the Gas Phase	71
4.2.1.1	Gas Cuvette and Monochromatic Radiation	71
4.2.1.2	Smog-chamber Method	74
4.2.2	Outlook	74
4.3	Degradation in the Adsorbed State	75
4.3.1	Introduction	75
4.3.2	Aerosol Chambers	76
4.3.3	Alternative Measurements of $k_{OH,ads}$	77
5	Additional Information Necessary for Calculating Lifetimes	78
5.1	Atmospheric Lifetimes	78
5.2	Indirect Photochemical Degradation	82
5.2.1	Average OH Concentration in the Troposphere	82
5.2.2	Average NO_3 Concentration in the Troposphere	86
5.2.3	Average O_3 Concentration in the Troposphere	87
5.3	Direct Photochemical Degradation	89
5.3.1	Introduction	89
5.3.2	Absorption Spectrum	89
5.3.3	Spectral Photon Irradiance	91
5.3.4	Final Comments on Direct and Indirect Photochemical Transformation	93
	References	94

Chapter 3 Table of Reaction Rate Constants of Photo-Degradation Processes *107*
1   Content of the Table   *107*
2   Explanation of the Column Headings   *108*
3   Content of the Footnotes   *109*
4   Completeness and Accuracy   *109*
5   Atmospheric Half-lives   *109*
6   Selection of Substances   *110*
7   Quality Index (QI)   *111*
8   Temperature Dependence of the Rate Constant   *111*
9   Pressure Dependence of the Rate Constant   *112*
10   Direct Photolysis   *113*

Table: Reaction Rate Constants and Quantum Efficiencies for Atmospheric Photo-degradation of Chemicals   *115*
Footnotes to the Table   *170*
References to the Table   *180*

Appendix: CAS Register   *209*

Subject Index   *237*

Chapter 1
Significance of Photo-degradation in Environmental Risk Assessment

1
Introduction

Photo-degradation (or transformation) occurs under the influence of solar radiation mainly in the atmosphere, and to a lesser extent in the hydrosphere and on soil surfaces. New developments in environmental risk assessment have given photo-degradation new significance, which will be described in this introduction. The data collected in this book refer to gas-phase photo-degradation in the atmosphere, and so does this introduction. Photo degradability is an intrinsic property of a chemical substance and must be measured. Quantitative structure activity relationships (QSARs) for estimating gas-phase rate constants of the indirect photo-degradation of organic chemicals are available and will be discussed in Section 5. Reactive species, which degrade a chemical substance in the atmosphere, are the hydroxyl radical, ozone and the nitrate radical. As these species are produced via solar radiation, this mechanism of degradation is known as "indirect", in order to distinguish it from the direct photolysis by solar radiation.

Photochemistry is a discipline within physical chemistry, and more specifically, atmospheric chemistry that is dealt with in the text books by Finlayson-Pitts and Pitts [1, 2]. Photo-degradation of organic substances is the subject of Chapter 2. The subject of interest to environmental chemists and administrators is the capacity of solar radiation to degrade, destroy and finally eliminate man-made chemical substances from the atmosphere. Atmospheric distribution of a chemical substance is critical because it can potentially lead to world-wide dissemination, if it is long-lived (persistent) and not destroyed. If this elimination mechanism did not occur, mankind would have suffered from severe air pollution and respiratory health problems, which are still the case in urban agglomerations as a result of air pollution, where the atmospheric elimination processes are not sufficiently efficient.

In the early 1980s Atkinson [3] and Becker et al. [4] began the systematic testing of volatile organic substances on gas-phase photo-degradation with the hydroxyl radical. At the time that the regulation of chemicals was drawn up, "photo-degradation" under the term of "abiotic degradation" was not considered important. Photo-degradation was not part of the OECD pre-minimum set of data in the assessment of chemicals [5], which was the example for the European legislation in the 6th Amendment of the 67/548/EEC Directive (1979).

This book, with the data collected in it, underlines the role of photo degradability as one of the important intrinsic properties that steers the atmospheric fate of chemicals and thereby contributes to "persistence" and "long-range transport potential". With the acceptance of multimedia environmental fate models in legal exposure assessment, photo degradability underwent a renaissance of importance as shown below.

2
Persistence and Long-range Transport Potential in Chemicals Regulation

Environmental persistence of organic substances was heralded by two publications that shook the scientific community as well as the politicians: Rachel Carson's *Silent Spring* [6] in 1962, and the discovery by Rowland and Molina [7] in 1975 that fluorochlorocarbons are stable (persistent) molecules in the troposphere (the lower 10 to 15 km of the atmosphere) and may deplete the ozone in the next layer of the atmosphere, the stratosphere. Persistence implies the absence of chemical, biological and physical degradation processes in the environment so that organic molecules, once emitted from the technosphere by anthropogenic activity, remain, distribute and accumulate in the worldwide environment. Persistence is an environmental (negative) term and should be distinguished from the (positive) term "durability" of a chemical product during use. In the following we will draw attention to the interaction between the scientific and the administrative communities with respect to the slow change in perception of the term "persistence".

"Persistence" first surfaced in the OECD in 1966, when the OECD Committee for Scientific Research held a conference on "Research on the Unintended Occurrence of Pesticides in the Environment". At that time the OECD established a study group on this problem, and its report led, in 1971, to the creation of the OECD Environment Committee with the following recommendations:

> "The Group wishes to stress the need internationally for means to make comprehensive investigations of the consequences of use, and limitations in use, of those chemicals which could be regarded as having unacceptable effects on man and his environment resulting from, either
> a) their undue persistence in natural conditions in biologically active form, or
> b) their wide distribution through water and air, or
> c) their accumulation which may lead to biological effective levels in organisms exposed to even low concentrations." [8, p. 11]

At the same time the United Nations Conference on the Human Environment (1972) recognised persistence as a negative environmental property in Recommendation 71:

"It is recommended that Governments use the best practical means available to minimize the release to the environment of toxic or dangerous substances, especially if they are persistent substances such as heavy metals and organochlorine compounds, until it has been demonstrated that their release will not give rise to unacceptable risks or unless their use is essential to human health or food production, in which case appropriate control measures should be applied." [9]

These two recommendations have not changed with the years and read like a contemporary political mandate. However, the environmental regulatory discussion treated "persistence" only as a sub-element of environmental effects assessment, and it took several years and arguments before "persistence" was accepted as an environmental criterion in its own right and designated as an "endpoint" in environmental exposure assessment equivalent to "ecotoxicity". In 1977 Stephenson foresaw the future problem of persistent chemicals, when he wrote:

"Persistent materials, because of this property, will accumulate in the environment for as long as they are released. Since the environment is not effective at cleansing itself of these materials, they will remain for indefinite periods, which were not recognized at the time of their original release. The problem could become entirely out of control and it would be extremely difficult if not impossible to do anything about it. Materials which are strongly persistent can accumulate to rather high levels in the environment and effects which would not otherwise be important could become so." [10, p. 48]

Frische et al. (1982) [11] and Klöpffer (1994) [12, 13] advocated that *"persistence" is the "central and most important environmental criterion"* often replacing ecotoxicity, which can never be determined with acceptable certainty.

At the beginning of the 1990s, semi-volatile organic chemicals (SOCs) came into focus for analytical chemists because of the worldwide distribution, ubiquitous occurrence and geo accumulation in remote areas (Ballschmiter and Wittlinger [14], Ballschmiter [15], Ockenden et al. [16], and AMAP [17]). This was a disturbing signal for sustainable chemical production and use, and alarmed the regulatory community, after "sustainable development" had been established as the key policy environmental term at the Earth Summit in Rio de Janeiro in 1992. In fact, Agenda 21 formulated the future chemicals risk assessment policy, by including:

"Governments, through the cooperation of relevant international organisations and industry, where appropriate, should adopt policies and regulatory and non-regulatory measures to identify, and minimize exposure to toxic chemicals by replacing them with less toxic substitutes and ultimately phasing out the chemicals that pose unreasonable and otherwise unmanageable risk to humans and the environment and those that are

toxic, persistent and bio-accumulative and whose use cannot be adequately controlled."
[18, Chap. 19.49 (c)]

Another milestone in this policy discussion was the Esbjerg Declaration, which nine European countries neighbouring the North Sea and the European Commission adopted in Esbjerg, Denmark, 8th–9th June 1995, at the Fourth International Conference on the Protection of the North Sea. The statement of zero concentrations for man-made synthetic substances in the North Sea was revolutionary and was received with scepticism, but did not miss its policy objective.

> *17. The Ministers AGREE that the objective is to ensure a sustainable, sound and healthy North Sea ecosystem. The guiding principle for achieving this objective is the precautionary principle.*
> *This implies the prevention of the pollution of the North Sea by continuously reducing discharges, emissions and losses of hazardous substances thereby moving towards the target of their cessation within one generation (25 years) with the ultimate aim of concentrations in the environment near background values for naturally occurring substances and close to zero concentrations for man-made synthetic substances. Esbjerg Declaration 1995.* [19]

In 1997 the Chemicals Policy Committee of the Swedish Ministry of the Environment outlined specific sustainability goals: "Substances that are persistent and liable to bioaccumulation should be banned, even if they are not known to have toxic effects." The Committee argued:

> *"Experience tells us, that new unexpected forms of toxicity may be uncovered in the future. For substances that are persistent and liable to bioaccumulate that knowledge will come too late. To act only when the knowledge of the hazard becomes available is not prevention. We therefore conclude that known or suspected toxicity is not a necessary criterion for measures against organic man-made substances that are persistent and liable to bioaccumulate. Such substances should in the future not be used at all."* [20]

One year later Martin Scheringer et al. [21], [22, Chap. 1–3], [23, Chap. 1–3] advocated a change in the paradigm of environmental risk assessment, in short a shift from the effects-based to the exposure-based assessment. Scheringer and Berg [24, 25] had prepared this change by introducing the following three indicators for measuring environmental threat:

- spatial range (potential for long-range transport)
- temporal range (persistence)
- bioaccumulation potential

Scheringer and Hungerbühler [26, p. 176] concluded:

"An exposure-based assessment requires different (and usually less) data than effect-based assessments and is (usually) performed faster than the various toxicity tests required for an effect-based assessment.

It should be noted that the combination of persistence and bioaccumulation, although relevant for many semi-volatile organic substances, is "narrower" than the concepts of Klöpffer and Scheringer and would not include, for example, new types of freons contributing to global warming or other, hitherto not recognized effects. The same is true for persistent water-soluble substances and thus not bioaccumulating substances.

Subsequently, in 2001, the German Umweltbundesamt [27] and Steinhäuser [28] argued similarly, when they published five policy principles on sustainability, two of which are concerned with persistence:

- *"The irreversible release of persistent and bioaccumulative or persistent and highly mobile pollutants (xenobiota) in the environment must totally be avoided regardless of their toxicity. This also holds for metabolites with the same properties.*
- *The increase of releases must be avoided independently of known adverse effects and other intrinsic properties, if it is practically impossible to recollect the substance from the environment because of its high mobility and/or its significant partitioning."*
[27, pp. 7–8]

In Canada, indigenous people, such as the Inuit, complained that they were not users of persistent chemicals, but suffered from the air-borne fallout and the contamination of their grounds living in the Arctic.

Important milestones in this policy discussion on persistence were the 1985 Vienna Convention for the Protection of the Ozone Layer [29] and the Montreal Protocol on Substances that deplete the Ozone Layer [30]. This Convention and its Protocol triggered off an intensive research on the gas-phase photo-degradation and the global warming potential of hydrofluorocarbons (HFCs), hydrofluoro-ethers (HFEs) [31] and polyfluoroethers [32], because such fluorated chemicals have valuable uses and are of high economic interest.

"Persistence," "bioaccumulation" and "long-range transport" of industrial chemicals and pesticides came into focus when Persistent Organic Pollutants (POPs) were detected in the remote and assumedly pristine areas of the Arctic. Analytical chemists were interested in measuring concentrations (Ballschmiter and Wittlinger [14], Ballschmiter [15], AMAP [17]), and environmental modellers tried to simulate the environmental movement of chemicals in multimedia fate models (Wania and Mackay [33, 34], Cowen et al. [35], Scheringer [22, 23, 36, 37], Wania [38]). Persistence and long-range transport potential were identified as the *intrinsic chemical properties* responsible for environmental migration. Transportation over the atmosphere is most probably the fastest route to worldwide distribution, and photo-degradation is the most likely elimination process of air-borne chemicals. Rivers and ocean currents transport chemicals much more slowly, particularly hydrophilic substances that are less volatile. Wash-out, especially of aerosol-bound chemicals, temporarily removes persistent chemicals from the atmosphere, but not from the environment. They may re-enter the troposphere by re-volatilisation from soil and surface waters (grasshopper-effect). Physical environmental sinks such as river or ocean sediments do not reduce the persistence because they do not eliminate the chemical from the environment.

How did the administrations react to this new environmental threat?

The administrative community had already reacted in November 1979, when the United Nations Economic Commission for Europe convened a High-level Meeting in Geneva in response to acute problems of transboundary air pollution through acidification. It resulted in the signature of the Convention on Long-range Transboundary Air Pollution by 34 Governments and the European Community [39]. The Convention was the first international legally binding instrument to deal with problems of air pollution on a regional basis. The Convention set up an institutional framework bringing together research and policy. Under this Convention, on 24th June 1998 in Aarhus, Denmark, the UN ECE adopted the Protocol on Persistent Organic Pollutants (POPs) [40]. It focuses on a list of 16 substances that have been singled out according to agreed risk criteria. The substances comprise 11 pesticides, two industrial chemicals and three by-products/contaminants. The ultimate objective is to eliminate any discharges, emissions and losses of POPs. The Protocol bans the production and use of some products outright (aldrin, chlordane, chlordecone, dieldrin, endrin, hexabromobiphenyl, mirex and toxaphene). Others are scheduled for elimination at a later stage (DDT, heptachlor, hexachlorobenzene, PCBs). Finally, the Protocol severely restricts the use of DDT, HCH (including lindane) and PCBs. The Protocol includes provisions for dealing with the wastes of products that are banned. It also obliges Parties to reduce their emissions of dioxins, furans, PAHs and HCB below their 1990 levels (or an alternative year between 1985 and 1995). For the incineration of municipal, hazardous and medical waste, it lays down specific limit values. Since it was enforced in October 2003 several "new" POP candidates have been added to the Protocol (http://www.unece.org/env/popsxg, May 2006).

In May 1995 the UNEP Governing Council was "aware that persistent organic pollutants pose major and increasing threats to human health and the environment" and adopted Decision 18/32 on Persistent Organic Pollutants and

> "*Invited, ... the Intergovernmental Forum on Chemical Safety to develop recommendations and information on international action, including such information as would be needed for a possible decision regarding an appropriate international legal mechanism on persistent organic pollutants, to be considered by the Governing Council and the World Health Assembly no later than in 1997."* [41]

Regulatory action on hazardous properties such as toxicity, carcinogenicity, mutagenicity or reproductive toxicity was straightforward; however, the fact that a chemical substance is recalcitrant to breakdown processes in the environment was not linked to these effects. This attitude changed when POPs were detected, and the voice of the Inuit populations in Canada and Greenland requested fairness and equity in regions where the benefits of POPs were not achieved, because these chemicals had not been used there. The international scientific and administrative community recognised these arguments and gave "persistence" and "long-range transport potential" the status of exposure assessment "criteria" in the Stockholm Convention [42, Annex D] in addition to "bioaccumulation" and "adverse effects." In 2004 the European Community transferred the Stockholm Convention and the UN ECE POP Protocol into European legislation by Regulation (EC) 850/2004 [43]. Industry's contribution to persistence is summarised in an ECETOC Monograph [44].

Persistence and long-range transport are not considered in isolation, but assessed in combination in two assessment schemes that reflected the policy discussions of the previous years:

- vPvB assessment = very persistent and very bioaccumulative
- PBT assessment = persistent, bioaccumulative and toxic

The Stockholm Convention set the stage in Annex D [42] for national or regional legislations by specifying values to these criteria[1]:

- *Persistence*
  - half-life in water > 2 months or
  - half-life in soil > 6 months or
  - half-life in sediment > 6 months or
  - evidence that the chemical is otherwise sufficiently persistent

- *Potential for long-range environmental transport*
  - measured levels in locations distant from the source of release
  - monitoring data showing long-range environmental transport
  - environmental fate properties and/or model results
  - half-life in air > 2 days

The Commission of the European Union and other Governments had already begun programmes to identify persistent and bioaccumulative chemicals:

- European Commission (2003): Technical Guidance Document, Part II: vPvB and PBT [45]
- US EPA: POP-Profiler [46]; Rodan et al. 1999 [47]
- Environment Canada (1995): Persistence and Bioaccumulation Criteria [48]
- Japan: Ikeda et al. 2001 [49]

---

1) Only these two criteria relevant for photo degradation are presented.

It is the intention of these programmes to screen new and existing commercial chemicals and pesticides against the Stockholm criteria. In fact, the Stockholm Convention stipulates in Article 3:

(3) that "each party (to the Convention) that has one or more regulatory and assessment schemes for new pesticides or new industrial chemicals shall take measures to regulate with the aim of preventing the production and use of new pesticides or new industrial chemicals which, taking into consideration the criteria in paragraph 1 of Annex D, exhibit the characteristics of persistent organic pollutants."

and

(4) that "each party that has one or more regulatory and assessment schemes for pesticides and industrial chemicals shall, where appropriate, take into consideration within these schemes the criteria in paragraph 1 of Annex D when conducting assessments of pesticides or industrial chemicals currently in use." [42]

The draft REACH regulation of 29th October 2003 [50] incorporates this task into Articles 54 and 55 with the criteria of Annex XII.[2] It is estimated that in the European Union about 30 000 existing chemicals will be screened within the next 15 years against the Stockholm Convention criteria. This legislation, in combination with the Precautionary Principle, should prevent POP-like chemical substances being produced and put on the market.

3
Multimedia Models as Tools to Estimate Persistence and Long-range Transport Potential

The above mentioned regulatory approaches in the PBT and vPvB assessment to measure persistence via the single media transformation (degradation) rate constants in the environmental compartments water and soil may have shortcomings because they do not consider the environmental partitioning of the chemical. Evidently, the persistence of a chemical partitioning, mostly to air, will not be adequately described by only measuring biodegradability in water. Environmental modellers took up this critique and created an integrated approach to persistence by introducing environmental fate models that were based on both environmental partitioning and transformation rate constants (Cowan et al. [35], Scheringer [22, 23], Webster et al. [51], Bennett et al. [52], Beyer et al. [53]).

This controversy became evident, when administrators and environmental modellers met in October 2001 in Ottawa in the OECD workshop on the use of

---

2) The REACH draft [50] presents the PBT and vPvB assessment criteria in Annex XII. They are not in full conformity with the concept of the Stockholm Convention [42, Annex D], because the criterion "potential for long-range environmental transport" is missing, which includes the half-life in air. The EU Council adopted the Common Position of the REACH draft in the 27th June 2006 version. Annex XII became Annex XIII and was not changed.

multimedia models for estimating overall persistence and long-range transport in the context of PBT/POPs assessment. This workshop's objective was to explore common ground for the use of multimedia models in screening chemicals against the two Stockholm criteria, "persistence" and "long-range transport potential" [54]. Upon recommendation of this workshop OECD established an expert group of modellers, who elaborated a Technical Guidance Document (OECD [55]) and two scientific articles (Fenner et al. [56], Klasmeier et al. [57]) and a consensus multimedia model for screening chemicals for high persistence and long-rang transport potential (Wegmann et al. [58]). OECD performed three workshops for the application of this screening software in Zürich, Switzerland (August 2005), Ottawa, Canada (May 2006) and Tsukuba, Japan (June 2006). This multimedia approach and its underlying philosophy slowly found its way into regulatory science. It can support industry and administrators in the task of screening industrial chemicals and currently used pesticides for the two criteria "overall persistence (Pov)" and "long-range transport potential (LRTP)". Overall persistence differs from the above mentioned single media persistences in water, soil, sediment and air. It constitutes a weighted persistence in the four environmental compartments: water, sediment, soil and air, in which a chemical partitions as a consequence of its physicochemical properties. Green and Bergman [59] reflected this multimedia understanding by proposing the following definition of "persistence":

"The persistence of a chemical is its longevity in the integrated background environment as estimated from its chemical and physicochemical properties within a defined model of the environment." [59]

A chemical that partitions to air is liable to atmospheric transport. This new concept of overall persistence, expressed in half-life [days], is the output of the multimedia fate model and is considered an adequate representation of the Stockholm persistence criterion. The long-range transport potential is the other output expressed in the "half-distance" [km], that the chemical may migrate with a certain wind speed. The plot of Pov versus LRTP (Fig. 1) is divided into four sectors [37, 57] and provides an indication for the chemical in question with respect to both criteria relative to a set of benchmark chemicals, amongst which are some of the agreed POPs. With this multimedia model the chemical industry and administrators will have a tool to hand that allows these two criteria to be determined in advance, so that persistent new chemicals will no longer be placed on the market and existing chemicals can be screened, as Schmidt-Bleek and Hamann [60] had called for in 1985 in their appeal to systematically screen chemicals for early warnings:

"Early recognition is understood to mean the timely observation of deleterious effects and changing exposure situations before significant damage manifest themselves. From an environmental protection point of view it would seem reasonable to approach a systematic solution from the exposure potential side for reasons of logic as well as saving time and resources:

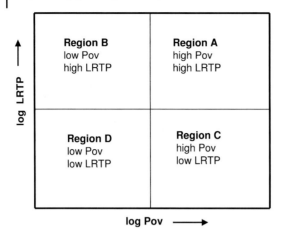

Fig. 1 Classification of chemicals in the Pov/LRTP space
(adapted from Klasmeier et al. 2006 [57], copyright of American Chemical Society).

– *if significant quantities can reach the environment*
– *if the environmental persistence of a chemical is long-term, and*
– *if its environmental mobility is high, such a chemical can be regarded as a threat to the environment, even in the absence of knowledge on its effect potentials."*
[60, p. 462–463]

4
Data Requirements for Multimedia Models

Klöpffer [61] identified the following data required in the application of multimedia models in screening chemicals for Pov and LRTP:

- *Physicochemical properties (partitioning)*
  – chemical identity
  – molecular weight
  – water solubility
  – vapour pressure
  – n-octanol/water partition coefficient

- *Transformation (degradation) half-lives*
  – biodegradation in water
  – biodegradation in soil
  – biodegradation in sediment
  – abiotic degradation
    · hydrolysis
    · indirect photo-degradation by hydroxyl radicals, ozone, nitrate radical
    · direct photolysis in air

Hydrolysis and direct photolysis in air are believed to be of subordinate importance for the persistence of most substances, as special structures are needed (hydrolysable groups, chromophores) to be effective. The reaction with hydroxyl radicals is by far the most important and predominant atmospheric elimination process (Bidleman [62]), so that the abiotic degradation in the *screening phase* may be expressed, by the rule of thumb, through the indirect photo-degradation with the hydroxyl radical. A more detailed discussion of the photo transformation processes, as required for a deeper understanding and refined assessment, is presented in Chapter 2.

The indirect photo-degradation is a decisive transformation rate constant that accounts for persistence and long-range transport potential of volatile and semi-volatile substances in multimedia fate models. The availability of literature data on reactions of organic substances with hydroxyl radicals has been the subject of several international workshops:

- Driebergen, The Netherlands 22–24 April 1998 (Bidlemann [62], Atkinson et al. [63])
- SETAC Pellston Workshop in Fairmont Hot Springs, British Columbia, Canada 12th–19th July 1998 (Franklin et al. [64], Chap. 2)
- OECD/UNEP Workshop in Ottawa, Canada 30th October – 1st November 2001 [54, Annex 7]

All three workshops concluded that the availability of photo-degradation data for semi-volatile substances is not sufficient. This book presents the rate constants for indirect photo-degradation of 1081 chemicals. The majority of these data result from volatile substances (vapour pressure > 10^{-3} Pa).

There is, however, a lack of rate constants for semi-volatile substances (vapour pressure < 10^{-3} to 10^{-6} Pa) because of experimental difficulties. Palm demonstrated (as quoted by Franklin et al. [64], Fig. 2-2) that the majority of currently used pesticides exhibit a vapour pressure of < 10^{-3} Pa and therefore constitute semi-volatile and non-volatile substances. A similar survey of the vapour pressure of existing industrial chemicals can probably not be made because of missing data.

5
Estimation of the Rate Constant of Organic Substances with Hydroxyl Radicals

The experimental rate constants collected from the literature, as given in this book, represent only a minority of chemicals *vis-à-vis* the 30 000 industrial chemical substances to be evaluated in Europe under the new legislation [50]. In the 1990s Meylan and Howard [65, 66] developed the empirical Atmospheric Oxidation Programme (AOP) based upon Atkinson's experimental data [cf. 67]. This QSAR software programme allows estimation of the gas-phase hydroxyl rate constant at 298 K from increments of the chemical structure. Its use is well established in industry and accepted in the administrative communities (Sabljic and Peijnenburg [68], Meylan and Howard [69], and Schüürmann [70]), but is often used in regulatory applications beyond its limitations. The SETAC Pellston Workshop expressed these limitations (Franklin et al. [64]):

> *"The empirical estimation method of Atkinson allows the OH rate constants at 298 K of 90% of approximately 500 to 600 organic compounds used in its development to be predicted within a factor of 2 of the experimental values.*
> *The limitations of present OH rate constant estimation methods must be recognised. In particular, most estimation methods cannot be used with any degree of liability for organic compounds outside of those compounds actually used in the development of the estimation method. Compound classes for which experimental data are not available for at least one member of the homologous series could have uncertainties in the estimated rate constants for reactions with the OH radical of a factor of 5 or greater. In such cases expert judgment should be sought. Atkinson's estimation technique has not been tested sufficiently to warrant its use for most N-, S- and P-containing compounds.*
> *Further studies need to be carried out to extend Atkinson's estimation method to additional compound classes represented by POP-type compounds and to develop more direct methods for the calculation of OH radical reaction rate constants."*
> [64, pp. 52–53]

These qualifying conclusions are particularly true for semi-volatile, persistent, particle-bound substances, which represent many existing organic substances and for which experimental hydroxyl rate constants are missing. The major obstacle to measuring the hydroxyl rate constant for semi-volatile substances lies in the low vapour pressure at room temperature so that traditional methods can not be used. In addition, semi-volatile substances ad- or absorb to atmospheric aerosols, thereby shielding the access of the hydroxyl radical.

Junge [71], and later Bidleman and Harner [72], Goss [73] and Goss and Schwarzenbach [74] created approximation models to predict the adsorption of semi-volatile organic substances to solid (aerosol) particles. The SETAC Pellston Workshop described (Franklin et al. [64]):

> *"Absorption of a gaseous species into an outer organic layer deposited on an inorganic particle may isolate the substance from gas-phase OH and hence impede its degradation. In a similar manner, adsorption onto the surface of pores whose particular shape hinders access by gaseous OH may lead to a reduced rate of degradation."*
> [64, p. 53]

These conclusions still hold. Scheringer [23, pp. 174–176] and [37, 75] discussed this dilemma and advocated introducing an effective atmospheric degradation rate constant in multimedia model calculations that is smaller than the gas-phase rate constant and accounts for a particle-bound fraction.

$$k_{eff} = (1 - \Phi) k_{OH} + \Phi k_{part} \tag{1}$$

where k_{eff} = effective atmospheric degradation rate constant; Φ = fraction of the particle-bound substance; k_{OH} = atmospheric rate constant for the gas-phase reaction with the hydroxyl radical; k_{part} = atmospheric rate constant of the particle-bound fraction. This latter rate constant is set to zero in most multimedia fate models.

He also pointed out that the multimedia model long-range transport results using AOP predicted hydroxyl rate constants for POPs are much too short and disagree with the findings of POPs in remote areas. The argument given was that either the hydroxyl rate constant is predicted to be too high or the model that assumes the sheltering of particle-bound substances from the hydroxyl radical reaction is incorrect. Wania and Daly [76] explained these inconsistencies with calculations of their Globo-POP multimedia model in a plausible manner. They adapted the hydroxyl radical concentration to tropospheric vertical and temporal conditions and the temperature dependent gas/aerosol partitioning to the tropospheric average temperature of –18 °C. By varying these two environmental parameters they showed that the atmospheric half-lives of PCBs, derived from laboratory hydroxyl rate constants increased substantially. They concluded:

"Reaction rate constants of the PCBs with OH radicals derived in the laboratory do not disagree with the observed global fate of these contaminants. However, the real atmospheric lifetimes of POPs are longer than may be expected from a cursory inspection of laboratory-derived reaction rate constants, as a result of low temperatures, variable OH radical concentrations and partitioning of semi-volatile chemicals onto atmospheric particles. ... Even a chemical with a seemingly fast atmospheric degradation rate constant can be subject to long-range atmospheric transport, especially if the chemical is semi-volatile and the atmosphere is cold and/or dark." [76]

6
Research Requirements for Photo-degradation of Semi-volatile Substances

Wania and Daly advocated further research for improving the quantitative knowledge on the reactions that PCBs may undergo when adsorbed to atmospheric particles, especially the kinetics of such reactions [76]. Researchers, therefore, are encouraged to engage in new approaches to kinetic photo-degradation studies of semi-volatile substances particularly when adsorbed on aerosols (cf. Wagner [77, 78], Scheringer [75, 79]).

Indeed, this issue has been the subject of research ever since and various approaches have been tried. The low vapour pressure of semi-volatile substances caused experimental difficulties, which have not yet been solved satisfactorily. Since the late 1980s Zetzsch and his group (Behnke et al. [80, 81]) have devoted research to heterogeneous photo-degradation of semi-volatile substances adsorbed on particles in smog chambers of different sizes. Palm et al. obtained rate constants for two semi-volatile pesticides (terbutylazine and pyrifenox) adsorbed on particles [82–84]. The group of Hites (Anderson and Hites [85, 86] and Brubaker and Hites [87, 88]) used a heated system (75–159 °C) to measure gas-phase hydroxyl reaction

Table 1 Reaction rate constants and half-lives for lindane coated on Aerosil®.

Substance CAS No.	$k_{OH} \times 10^{-12}$ [cm³ molecule⁻¹ s⁻¹]	Half-life [d][a]	Experimental conditions	Reference
Lindane 58-89-9	0.60 ± 0.3	26.7	Smog chamber 27 °C 50% humidity	Behnke and Zetzsch 1989 [94], Zetzsch 1991 [95]
Lindane	0.345 ± 0.3	46.5	Eletrodynamic trap room temperature	Rühl 2004 [89]
Lindane	3.0 ± 0.3	5.3	Smog chamber 6.5 °C	Krüger et al. 2005 [90]
Lindane	0.19 ± 0.06	84.4	Measured at 73–113 °C and extrapolated to room temperature: 25 °C	Brubaker and Hites 1998 [88]

a) Calculated with 5×10^5 OH radicals per cm³.

rate constants of polychlorinated dibenzo-*p*-dioxins and furans in addition to PCBs and α- and γ-hexachlorocyclohexanes and hexachlorobenzene and extrapolated them to room temperature through the Arrhenius equation. More recently two research groups provided particle-bound rate constants for lindane (Table 1). Rühl [89] used an electrodynamic trap to expose lindane coated particles and analysed its decay by infrared spectroscopy. Krüger et al. [90] used a smog chamber at 6.5 °C with conventional gas chromatographic analysis. Both experiments were performed with artificial inorganic aerosols (Aerosil®, silica flocks from Degussa) mimicking desert sand aerosols, i.e., particles containing no organic matter. The situation with an organic matter surface that covers the inorganic nucleus of an aerosol may be different [cf. 76]. The other extreme is particles covered by a water layer [cf. 74]. All variations may occur in nature, depending on the origin and chemical composition of the aerosol particle and the atmospheric conditions (humidity, temperature). Recently, Sun et al. [91] measured the gas-phase hydroxyl reaction rate constant of four semi-volatile pesticides at 296 K (dichlorvos, carbaryl, chlordimeform and 2,4-D-butyl ester) with a Fourier transform infrared spectrometer (FTIR). Feigenbrugel et al. [92] also measured dichlorvos with an FTIR technique at room temperature.[3] Mandalakis et al. [93] determined the hydroxyl reaction rate constant of PCBs in a field experiment in Crete (Greece) by monitoring the gas concentrations over one day. The known rate constant for PCB 28 served as the reference. The field rate constants were notably coherent with the laboratory measured values. The research aim may still be the conclusion of the Pellston Workshop (Franklin et al. [64]), which recommended that

---

3) The rate constants of the quoted papers are summarised in Chapter 3.

"consistent data from at least two independent studies (preferably using different techniques) are needed before a rate constant can be considered to be well established." [64, p. 14]

7
Conclusions

The international scientific and administrative communities now agree that persistence has a central role in environmental exposure and risk assessment. Screening multimedia fate models have now been developed and tested to such an extent that regulatory science can rely on their results for persistence and long-range transport potential. This multimedia approach realises the change in paradigm from effects to exposure based environmental risk assessment, as postulated above by several workers and institutions.

Multimedia models require two classes of intrinsic substance properties: partitioning and degradation (transformation) data. Environmental partitioning data are reliably calculated on the basis of measured physicochemical properties. The weak and salient point is still the experimental determination of transformation half-lives in water, soil, sediment and air. In particular, the last half-life in air is responsible for assessing the world-wide distribution of a chemical. This book presents the state-of-the-art methodology for measuring indirect and direct gas-phase photo-degradation (transformation) and includes quality checked photo-degradation rate constants for 1081 chemicals published in the literature up to July 2006. Research is needed to develop new methods for measuring the photo-degradation of semi-volatile chemical substances, which, at environmental temperatures, adsorb partially or totally to aerosols [72]. Such data could extend the database of the Atmospheric Oxidation Programme (AOP) and then make this QSAR programme more valuable.

References

1. Finlayson-Pitts, B. J.; Pitts, Jr., J. N.: Atmospheric Chemistry – Fundamentals and Experimental Techniques. Wiley, New York 1986.
2. Finlayson-Pitts, B. J.; Pitts, Jr., J. N.: Chemistry of the Upper and Lower Atmosphere. Theory, Experiment, and Application. Academic Press, San Diego 2000.
3. Atkinson, R: Kinetics and Mechanisms of the Gas-phase Reaction of the Hydroxyl Radical with Organic Compounds under Atmospheric Conditions, Chem. Rev. 85 (1985) 69–201.
4. Becker, K. H.; Biehl, H. M.; Bruckmann, P.; Fink, E. H.; Führ, F.; Klöpffer, W.; Zellner, R.; Zetzsch, C. (Eds.): Methods of the Ecotoxicological Evaluation of Chemicals – Photo Degradation in the Gas Phase – Vol. 6: OH Reaction Rate Constants and Tropospheric Lifetimes of Selected Environmental Chemicals – Report 1980–1983. Kernforschungsanlage Jülich GmbH – Projektträger Umweltchemikalien. Jül-Spez-279. Jülich 1984.
5. OECD: Decision of the Council Concerning the Minimum Pre-market Set of Data in the Assessment of Chemicals. C(82)196 (Final). Paris 1982.

6 Carson, R.: The Silent Spring. Greenwich (Connecticut) 1962. http://www.rachelcarson.org (June 2006).
7 Rowland, F. S.; Molina, M. J.: Chlorofluoromethanes in the Environment. Rev. Geophy. Space Phys. 13 (1975) 1.
8 OECD: The Problems of Persistent Chemicals – Implications of Pesticides and other Chemicals in the Environment. Report of the Study Group on Unintended Occurrence of Pesticides. Paris 1971.
9 United Nations: The Results of Stockholm. General Assembly Document A/CONF.48.14: Beiträge zur Umweltgestaltung. Heft A 10. Erich Schmidt Verlag, Berlin 1973.
10 Stephenson, M. E.: An Approach to the Identification of Organic Compounds Hazardous to the Environment and Human Health. Ecotox. Environ. Safety 1 (1977) 39–48.
11 Frische, R.; Klöpffer, W.; Esser, G.; Schönborn, W.: Criteria for Assessing the Environmental Behaviour of Chemicals: Selection and Preliminary Quantification. Ecotox. Environ. Safety 6 (1982) 283–293.
12 Klöpffer, W.: Kriterien zur Umweltbewertung von Einzelstoffen und Stoffgruppen. UWSF – Z. Umweltchem. Ökotox. 6 (1994) 61–63.
13 Klöpffer, W.: Environmental Hazard Assessment of Chemicals and Products. Part II. Persistence and Degradability. ESPR – Environ. Sci. Pollut. Res. 1 (1994) 108–116.
14 Ballschmiter, K.; Wittlinger, R.: Interhemispheric Exchange of Hexachlorocyclohexanes, Hexachlorobenzene, Polychlorobiphenyls, and 1,1,1-Trichloro-2,2-bis(p-chlorophenyl)-ethane in the Lower Troposphere. Environ. Sci. Technol. 25 (1991) 1103–1111.
15 Ballschmiter, K.: Transport and Fate of Organic Compounds in the Global Environment. Angew. Chem. Int. Ed. Engl. 31 (1992) 487–515.
16 Ockenden, W. A.; Sweetman, A. J.; Prest, H. F.; Steinnes, E.; Jones, K. C.: Toward an Understanding of the Global Atmospheric Distribution of Persistent Organic Pollutants: The use of Semipermeable Membranes Devices as Time-integrated Passive Samplers. Environ. Sci. Technol. 32 (1998) 2795–2803.
17 AMAP Assessment 2002: Persistent Organic Pollutants in the Arctic. Arctic Monitoring and Assessment Programme (AMAP). Oslo 2004. http://www.amap.no (May 2006).
18 United Nations: Report of the United Nations Conference on Environment and Development – Rio de Janeiro, 3–14 June 1992 (Agenda 21). United Nations A/CONF.151/26/Rev. 1 (Vol. I). New York 1993. http://www.un.org/esa/sustdev/documents/agenda21/index.htm (May 2006).
19 Esbjerg Declaration – Ministerial Declaration of the Fourth International Conference on the Protection of the North Sea. Esbjerg 1995. http://www.odin.no/md/nsc/declaration/022001-990243/dok-bn.html (May 2006).
20 Swedish Chemicals Policy Committee. Towards a Sustainable Chemicals Policy. Ministry of the Environment: Government Official Report, Vol. 1997: 84. Stockholm 1998.
21 Scheringer, M.; Mathes, K.; Weidemann, G.; Winter, G.: Für einen Paradigmenwechsel bei der Bewertung ökologischer Risiken durch Chemikalien im Rahmen der staatlichen Chemikalienregulierung. Z. Angew. Umweltforsch. 11 (1998) 227–233.
22 Scheringer, M.: Persistenz und Reichweite von Umweltchemikalien. Wiley-VCH Verlag, Weinheim 1999.
23 Scheringer, M.: Persistence and Spatial Range of Environmental Chemicals – New Ethical and Scientific Concepts for Risk Assessment. Wiley-VCH Verlag GmbH, Weinheim 2002.
24 Berg, M.; Scheringer, M.: Problems in Environmental Risk Assessment and the Need for Proxy Measures. Fresenius Environ. Bull. 3 (1994) 487–492.
25 Scheringer, M.; Berg, M.: Spatial and Temporal Range as Measures of Environmental Threat. Fresenius Environ. Bull. 3 (1994) 493–498.
26 Scheringer, M.; Hungerbühler, K.: Exposure-based and Effect-based Environmental Risk Assessment for Chemicals: Two Complementary Approaches. Proc. ECO-INFORMA '97, Information and Communication in Environmental and Health Issues. Munich, October 6–9, 1997. Munich 1997. pp. 173–178.

27 Umweltbundesamt (Berlin): Nachhaltigkeit und Vorsorge bei der Risikobewertung und beim Risikomanagement von Chemikalien. Vol. I: Ahlers, J.; Beulshausen, T.; Bigalke, T.; Eggers, K. H.; Gies, A.; Greiner, P.; Henseling, K. O.; Mehlhorn, B.; Merkel, H.; Paulini, I.; Steinhäuser, K.; Stolzenberg, H. C.; Vormann, K.; Wiandt, S.: Neue Strategien zur ökologischen Risikobewertung und zum Risikomanagement von Stoffen. Umweltbundesamt (UBA) Texte 30/01. Berlin 2001, pp. 1–42.
28 Steinhäuser, K. G.: Hintergrund zu Ferntransport und Persistenz. Umweltbundesamt Texte 16/02. Berlin 2002, pp. 27–30.
29 The Vienna Convention on the Protection of the Ozone Layer. UNEP. Nairobi 1985. http://www.unep.ch/ozone/Treaties_and_Ratification/2A_vienna_convention.asp (May 2006).
30 The Montreal Protocol on Substances that Deplete the Ozone Layer. UNEP. Nairobi 1987. http://hq.unep.org/ozone/Montreal-Protocol/Montreal-Protocol2000.shtml (May 2006).
31 Orkin, V. L.; Villenave E.; Huie, R. E.; Kurylo, M. J.: Atmospheric Lifetimes and Global Warming Potentials of Hydrofluoroethers: Reactivity Toward OH, UV Spectra, and IR Adsorption Cross Sections. J. Phys. Chem. A 103 (1999) 9770–9779.
32 Young, C. J.; Hurley, M. D.; Wallington, T. J.; Mabury, S. A.: Atmospheric Lifetime and Global Warming Potential of a Perfluoropolyether. Environ. Sci. Technol. 40 (2006) 2242–2246.
33 Wania, F.; Mackay, D.: Global Fractionation and Cold Condensation of Low Volatility Organic Compounds in Polar Region. Ambio 22 (1993) 10–17.
34 Wania, F.; Mackay, D.: Tracking the Distribution of Persistent Organic Pollutants. Environ. Sci. Technol. 30 (1996) 390A–396A.
35 Cowan, C. E.; Mackay, D.; Feijtel, T. C. J.; van de Meent, D.; di Guardo, A.; Davies, J.; Mackay, N.: The Multimedia Fate Model: A Vital Tool for Predicting the Fate of Chemicals. Society of Environmental Toxicology and Chemistry (SETAC). Pensacola, FL, 1995.
36 Scheringer, M.: Persistent and Spatial Range as Endpoints of an Exposure-based Assessment of Organic Chemicals. Environ. Sci. Technol. 30 (1996) 1652–1659.
37 Scheringer, M.: Characterization of the Environmental Distribution Behaviour of Organic Chemicals by Means of Persistence and Spatial Range. Environ. Sci. Technol. 31 (1997) 2891–2897.
38 Wania, F.: Schadstoffe ohne Grenzen – Ferntransport persistenter organischer Umweltchemikalien in die Kälteregionen der Erde. GAIA 13 (2004) 176–185.
39 UN ECE: Convention on Long-range Transboundary Air Pollution (LRTP). Geneva 1979. http://www.unece.org/env/lrtap/welcome.html. (May 2006).
40 UN ECE: POP Protocol. Geneva 1998. http://www.unece.org/env/popsxg. (May 2006).
41 UNEP Governing Council (1995): Decision 18/3 on Persistent Organic Pollutants. http://www.pops.int/documents/background/gcdecision/18_32/gc1832en.html (May 2006).
42 UNEP: Stockholm Convention on Persistent Organic Pollutants. Geneva 2001. http://www.pops.int. (May 2006).
43 European Commission: Corrigendum to Regulation (EC) No. 850/2004 of the European Parliament and of the Council of 29 April 2004 on Persistent Organic Pollutants and Amending Directive 79/117/EEC (OJ L 158, 30.4.2004). In: OJ L229 (2004, 29 June), p. 5–22.
44 European Centre for Ecotoxicology and Toxicology of Chemicals (ECETOC): Persistence of Chemicals in the Environment. JACC No. 90. Brussels 2003.
45 European Commission: Technical Guidance Document on Risk Assessment, 2nd edition. Part II, Chapter 4.4. Joint Research Centre, Ispra, Italy 2003. http://ecb.jrc.it/existing-chemicals (May 2006).
46 U.S. EPA (1998): Proposed Category for Persistent, Bioaccumulative, and Toxic Chemicals. Federal Register 63 (192) (1998, 5 Oct.), p. 53417-53423.
See also: http://www.epa.gov/international/toxics/brochure.html (May 2006).

47 Rodan, B. D.; Pennington, D. W.; Eckley, N.; Boethling, R. S.: Screening for Persistent Organic Pollutants: Techniques to Provide a Scientific Basis for POPs Criteria in International Negotiations. Environ. Sci. Technol. 33 (1999) 3482–3488.
48 Environment Canada: Toxic Substances Management Policy. Persistence and Bioaccumulation Criteria – Final report. EN 40-499/2-1995E. Ottawa, Canada 1995. See also: Persistence and Bioaccumulation Regulation of 29 March 2000. Canada Gazette, Part II, Vol. 134, No. 7, pp. 607–612.
http://www.ec.gc.ca/CEPARegistry/regulations/detailReg.cfm?intReg=35 (August 2006).
49 Ikeda, M.; Takatsuki, M.; Yakabe, Y.; Arimoto, Y.; Fukuma, T.; Higeshikawa, K.: Experience on Persistent Organic Pollutants under the Law Concerning the Examination and Regulation of Manufacture, etc. of Chemical Substances, Japan, with Reference to Bio-degradation and Bioaccumulation. Int. Arch. Occup. Environ. Health 74 (2001) 295–301.
50 European Commission: Proposal for a Regulation of the European Parliament and of the Council Concerning the Registration, Evaluation, Authorisation and Restriction of Chemicals (REACH), Establishing a European Chemicals Agency and Amending the Directive 1999/45/EC and Regulation (EC) on Persistent Organic Pollutants. COM (2003) 644 final. Brussels, 29 Oct. 2003.
http://ec.europa.eu/enterprise/reach/index_en.htm (August 2006).
The EU Council adopted the REACH draft in the version of the Common Position of 27 June 2006 (Document 7524/06).
51 Webster, E.; Mackay, D.; Wania, F.: Evaluating Environmental Persistence. Environ. Toxicol. Chem. 17 (1998) 2148–2158.
52 Bennett, D. B.; McKone, T. E.; Matthies, M.; Kastenberg, W. E.: General Formation of Characteristic Travel Distance for Semivolatile Organic Chemicals in a Multimedia Environment. Environ. Sci. Technol. 32 (1998) 4023–4030.
53 Beyer, A.; Mackay, D.; Matthies, M.; Wania, F.; Webster, E.: Assessing Longrange Transport Potential of Persistent Organic Pollutants. Environ. Sci. Technol. 34 (2000) 699–703.
54 OECD: Report of the OECD/UNEP Workshop on the Use of Multimedia Models for Estimating Overall Environmental Persistence and Long-range Transport in the Context of PBT/POPs Assessment. Series on Testing and Assessment. Vol. 36. Paris 2002. http://www.oecd.org/ehs (May 2006).
55 OECD: Guidance Document on the Use of Multimedia Models for Estimating Overall Environmental Persistence and Long-range Transport. Series on Testing and Assessment. Vol. 45. Paris 2004. http://www.oecd.org/ehs (May 2006).
56 Fenner, K.; Scheringer, M.; MacLeod, M.; Matthies, M.; McKone, T.; Stroebe, M.; Beyer, A.; Bonnell, M.; Le Gall, A. C.; Klasmeier, J.; Mackay, D.; van de Meent, D.; Pennington, B.; Scharenberg, B.; Suzuki, N.; Wania, F.: Comparing Estimates of Persistence and Long-range Transport Potential among Multimedia Models. Environ. Sci. Technol. 39 (2005) 1932–1942.
57 Klasmeier, J.; Matthies, M.; Fenner, K.; Scheringer, M.; Stroebe, M.; le Gall, A. C.; McKone, T.; van de Meent, D.; Wania, F.: Application of Multimedia Models for Screening Assessment of Long-range Transport Potential and Overall Persistence. Environ. Sci. Technol. 40 (2006) 53–60.
58 Wegmann, F.; Cavin, L.; MacLeod, M.; Scheringer, M.; Hungerbühler, K.: The OECD Pov and LRTP Screening Model. Environmental Modelling and Software (2006), in preparation.
59 Green, N.; Bergman A.: Chemical Reactivity as a Tool for Estimating Persistence – A Proposed Experimental Approach for Measuring this Key Environmental Factor. Environ. Sci. Technol. 39 (2005) 480A–486A.
60 Schmidt-Bleek, F.; Hamann, H. J.: Priority Setting Among Existing Chemicals for Early Warning. Gesellschaft für Strahlen- und Umweltforschung mbH – Projektgruppe "Umweltgefährdungspotentiale von Chemikalien": Environmental Modelling for Priority Setting among Existing Chemicals – Workshop 11–13. Nov. 1985 München-Neuherberg

- GSF Proceedings. GFS-Bericht 40/85. ecomed verlagsgesellschaft mbH, Landsberg, Lech 1986, pp. 455–464.
61 Klöpffer, W.: Physikalisch-chemische Kenngrößen von Stoffen zur Bewertung ihres atmosphärisch-chemischen Verhaltens: Datenqualität und Datenverfügbarkeit. Beratergremium für Altstoffe (BUA): Stofftransport und Transformation in der Atmosphäre – Ein Beitrag der Atmosphärenwissenschaften für die Expositions-abschätzung – 10. BUA-Colloquium 25. November 2003. GDCh-Monographie, Bd. 28. Gesellschaft Deutscher Chemiker, Frankfurt am Main 2004, pp. 135–136.
62 Bidleman, T. F.: Atmospheric Transport and Air-surface Exchange of Pesticides. Water, Air, Soil Pollut. 115 (1999) 115–166.
63 Atkinson, R.; Guicherit, R.; Hites, R. A.; Palm, W. U.; Seiber, J. N.; de Voogt, P.: Transformation of Pesticides in the Atmosphere: A State of the Art. Water, Air, Soil Pollut. 115 (1999) 218–243.
64 Franklin, J.; Atkinson, R.; Howard, P. H.; Orlando, J. J.; Seigneur, C.; Wallington, T. J.; Zetzsch, C.: Quantitative Determination of Persistence in Air. In Klecka, G.; Boethling, B.; Franklin, J.; Grady, L.; Graham, D.; Howard, P. H.; Kannan, K.; Larson, R. J.; Mackay, D.; Muir, D.; van der Meent D. (Eds.): Evaluation of Persistence and Longrange Transport of Organic Chemicals on the Environment. SETAC Publication, Pensacola 2000, Chapter 2, pp. 7–62.
65 Meylan, W. M.; Howard, P. H.: Computer Estimation of the Atmospheric Gas-phase Reaction Rate of Organic Compounds with Hydroxyl Radicals and Ozone. Chemosphere 26 (1993) 2293–2299.
66 Meylan, W. M.; Howard, P. H.: Atmospheric Oxidation Programme (AOP) version 1.9. US EPA: EPISUIT Software: http://www.epa.gov/opptintr/exposure/docs/episuite.htm. Washington 1999.
67 Atkinson, R.: Atmospheric Oxidation. In: Boethling, R. S.; Mackay, D.: Handbook of Property Estimation Methods for Chemicals – Environmental and Health Sciences. Lewis Publishers, Boca Raton 2000, pp. 335–354.
68 Sabljic, A.; Peijnenburg, W.: Modelling Lifetime and Degradability of Organic Compounds in Air, Soil, and Waters Systems (IUPAC Technical Report). Pure Appl. Chem. 73 (2001) 1331–1348.
69 Meylan, W. M.; Howard, P. H.: A Review of Quantitative Structure-activity Relationship Methods for the Prediction of Atmospheric Oxidation of Organic Chemicals. Environ. Toxicol. Chem. 22 (2003) 1724–1732.
70 Schüürmann, G.: Modellierung der Lebensdauer und Abbaubarkeit organischer Verbindungen in Luft, Boden und Wasser. Ang. Chemie 117 (2005) 834–845.
71 Junge, C. E.: Basic Consideration about Trace Constituents in the Atmosphere as Related to the Fate of Global Pollution. In: Suffet, I. H.: Fate of Pollutants in the Air and Water Environments. John Wiley & Sons, New York 1977, pp. 7–26.
72 Bidleman, T. F.; Harner, T.: Sorption to Aerosols. In: Boethling, R. S.; Mackay, D.: Handbook of Property Estimation Methods for Chemicals – Environmental and Health Sciences. Lewis Publishers, Boca Raton 2000, pp. 233–260.
73 Goss, K. U.: The Air/Surface Adsorption Equilibrium of Organic Compounds under Ambient Conditions. Crit. Rev. Environ. Sci. Technol. 34 (2004) 339–389.
74 Goss, K. U.; Schwarzenbach RP.: Adsorption of a Divers Set of Organic Vapours on Quartz, $CaCO_3$ and α-Al_2O_3 at Different Relative Humidities. J. Colloid Interface Sci. 252 (2002) 31–41.
75 Scheringer, M.: Persistent Organic Pollutants (POPs) in the Focus of Science and Politics – Editorial. ESPR – Environ. Sci. Pollut. Res. 11 (2004) 1–2.
76 Wania, F.; Daly, G. L.: Estimating the Contribution of Degradation in Air and Deposition to the Deep Sea to Global Loss of PCBs. Atmos. Environ. 36 (2002) 5581–5593.
77 Wagner, B. O.: Zweites UBA-Fachgespräch über Persistenz und Ferntransport von POP-Stoffen, Berlin 6 and 7 September 2001. UWSF – Z. Umweltchem. Ökotox. 14 (2002) 268–270.

78 Wagner, B. O.: Zweites Fachgespräch zu Persistenz und Ferntransport von POP-Stoffen. Umweltbundesamt: UBA Texte 16/02. Berlin 2002.
79 Scheringer, M.: Part 2: Challenges for Environmental Sciences – Editorial. ESPR – Environ. Sci. Pollut. Res. 12 (2005) 186–187.
80 Behnke, W.; Hollander, W.; Koch, W.; Nolting, F.; Zetzsch, C.: A Smog Chamber for Studies of the Photochemical Degradation of Chemicals in the Presence of Aerosols. Atmos. Environ. 22 (1988) 1113–1120.
81 Behnke, W.; Nolting, F.; Zetzsch, C.: The Atmospheric Fate of Di(2-ethylhexyl-)phthalate, Adsorbed on Various Metal Oxide Model Aerosols and on Coal Fly Ash. J. Aerosol. Sci. 18 (1987) 849–852.
82 Palm, W. U.; Elend, M.; Krüger, H. U.; Zetzsch, C.: OH Radical Reactivity of Airborne Terbutylazine Adsorbed on Inert Aerosol. Environ. Sci. Technol. 31 (1997) 3389–3396.
83 Palm, W. U.; Elend, M.; Krüger, H. U.; Zetzsch, C.: Atmospheric Degradation of Semivolatile Aerosol-borne Pesticide: Reaction of OH with Pyrifenox (an Oxime-Ether), Absorbed on SiO_2. Chemosphere 38 (1999) 1241–1252.
84 Palm, W. U.; Millet, M.; Zetzsch, C.: OH Radical Reactivity of Pesticides Adsorbed on Aerosol Materials: First Results of Experiments with Filter Samples. Ecotox. Environ. Saftey 41 (1998) 36–43.
85 Anderson, P. H.; Hites, R. A.: System to Measure Relative Rate Constants of Semi Volatile Organic Compounds with Hydroxyl Radicals. Environ. Sci. Technol. 30 (1996) 301–306.
86 Anderson, P. A.; Hites, R. A.: OH Radical Reactions: The Major Removal Pathway for Polychlorinated Biphenyls from the Atmosphere. Environ. Sci. Technol. 30 (1996) 1756–1763.
87 Brubaker, Jr., W. W.; Hites, R. A.: Polychlorinated Dibenzo-p-dioxins and Dibenzofurans: Gas-phase Hydroxyl Reactions and Related Atmospheric Removal. Environ. Sci. Technol. 31 (1997) 1805–1810.
88 Brubaker, Jr., W. W.; Hites, R. A.: OH Reaction Kinetics of Gas-phase α- and γ-Hexachlorocyclohexane and Hexachlorobenzene. Environ. Sci. Technol. 32 (1998) 766–769.
89 Rühl, E.: Messung von Reaktionsgeschwindigkeitskonstanten zum Abbau von langlebigen, partikelgebundenen Substanzen durch indirekte Photooxidation. In: Umweltforschungsplan (UFOPLAN). FKZ 202 67 434. Umweltbundesamt, Berlin 2004, p. 59.
90 Krüger, H. U.; Gavrilov, R.; Lio Qing.; Zetzsch, C.: Entwicklung eines Persistenz-Messverfahrens für den troposphärischen Abbau von mittelflüchtigen Pflanzenschutzmitteln durch OH-Radikale. In: Umweltforschungsplan (UFOPLAN). FKZ 201 67 424/02. Umweltbundesamt, Berlin 2005, p. 143.
91 Sun, F.; Zhu, T.; Shang, J.; Han, L.: Gas-phase Reaction of Dichlorvos, Carbaryl, Chlordimeform, and 2,4-D-Butyl Ester with OH Radicals. Int. J. Chem. Kinet. 37 (2005) 755–762.
92 Feigenbrugel, V.; Le Person, A.; Le Calvé, S.; Mellouki, A.; Munoz, A.; Wirtz, K.: Atmospheric Fate of Dichlorvos Photolysis and OH-initiated Oxidation Studies. Environ. Sci. Technol. 40 (2006) 850–857.
93 Mandalakis, M.; Berresheim, H.; Stephanou, E. G.: Direct Evidence for Destruction of Polychlorobiphenyls by OH Radicals in the Subtropical Troposphere. Environ. Sci. Technol. 37 (2003) 542–547.
94 Behnke, W.; Zetzsch C.: Abschlussbericht zum 30.11.1989 über die Untersuchung der Reaktivität von Lindan gegenüber OH-Radikalen. Fraunhofer-Institut für Toxikologie und Aerosolforschung Hannover. Report for Umweltbundesamt, Berlin 1989.
95 Zetzsch, C.: Photochemischer Abbau in Aerosolphasen. UWSF – Z. Umweltchem. Ökotox. 3 (1991) 59–64.

Chapter 2
Abiotic Degradation in the Atmosphere

1
Introduction

In the atmosphere, organic molecules can only be degraded abiotically. With the exception of the droplet phase, where hydrolytic processes are likely to occur, only indirect and direct photochemical reactions are efficient sinks[1)] [1]. The emphasis of this chapter is on degradation in the gas phase of the troposphere, the lower (10–15 km) layer of the atmosphere. There is much experimental data available for the reactions dominant in this phase, which are presented in Chapter 3 (Table of degradation reaction rate constants and of photo-degradation processes). Abiotic degradation of molecules adsorbed onto aerosol particles has been investigated much less, but is discussed within the framework of the basis of the small amount of concrete knowledge that there is on these reactions. The situation is even worse for the droplet phase, and the quantitative importance of this phase for the degradation of organic molecules is not yet clear. It should be investigated, however, and taken into account in the fate and transport assessments of chemicals in the future.

Interest in the quantification of the relevant reactions originated from three main areas of environmental research and policy:

1. Analysis and computation of *photochemical smog formation* (tropospheric formation of ozone and other photo-oxidants in the boundary layer) [2, 14].
2. Quantification of the *abiotic degradation of chemicals in the atmosphere* (fate and transport modelling, assessment of persistent organic pollutants – POPs – and semi-volatile organic compounds – SOCs) [3, 4].
3. Estimation of the contribution of chemicals to global environmental processes that contribute to climate change (*stratospheric ozone depletion and global warming*) [5–7].

---

1) A sink is defined here as an irreversible chemical, photochemical or biochemical reaction which removes a molecule from the environment; a perfect sink leads to mineralization. Within this definition, the transfer to and the temporary fixation in another compartment (e.g., soil, vegetation or water) without chemical modification is not considered to be a sink [1].

Atmospheric Degradation of Organic Substances. W. Klöpffer and B. O. Wagner
Copyright © 2007 WILEY-VCH Verlag GmbH & Co. KGaA, Weinheim
ISBN: 978-3-527-31606-9

For these three areas of environmental concern, in general the same rate constants are required (specifically k_{OH}), but the assessment differs: whereas in smog formation (1) the highly reactive substances are the "bad guys" (strong photo-oxidant formation potential), the contrary is the case in the assessment of environmental chemicals (2) (persistence, long-range transport) and in stratospheric ozone depletion (3); in assessing the potential contribution to climate change (3), abiotic degradation – or rather the lack of it, i.e., persistence – contributes to global warming in combination with IR absorption in the spectral window between about 10 and 15 µm.

The abiotic degradation processes can be divided into:
- indirect photochemical processes, including dark reactions by reactive species formed in photochemical reaction cycles;
- direct photochemical transformations, often incorrectly called "photolysis"[2)]
- degradation in the adsorbed state, the special case of semi-volatile organic compounds;
- degradation in the droplet phase.

The availability of data is best for indirect photochemical processes in the gas phase, including the reactions with the hydroxyl radical (OH), the nitrate radical (NO_3) and ozone (O_3). These reactions and the direct photochemical reactions are discussed in Section 2, followed by the heterogeneous degradation (Section 3) and a short description of the measurement methods (Section 4). Additional information necessary for calculating lifetimes is presented in Section 5.

2
Photo-degradation in the Homogenous Gas Phase of the Troposphere

2.1
Indirect Photochemical Reactions

2.1.1 The Reaction with OH-Radicals

2.1.1.1 Sources and Sinks of the OH-Radical

The importance of the OH-radical as a very reactive intermediate – even in remote areas of the troposphere – was recognised in 1971 by Levy [8]. The role of OH in polluted air ("photochemical smog") had already been recognised 10 years earlier by Leighton [9].

The most important tropospheric *source* of OH-radicals is the photolysis of ozone, Eq. (2.1):

$$O_3 + h\nu \ (\lambda < 320 \text{ nm}) \rightarrow O^*(^1D) + O_2^*(^1\Delta_g) \tag{2.1}$$

---

2) Not all photochemical transformations are photolytic reactions, e.g., isomerisations.

Ozone is ubiquitous in the troposphere (see Section 5) and is photochemically split by visible light and by UV radiation. Radiation of low wavelengths, corresponding to high energy per photon, creates oxygen atoms in an excited state, which in turn react with water molecules always present in the atmosphere, Eq. (2.2); for reasons of spin conservation, the oxygen molecule is also formed in an excited state ("singlet oxygen"). Ground-state oxygen atoms, as formed by longer wavelength ozone splitting, cannot react with water.

$$O^*(^1D) + H_2O \rightarrow 2\ OH \tag{2.2}$$

The upper limit of the quantum efficiency of O^*-formation ($\Phi' = 0.9$), which is close to zero above 320 nm, is reached at about 305 nm [10]. Reaction (2.2) therefore strongly depends on the spectral distribution of the UV radiation at the stratospheric absorption edge close to 290 nm. This edge is due to the absorption of UV radiation by stratospheric ozone, which prevents shorter wavelengths reaching the troposphere and the surface of the Earth.

The second source of OH-radicals, photolysis of nitrous acid molecules, Eq. (2.3), is spectrally more favourable due to the HNO_2 absorption in the near-UV (300–400 nm); this type of OH-formation can play a role during the mornings until the reservoir of nitrous acid, built up during the night, is exhausted:

$$HO-NO + h\nu \rightarrow OH + NO \tag{2.3}$$

A third source of tropospheric OH is the reaction of photochemically formed hydroperoxyl (OOH) radicals with nitrogen monoxide [11a], Eq. (2.4):

$$OOH + NO \rightarrow OH + NO_2 \tag{2.4}$$

The most important *sink* of OH is carbon monoxide, as already recognised by Levy [8] in 1971, Eq. (2.5):

$$OH + CO \rightarrow CO_2 + H \tag{2.5}$$

Reaction (2.5) is relatively slow, the bimolecular reaction rate constant is

$$k_{OH}(CO) = 1.7 \times 10^{-13}\ cm^3\ molecule^{-1}\ s^{-1}\ ^{3)}$$

at 200 hPa and 296 K [12]; it depends both on pressure and temperature [13][4].

---

3) The most commonly used unit for second-order rate constants of gas-phase reactions is [$cm^3\ molecule^{-1}\ s^{-1}$], which is based on the (non-SI) unit of concentration [$molecule\ cm^{-3}$]; the SI base unit would be [$mol\ m^{-3}$], giving a unit of [$m^3\ mol^{-1}\ s^{-1}$], which is unusual in gas-phase kinetics. In liquid-phase kinetics, the derived unit [$L\ mol^{-1}\ s^{-1}$] is used for second-order rate constants.
4) The temperature dependence at low pressure ($p \leq 10$ hPa) under Ar is given by Frost et al. [13]: $k_{OH}(T) = 2.81\ E\text{-}13\ \exp.(-E_a/RT)\ cm^3\ molecule^{-1}\ s^{-1}$; $E_a = 1.5\ kJ\ mol^{-1}$.

If used together with an estimate of the total source strength, the sink strength can be used for an estimate of the globally and annually averaged tropospheric concentration of the OH-radicals. Several estimates of this type have produced values between 10^5 and 10^6 radicals cm^{-3} [6]. A more detailed discussion of this important concentration, which can be used to estimate tropospheric lifetimes of molecules, is given in Section 5. In the following, a conservative (i.e., not underestimating the persistence) value of

$$<[OH]> \text{ (global, annual average)} = 5 \times 10^5 \text{ cm}^{-3}$$

is used [1, 14–16, 194]. This value lies within the range reported by BUA [17]:

$$<[OH]> = (6 \pm 2) \times 10^5 \text{ cm}^{-3}$$

The OH-radical is very short-lived due to its high reactivity with many molecules present in the atmosphere. Even in pure air, the reaction with CO yields an average lifetime in the order of seconds, Eq. (2.6). Any other reaction reduces the OH lifetime further.

$$\tau_{OH} = 1/k^I_{OH} = 1/(k_{OH} <[CO]>) \approx 2 \text{ s} \tag{2.6}$$

τ_{OH} = average lifetime of an OH-radical; time after which the concentration of OH-radicals is decreased to 1/e (37%) after terminating the production of new radicals; reciprocal of the first-order rate constant of OH decay.

In Eq. (2.6), the average concentration of CO (mixing ratio ≈ 0.1 ppmv [15]) is taken to be $<[CO]> = 2.7 \times 10^{12}$ cm^{-3}. Owing to this short lifetime, the OH-induced reactions disappear during the night, except for traces of OH that may be formed even in the dark in the presence of ozone.

2.1.1.2 Reactions of OH with Organic Compounds

The high reactivity of the OH-radical is due to the high enthalpy of the H–O bond in water [formed in H-abstraction reactions, see Eq. (2.7)]:

$$\Delta H \text{ (HO–H)} = (498 \pm 4) \text{ kJ mol}^{-1} \text{ [18]}$$

and to the strong electrophilic behaviour of the radical. The most important reactions of OH are, therefore, the H-abstraction and the addition to π-electron systems (double bonds, aromatic molecules) [11, 19]. H-Abstraction by OH, Eq. (2.7), is more efficient the weaker the H-atom to be abstracted is bound [19].

$$RH + OH \rightarrow R + HOH \tag{2.7}$$

If an organic molecule has several types of H-atoms with different reactivities towards OH, it is the one that is bound weaker that reacts faster, e.g., in methanol, Eq. (2.7a):

Path a: ca. 90%

$$CH_3OH + {}^{\bullet}OH \rightarrow {}^{\bullet}CH_2OH + H_2O \qquad (2.7a)$$

Path b: ca. 10%

$$CH_3OH + {}^{\bullet}OH \rightarrow CH_3O^{\bullet} + H_2O \qquad (2.7b)$$

The *measured* reaction rate constant $k_{OH}(CH_3OH)$ results from an average of reactions (2.7a) and (2.7b) and amounts at room temperature[5] to

$$k_{OH}(CH_3OH) = 9.4 \times 10^{-13} \text{ cm}^3 \text{ mol}^{-1} \text{ s}^{-1}$$

The quantitative identification of the primary product ($CH_3O^{\bullet}$) allowed the conclusion that reaction (2.7b) contributes only 10% to the total reaction [21]. This finding is in accord with the fact that the enthalpy of CH_3O-H is 40 kJ mol^{-1} higher than that of $H-CH_2OH$ [18]; the reaction path (Eq. 2.7a) is therefore energetically favoured relative to Eq. (2.7b).

OH addition, Eq. (2.8), is especially important for aromatic compounds, e.g., for derivatives of benzene (for example, toluene) and polycyclic aromatic hydrocarbons (for example, anthracene).

$$ArH + OH \leftrightarrow ArHOH \qquad (2.8a)$$
$$\hookrightarrow Products \qquad (2.8b)$$

Ar = aryl, e.g., C_6H_5 (phenyl); ArH, e.g., C_6H_6 (benzene)

Frequently it is observed that the first step (OH addition; Eq. 2.8a) is reversible, so that the total reaction becomes complicated and may show unusual pressure and temperature profiles (not simple Arrhenius behaviour) [22, 23].

Aromatic compounds with aliphatic side groups (e.g., toluene, ethylbenzene) or cycloaliphatic rings (e.g., tetraline) can react according to the H-abstraction mechanism, Eq. (2.7), and according to the OH-addition mechanism (Eq. 2.8).

The general importance of OH as the "detergent" of the atmosphere relies on the fact that *all* compounds showing H-atoms or double bonds react with OH, albeit with reaction rate constants varying over a wide range from about 10^{-16} to 10^{-10} cm^3 molecule^{-1} s^{-1}. The upper limit roughly corresponds to the diffusion limit (each encounter leads to a reaction). Only saturated perhalogenated (F, Cl, Br) aliphatic hydrocarbons (and SF_6) seem to be totally persistent towards the attack of OH-radicals; for these compounds, only the upper limits of k_{OH} are given in the literature.

---

5) Throughout this chapter, rate constants without citation are taken from the Table of reaction rate constants in Chapter 3 (293–298 K, normally measured at 298 K).

According to Atkinson et al. [24], the upper limit of k_{OH} for Freon 11 (CCl_3F) at room temperature amounts to

$$k_{OH} (CCl_3F, 298\ K) \leq 5 \times 10^{-18}\ cm^3\ molecule^{-1}\ s^{-1}$$

The lower limit of the chemical lifetime in the troposphere, related to the reaction with OH, is calculated as follows [1] (see also Section 5):

$$\tau_{OH} (CCl_3F) = \frac{1}{k_{OH} \times <[OH]>} \geq \frac{10^{18}}{5 \times 5 \times 10^5} = 4 \times 10^{11}\ s \approx 13\,000\ a$$

Similar values are obtained for Freon 12 (CCl_2F_2). It should be noted that this is not the residence time, which in the case of highly persistent, not particularly water-soluble gases is controlled by the transfer time into the stratosphere (about 10 years)[6].

At the other end of the OH-reactivity scale there are, for example, the terpenes with α-terpinene as a very reactive example:

$$k_{OH} (\alpha\text{-terpinene}, 294\ K) = (3.6 \pm 0.4) \times 10^{-10}\ cm^3\ molecule^{-1}\ s^{-1}$$

The chemical lifetime during the day ($[OH] \approx 10^6\ cm^{-3}$) amounts to:

$$\tau_{OH,day} = \frac{1}{k_{OH} \times [OH]} = \frac{10^{10}}{3.6 \times 10^6} = 2780\ s \approx 46\ min$$

A similar very high reactivity is shown by the aromatic amines, e.g., N,N-dimethylaniline:

$$k_{OH}\ [C_6H_5N(CH_3)_2,\ 278\text{–}298\ K] = 1.5 \times 10^{-10}\ cm^3\ molecule^{-1}\ s^{-1}$$

Alkenes, e.g., propene and ethene, react with OH an order of magnitude slower. Typical reaction rate constants for these compounds are about $10^{-11}\ cm^3\ molecule^{-1}\ s^{-1}$. Using 5×10^5 radicals cm^{-3} as the average global OH concentration, this rate constant corresponds to an average lifetime of 200 000 s or approximately 2 days.

Naphthalene, initially measured in a smog chamber,

$$k_{OH} (naphthalene, 300\ K) = 2.0 \times 10^{-11}\ cm^3\ molecule^{-1}\ s^{-1}\ [25]$$

has about the same reactivity and, hence, chemical OH lifetime. The kinetics of this reaction are dominated by the reversible formation of an OH adduct according to Eq. (2.8a) and are similar to the kinetics of the reaction with benzene [22].

---

6) The characteristic time for mixing within one hemisphere is about 1 month, for interhemispheric mixing about 1 year and for troposphere into the stratosphere about 10 years (near the equator, there are processes that allow a rapid transfer, the average transfer is much slower).

Benzene reacts much slower, however:

$$k_{OH} \text{ (benzene)} = 1.4 \times 10^{-12} \text{ cm}^3 \text{ molecule}^{-1} \text{ s}^{-1}$$

This corresponds to an average (global) OH-lifetime of

$$\tau_{OH} = \frac{1}{k_{OH} \times \langle [OH] \rangle} = \frac{10^{12}}{1.4 \times 5 \times 10^5} \text{ s} \approx 17 \text{ d}$$

This value is typical for many substances of average reactivity and can, in the absence of other sinks, lead to long-range transport in the troposphere, at least within one hemisphere.

Hexafluorobenzene reacts slower than benzene by a factor of 10 and shows simple (Arrhenius-type) kinetics; it is thus concluded that no intermediate OH-adduct formation is involved in this reaction [26].

$$k_{OH} \text{ (C}_6\text{F}_6\text{)} = 1.7 \times 10^{-13} \text{ cm}^3 \text{ molecule}^{-1} \text{ s}^{-1}$$

Some reactions with OH show a pronounced dependence on the pressure used in measuring k_{OH}, e.g., the reaction with acetylene, Eq. (2.9):

$$\text{OH} + \text{CH} \equiv \text{CH} \rightarrow \text{products} \tag{2.9}$$

At normal pressure, e.g., in a smog chamber, this reaction gives

$$k_{OH} \text{ (C}_2\text{H}_2\text{, 300 K)} = 8.1 \times 10^{-13} \text{ cm}^3 \text{ molecule}^{-1} \text{ s}^{-1}$$

At low pressures, however, much lower rate constants are measured, suggesting much too high a persistence in the real atmosphere (troposphere).

Substances with low reactivity towards OH are often halogenated. Acetonitrile is an example of a non-halogenated, persistent compound [20] (CH$_3$C≡N; AcN):

$$k_{OH} \text{ (AcN, 298 K)} = 2.6 \times 10^{-14} \text{ cm}^3 \text{ molecule}^{-1} \text{ s}^{-1}$$

For this substance, the average OH lifetime amounts to

$$\tau_{OH} = \frac{1}{k_{OH} \times \langle [OH] \rangle} = \frac{10^{14}}{2.6 \times 5 \times 10^5} \text{ s} \approx 2.5 \text{ a}$$

Substances with such or even lower OH-reaction rate constants can be distributed on a global scale (both hemispheres in the absence of other sinks, or efficient physical removal from the troposphere) and may enter the stratosphere. As a final example of a non-fluorinated, non-perhalogenated solvent, 1,1,1-trichloroethane ("methylchloroform", MCF), k_{OH} is presented as a function of temperature for the range 243–379 K [27]:

$$k_{OH}(T) = (2.25 + 0.46 - 0.39) \times 10^{-18}\, T^2 \exp\{(-910 \pm 56)/T\}\ \text{cm}^3\ \text{molecule}^{-1}\ \text{s}^{-1}$$

$$k_{OH}(298\ \text{K}) = 9.43 \times 10^{-15}\ \text{cm}^3\ \text{molecule}^{-1}\ \text{s}^{-1}$$

More interesting for the average OH lifetime is k_{OH} at the (weighted or effective) average temperature of the troposphere. This temperature has been estimated by Warneck [15] in an analysis of the tropospheric $CH_4 + OH$ reaction as $<T>_{\text{eff}} = 280$ K. According to an earlier estimate, based on the lifetime of chlorinated hydrocarbons [34], this temperature may be as low as 260 K. Using the more recent estimate by Warneck, we obtain

$$k_{OH}(280\ \text{K}) = 6.84 \times 10^{-15}\ \text{cm}^3\ \text{molecule}^{-1}\ \text{s}^{-1}$$

The average tropospheric OH lifetimes of MCF is calculated to be

$$\tau_{OH} = 9.3\ \text{a}$$

This value is higher than the average residence times calculated for MCF, which are based on long-term measurements of MCF concentrations and estimates of the amounts emitted yearly (see Section 5).

MCF is photolysed at wavelengths around 200 nm [28] and can therefore act as a source of chlorine atoms in the stratosphere and contribute to ozone destruction. It has been phased out as a result of the Protocol of Montreal and its follow-up agreements.

An up-to-date data collection of OH-reaction rate constants can be found in Chapter 3. Earlier critical data collections are mostly attributable to Roger Atkinson, whose group also measured many substances for the first time [11a, 11b, 19, 20, 27, 147, 148].

2.1.2 The Reaction with NO_3-Radicals

2.1.2.1 Sources and Sinks of the NO_3-Radical

The tropospheric sink "reaction with NO_3-radical" has been studied since about 1980 [29, 30]. The sink is active only during the night, as the NO_3 radical absorbs light in the visible region of the spectrum and is photolytically unstable. A comparison with the other main sinks in the gas phase of the troposphere shows that this is a peculiarity of the nitrate radical:

Sink	Time of activity
$h\nu$ (direct) and OH:	Only during the day
O_3:	Day and night
NO_3:	Only during the night

NO$_3$ is produced by the reaction of NO$_2$ with ozone, Eq. (2.10):

$$NO_2 + O_3 \rightarrow NO_3 + O_2 \tag{2.10}$$

NO$_3$ was first measured in 1980 by Platt et al. [30] in the troposphere during the night. As NO is also oxidised by O$_3$ to NO$_2$ [in analogy with Eq. (2.10)], the nitrate radical can always be formed if NO$_x$ (= NO + NO$_2$) and ozone are present in the atmosphere. This is the case particularly during smog events, but also must not be neglected in moderately polluted air masses. Thus, NO$_3$ is an important component in smog during the night, albeit not in an absolute sense (mixing ratio, concentration), but rather with regard to its high reactivity [2, 29, 31, 32]. NO$_3$ is a reactive radical as is OH, and shows several similarities with the dominant daytime radical. It has therefore been nicknamed *"the OH of the night"* [33a].

Early measurements of the mixing ratios of NO$_3$ were in the range of 10–100 pptv[7]. At sea level and 0 °C this corresponds to a range of concentrations

$$[NO_3] = 2.7 \times 10^8 - 2.7 \times 10^9 \text{ NO}_3 \text{ radicals cm}^{-3}$$

In strongly polluted air, [NO$_3$] as a function of time shows a sharp maximum in the early hours of the night, whereas in relatively clean air a more constant trend at a lower level is observed [32]. During the MINOS campaign, Crete, summer 2001 (see Section 5.2.3) [213], the maximum was observed after midnight, the average NO$_3$ concentration measured was about 10^8 cm^{-3}.

Nitrogen trioxide and nitrogen dioxide form an equilibrium in the absence of light (Eq. 2.11):

$$NO_2 + NO_3 + M \leftrightarrow N_2O_5 + M \tag{2.11}$$

M = inactive molecule.

This equilibrium (Eqs. 2.11 and 2.12) strongly depends on temperature.

$$K(T) = \frac{[N_2O_5]}{[NO_2] \times [NO_3]} \tag{2.12}$$

---

7) The mixing ratio pptv = parts per trillion (per volume), 10^{-12}; other commonly used volume mixing ratio units are volume percent (10^{-2}), ppmv = parts per million (10^{-6}) and ppbv = parts per billion (10^{-9}). The mixing ratios are very useful in atmospheric chemistry, as they do not change with pressure (and height); they cannot be used in kinetics, however, because the actual concentrations required in kinetic equations depend on the pressure. When assuming ideal gas behaviour, for given pressure (p) and temperature (t), mixing ratios are proportional to partial pressures and mole fractions. They can easily be converted into gas-phase concentrations [molecule cm^{-3}]: conc. [molecule cm^{-3}] = mixing ratio $N_Lp/(RT)$. Thus, at 298 K and $p = 10^5$ Pa, atmospheric concentrations and mixing rations are related by conc. [molecule cm^{-3}] = $2.43 \times 10^{19} \times$ mixing ratio [146b, p. 673].

Chapter 2: Abiotic Degradation in the Atmosphere

Table 1 Dependence of K ($NO_2 + NO_3 = N_2O_5$) on temperature.

T [K]	T [°C]	Equilibrium constant K [cm^3 molecule^{-1}]
260	−13	7.38×10^{-9}
273	0	9.85×10^{-10}
280[a)]	7	3.60×10^{-10}
293	20	6.31×10^{-11}
298	25	3.36×10^{-11}
310	37	8.07×10^{-12}
320	47	2.67×10^{-12}

a) Weighted average temperature of the troposphere [15].

The quantitative relationship between K and the absolute temperature T is given by the empirical equation (2.13) [36]

$$K = 2.4 \times 10^{-27} \, (T/300)^{0.32} \, e^{11080/T} \, cm^3 \, molecule^{-1} \tag{2.13}$$

Table 1 shows the values of K (Eq. 2.12), calculated according to Eq. (2.13) for selected temperatures.

The data in Table 1 show that high temperature favours the formation of NO_3 radicals so that the sink is (more) effective in the lower layer of the troposphere and in warm regions.

Owing to the equilibrium in Eq. (2.13) and the hydrolysis in Eq. (2.14), N_2O_5 is linked to nitric acid, so that water is indirectly – adding to the daytime photolysis – an important sink for the nitrate radical.

$$N_2O_5 + H_2O \rightarrow 2 \, HNO_3 \tag{2.14}$$

In addition, many organic substances, including naturally occurring ones (e.g., terpenes) act as sinks for this highly reactive species.

As in the case of OH, the average concentration <[NO_3]> cannot be determined easily (see Section 5.2.2). In the following examples we adopt a preliminary night-average of

$$<[NO_3]>_{night} \approx 10^8 \, cm^{-3}$$

corresponding to an average mixing ratio at the surface of the earth of 5 pptv. The 24-hour average can thus be assumed to be about

$$<[NO_3]> \, (global, annual \, average) \approx <[NO_3]>_{night} / 2 \approx 5 \times 10^7 \, cm^{-3}$$

2.1.2.2 Reactions of NO_3 with Organic Compounds

The reactions of NO_3 follow, in principle, similar mechanisms to those of the OH-radical:

- addition to double bonds (unsaturated compounds and aromatics),
- H-abstraction under formation of nitric acid in the case of saturated compounds, Eq. (2.15).

$$RH + NO_3 \rightarrow R + HNO_3 \qquad (2.15)$$

NO_3 is *not* a nitration agent – in contrast to N_2O_5; hence, no increase in the persistence due to the formation of nitro-aromatic compounds is to be expected, as observed in the reaction of dinitrogen pentoxide with PAH [14]. *The atmospheric species NO_3 is – as is OH – a highly reactive electrophilic radical.* It can thus contribute significantly to the degradation of organic compounds in the troposphere, despite its very low average concentration or mixing ratio.

The reaction of NO_3 with aromatic compounds frequently leads, also similar to the case of OH, in the first step to the formation of loose complexes, which may dissociate without permanent change to the molecule. This behaviour has been observed, for instance, in the case of naphthalene + NO_3 [38].

As with OH and O_3, terpenes belong to the most reactive substances, e.g., d-limonene:

$$k_{NO_3}\text{ (d-limonene)} = 1.2 \times 10^{-11} \text{ cm}^3 \text{ molecule}^{-1} \text{ s}^{-1}$$

Using the "night-average" $<[NO_3]>_{night} \approx 10^8 \text{ cm}^{-3}$, the following lifetime (night-hours only) can be estimated:

$$\tau_{NO_3} = \frac{1}{k_{NO_3} \times <[NO_3]>_{night}} = \frac{10^{11}}{1.2 \times 10^8} \text{ s} \approx 14 \text{ min}$$

This chemical lifetime is shorter than those calculated for ozone (40 min) and OH (2 h daytime) by Atkinson et al. [36].

The reaction of NO_3 with azulene is even faster [39]:

$$k_{NO_3}\text{ (azulene)} = 3.9 \times 10^{-10} \text{ cm}^3 \text{ molecule}^{-1} \text{ s}^{-1}$$

corresponding to τ_{NO_3} (night) ≈ 25 s

Aliphatic olefins are also very reactive towards NO_3, e.g., isobutene (2-methylpropene):

$$k_{NO_3}\text{ (isobutene)} = 3.3 \times 10^{-13} \text{ cm}^3 \text{ molecule}^{-1} \text{ s}^{-1}$$

This value was measured by means of an absolute method by Canosa-Mas et al. [40], and is in very good agreement with the relative value of Atkinson et al. [41].

τ_{NO_3} (night) = 30 300 s ≈ 8.4 h

Alkanes ($k_{NO_3} \approx 10^{-16}$ cm^3 molecule^{-1} s^{-1} [41]) and benzene-type aromatic hydrocarbons, e.g., toluene, are much less reactive towards NO_3:

$$k_{NO_3} \text{(toluene)} = 6.8 \times 10^{-17} \text{ cm}^3 \text{ molecule}^{-1} \text{ s}^{-1}$$

The corresponding chemical lifetime, calculated with the global day/night average, amounts to about 9 years. A similar reaction rate constant has been measured for acetylene [40].

Organic sulphides [40, 42], e.g., the atmospheric species dimethyl sulphide (DMS), are fairly reactive towards NO_3:

$$k_{NO_3} \text{(DMS)} = 7.0 \times 10^{-13} \text{ cm}^3 \text{ molecule}^{-1} \text{ s}^{-1}$$

This corresponds, using the night average, to an NO_3-related lifetime of about 4 hours. This was one of the research problems to be solved by the MINOS-experiment [213]: what role does NO_3 play in the tropospheric DMS budget?

An up-to-date collection of NO_3 reaction rate constants can be found in Chapter 3; earlier data collections can be found in Wayne et al. [29] and Atkinson [27, 43].

2.1.3 The Reaction with Ozone

2.1.3.1 Sources and Sinks of O_3 in the Troposphere

Ozone is a ubiquitous, oxidizing trace compound in the troposphere. Two *sources* of ozone have been known for a long time [2, 9, 14, 44]:

(1) Formation of ozone in the stratosphere with short wavelength UV (Eqs. 2.16 and 2.17), followed by inflow down into the troposphere, predominantly through the "crevices" of the tropopause.

$$O_2 + h\nu \rightarrow 2\,O \qquad (2.16)$$

$$O + O_2 + M \rightarrow O_3 + M \qquad (2.17)$$

The surface of the earth acts as a sink for ozone, soil being more effective than water and snow. This can be recognized in the deposition velocities [44] to soil ($v = 0.6$ cm s^{-1}), water ($v = 0.04$ cm s^{-1}) and snow ($v = 0.02$ cm s^{-1}).

(2) Photochemical smog in the boundary layer, especially near the surface of the earth in polluted air, contains ozone. It requires the simultaneous presence of NO_x and volatile organic substances (VOC), especially unsaturated hydrocarbons. In this formation pathway, the necessary oxygen atoms, Eq. (2.17), are formed by photolysis of the red gas nitrogen dioxide, Eq. (2.18):

$$NO_2 + h\nu \rightarrow NO + O \tag{2.18}$$

The oxygen atoms react with O_2 to yield O_3, Eq. (2.17). The ozone thus formed is, however, reduced by NO, unless NO is removed from the reaction mixture. It is a central requirement of the smog chemistry that NO is oxidized to NO_2 by peroxy radicals, Eqs. (2.19) and OOH (2.20). These species originate from complex reaction cycles driven by VOCs, NO_2 and light [2, 14].

$$ROO + NO \rightarrow RO + NO_2 \tag{2.19}$$

$$OOH + NO \rightarrow OH + NO_2 \tag{2.20}$$

In the early phase of smog formation, the reactive mixture is composed of NO_x (mostly in the form of NO) and VOCs. Mature smog is characterized by NO_2 and O_3 until NO_2 is finally converted into HNO_3 and numerous other photo-oxidants (e.g., peroxyacetyl nitrate). In these later stages, the chemistries of ozone and the nitrate radical (during the night) are interwoven (see Section 2.1.2.1).

A third source or formation mechanism of ozone has been detected more recently:

(3) The asymmetry in the ozone mixing ratios observed between the northern and the southern hemispheres could not be explained with the two sources (1 and 2); Fishman and Crutzen [44] proposed that *ozone formation also occurs in remote areas*. As there are not enough ROO radicals present in the pure troposphere, OOH has to carry out the oxidation alone, Eq. (2.20). According to this theory, OOH stems from the reaction CO + OH (2.5), followed by the reaction $H + O_2 = OOH$; see also [45].

Sinks

The removal of ozone in the troposphere occurs (in addition to NO and other reductive agents) photochemically [46]. The photolysis of ozone (Eq. 2.21) does not require much energy $\{\Delta H (O_2 - O) = 101$ kJ mol^{-1} [46]$\}$, corresponding to about 1 eV molecule^{-1} or near-IR[8] radiation.

$$O_3 + h\nu \ (\lambda \leq 1.2 \ \mu m) \rightarrow O_2 + O \tag{2.21}$$

The (average) ozone concentrations vary greatly depending on the relative strength of the formation and degradation processes. Atkinson and Carter [47] proposed an average tropospheric mixing ratio of 30 ppbv, corresponding to a concentration near the surface of the earth (see also the detailed discussion in Section 5.2.3):

$$<[O_3]>_{global} \approx 7 \times 10^{11} \text{ molecule cm}^{-3}$$

---

8) 1 eV molecule^{-1} (96.487 kJ mol^{-1}) corresponds to λ = 1240 nm = 1.24 μm.

This value is much lower than the ozone concentration in a mature photochemical smog after the disappearance of NO (Section 5.2.3). Furthermore, it is about a million times higher than the average (global) concentration of OH. If the significant range of the reaction rate constants (OH) is in the region of

$$k_{OH} \approx 10^{-10} - 10^{-14} \text{ cm}^3 \text{ molecule}^{-1} \text{ s}^{-1}$$

the range of relevant reaction rate constants for ozone is, accordingly,

$$k_{O_3} \approx 10^{-16} - 10^{-20} \text{ cm}^3 \text{ molecule}^{-1} \text{ s}^{-1}$$

Bimolecular reaction rate constants $k_{O_3} \leq 10^{-20}$ [cm^3 molecule^{-1} s^{-1}] have no significance for the degradation of organic molecules in the troposphere [47], i.e., such compounds can be considered to be persistent with regard to ozone. Another consequence of the much lower reactivity of ozone relative to OH and NO$_3$ is the absence of a pronounced day/night variation.

2.1.3.2 Reactions of O₃ with Organic Compounds

In contrast to OH and NO$_3$, ozone is not a highly reactive and strongly electrophilic radical. Rather, it is a moderately reactive oxidizing molecule in which one oxygen atom can split off and be transferred to other molecules.

The most important primary reaction of ozone with organic molecules is the addition of oxygen to double bonds [9, 14, 47]. This reaction, discovered by Harries in 1904, is a key reaction in organic chemistry. The reaction mechanism was elucidated by Criegee, Staudinger and Rieche as cited in Ref. [14] for the reaction in water-free solvents. According to this theory, the primary reaction leads to an addition product (primary ozonide), which is split in a secondary reaction into carbonyl compounds (Criegee's intermediate) and finally splits off into aldehydes and ketones, depending on the structure of the compound reacting with O$_3$. In the gas phase, the reaction is expected to follow a similar path.

The quantitative product spectrum of the reaction isoprene + ozone in the gas phase was investigated by Paulson, Flagan and Seinfeld [48, 49].

An example will be given for a substance very reactive towards ozone, d-limonene (see also Section 2.1.2.2) [25, 47]:

$$k_{O_3} = 2.0 \times 10^{-16} \text{ cm}^3 \text{ molecule}^{-1} \text{ s}^{-1}$$

The ozone-related chemical lifetime of this natural chemical is calculated to be (using the global average)

$$\tau_{O_3} = \frac{1}{k_{O_3} \times <[O_3]>_{global}} = \frac{10^{16}}{2.0 \times 7 \times 10^{11}} \text{ s} \approx 2 \text{ h}$$

Limonene and other naturally occurring terpenes are – in the presence of NO$_x$ – very probably involved in smog formation. As the actual concentration of ozone during smog episodes is much higher than the average value taken above (ca. 300 ppbv ≈ 7×10^{12} molecule cm^{-3}) [14], the actual, ozone-related lifetime of limonene shortens to τ_{O_3} (smog) ≈ 10 min.

Among the technical chemicals, propene has to be mentioned as being very reactive towards ozone. As shown in smog-chamber experiments, the ozone reaction may actually dominate over the OH reaction [25] at elevated values of [O_3]:

$$k_{O_3} \text{(propene, 298 K)} = 1.0 \times 10^{-17} \text{ cm}^3 \text{ molecule}^{-1} \text{ s}^{-1}$$

The resulting ozone-related lifetime (at <[O_3]>$_{global}$) amounts to 1.6 days. As in all other examples, it would be better to use the reaction rate constant at 280 K (average temperature of the troposphere [15]). As the temperature dependence of the rate constants is not known in all instances, the room temperature values are used instead. For propene, k_{O_3} (T) is known for the range 230–370 K and can be fitted by means of an Arrhenius-function[9)] with $A = 5.5 \times 10^{-15}$ cm^3 molecule^{-1} s^{-1} and E_a = 15.6 kJ. The reaction rate constant for 280 K is calculated to be 5.9×10^{-18} cm^3 molecule^{-1} s^{-1} and the corresponding lifetime as 2.8 days.

Another important industrial (carcinogenic) chemical is vinyl chloride (VC) [25, 47]:

$$k_{O_3} \text{(VC)} = 2.4 \times 10^{-19} \text{ cm}^3 \text{ molecule}^{-1} \text{ s}^{-1}$$

corresponding to an ozone-related lifetime (at <[O_3]>$_{global}$) of about 70 days. This shows a long-range transport potential, otherwise, OH and/or NO$_3$ reduce the overall lifetime (Section 5.1).

Chemicals without aliphatic double bonds, in general, do not show reaction rate constants in the relevant range above 10^{-20} cm^3 molecule^{-1} s^{-1} [47]. Exceptions are the phenols, nitrites and amines.

The actual degradation in the troposphere, especially above the boundary layer, depends on the thermal activation of the reactions, as shown in the example of propene. This should be taken into account in assessing slowly degrading compounds that are likely to be distributed throughout the whole troposphere.

A list of measured k_{O_3} data is given in Chapter 3. Earlier compilations have been published by Atkinson and Carter [47] and Atkinson [27].

2.2
Direct Photochemical Reactions

2.2.1 Quantum Efficiency

Photochemical reactions of molecules (A) occur in general in two steps, Eqs. (2.22a) and (2.22b):

[9)] $k(T) = A \exp(-E_a/RT)$: A = pre-exponential factor = hypothetical rate constant at infinite temperature; E_a = activation energy.

$$A + h\nu \to A^* \qquad (2.22a)$$

$$A^* \to B, C, \ldots \text{ (products)} \qquad (2.22b)$$

In most instances A^* is an excited singlet or triplet state [50, 51]. It is a prerequisite that the absorption (2.22a) actually occurs, i.e., that the spectrum of the radiation (in the environment: solar) overlaps with the (UV, VIS) absorption spectrum of A. This sounds trivial today, but it was not the case in the early 19th century when this "First Law of Photochemistry" was formulated by Grotthuß [52]. In the environment this means that absorption has to take place at wavelengths $\lambda \geq 295$ nm, the short wavelength limit of tropospheric solar radiation. This is true for photochemical reactions in the free gas phase of the troposphere, and also for molecules adsorbed at the surface of aerosol particles or dissolved in water droplets or in surface water near the surface[10].

In order to calculate the first-order rate constants $k_{h\nu}$ of the direct phototransformation of molecules, the quantum efficiency (Φ) of reaction (2.22) is required (2.23):

$$\Phi = \frac{\text{Number of \textbf{transformed} molecules (or moles) A}}{\text{Number of absorbed photons (or Einstein)}} \qquad (2.23)$$

$$= N_r / N_a \leq 1$$

1 Einstein = 1 mol photons

In this definition, the quantum efficiency cannot be higher than 1, as a molecule cannot be transformed and thus disappear more than once. If, however, the quantum efficiency is to be related to the number of molecules or moles formed, e.g., if the primary reaction is followed by a chain reaction, another definition of the quantum efficiency, Eq. (2.24), is used together with another symbol (Φ'). This is often neglected in the literature and is a frequent source of errors.

$$\Phi' = \frac{\text{Number of \textbf{formed} molecules (or moles) B, C} \ldots}{\text{Number of absorbed photons (or Einstein)}} \qquad (2.24)$$

Φ' may be < 1, $= 1$, or > 1. If only one molecule (B) is formed per lost A, Φ equals Φ'.

The quantum efficiency Φ can, but does not have to, be independent of the wavelength of irradiation. In the condensed phase (e.g., in water), the independence of Φ from the wavelength is the rule due to the fast relaxation processes leading to the first excited singlet and triplet states (S_1, T_1), which are the reacting states

---

[10] The penetration depth for UV radiation strongly depends on the presence of humic acids and other UV-absorbing substances in the lakes, rivers and oceans. In clean water, as often found in the open sea, the penetration depth may amount to 30 m or more [1]. In polluted rivers, only the uppermost few cm are available for UV radiation. In quantitative assessments, the UV spectrum of the water has to be measured [231].

[50]. The exceptions from this rule point to higher excited states and extremely fast reactions from these states – before equilibrium with the surrounding matrix. In the gas phase, there is no surrounding matrix and the excited molecules can get rid of the excess energy only by radiation (emission of photons) and collisions with other molecules present in the gas (in the troposphere mainly N_2 and O_2). Therefore, deviations from the rule of wavelength-independence are much more likely in the gas phase compared with the condensed phases.

This question of wavelength dependence or independence is of paramount importance in the calculation of the rate of direct photo-transformation according to Eq. (2.25), which is valid for small absorbance ($E \leq 0.02$) and polychromatic irradiation, as in the case of solar radiation:

$$k_{h\nu} = \int \sigma(\lambda)\, \Phi(\lambda)\, I(\lambda)\, d\lambda \quad [s^{-1}] \tag{2.25}$$

where σ = absorption cross section [cm^2 molecule^{-1}]; Φ = quantum efficiency, defined according to Eq. (2.23); I = spectral photon irradiance [53] in [photons cm^{-2} s^{-1} nm^{-1}]; and λ = wavelength [nm].

The absorption cross section used in Eq. (2.25) is defined by the Lambert–Beer law in the formulation given in Eq. (2.26):

$$I = I_0\, e^{-\sigma C' d} \tag{2.26a}$$

$$\ln I = \ln I_0 - \sigma\, C' d \tag{2.26b}$$

where σ = absorption cross section [cm^2 molecule^{-1}]; C' = concentration [A] in [molecule cm^{-3}]; and d = irradiated layer [cm].

The conversion into the molar decadal absorption coefficient, ε, used in chemistry is achieved using Eq. (2.27):

$$\sigma = 1000 \ln 10\, \varepsilon / N_L \tag{2.27}$$

where N_L = Avogadro constant ($6.022\,136\,7 \times 10^{23}$ mol^{-1}); $\sigma = 3.8235 \times 10^{-21}\, \varepsilon$ [cm^2 molecule^{-1}]; $\varepsilon = 2.6154 \times 10^{20}\, \sigma$ [L mol^{-1} cm^{-1}].

One mole of photons is often called "1 Einstein" in the photochemical literature, although this name is not recognized officially by IUPAC [53].

If the quantum efficiency Φ is independent of the wavelength of the radiation, it can be taken out of the integral in Eq. (2.25), which is then replaced by a summation for practical purposes (Eq. 2.28):

$$k_{h\nu} \approx \Phi \sum \sigma(\lambda)\, I(\lambda)\, \Delta\lambda \quad [s^{-1}] \tag{2.28}$$

σ and I in Eq. (2.28) are average values of the selected wavelength intervals ($\Delta\lambda$), e.g., 2.5 or 10 nm for relatively broad absorption bands.

Equation (2.28) is suitable for calculating the rate of direct photo-transformation both in water and in the gas phase, provided that Φ is independent of the wavelength. To be sure that this is indeed the case, Φ should be measured at two or more wavelengths in the overlap region of UV/VIS absorption and solar radiation. This is not necessary in the frequent situation of a small spectral overlap near to 300 nm. In this instance the tail of the absorption spectrum (maximum below 300 nm) overlaps weakly with the edge of the solar radiation between about 295 and 310 nm (for details see Section 5.3).

Besides Φ, only the absorption spectrum $\sigma(\lambda)$ in the near-UV (and in the visible range, "VIS", for coloured substances) has to be known. The average solar photon irradiance $I(\lambda)$ can be taken from the literature [54–58]. Taking Φ = 1, an upper limit of k_{hv} is obtained, corresponding to a *lower limit* of the photo-transformation lifetime (Eq. 2.29) of the substance investigated. This estimate is, of course, not at all conservative and should therefore not be used for any hazard or risk assessment; it may serve, however, for estimating whether or not a compound can be degraded photochemically in a realistic time period. If not, any further work on direct photo-degradation is a waste of time.

The photochemical half-life $t_{1/2(hv)}$, the first-order rate constant k_{hv} (Eqs. 2.25 and 2.29) and the lifetime τ_{hv} are related in first-order kinetics by Eq. (2.29), see also Section 5.1:

$$t_{1/2(hv)} = \ln 2 / k_{hv} = \ln 2\ \tau_{hv} \tag{2.29}$$

where τ_{hv} = photochemical lifetime, related to direct photo-transformation.

In the gas phase, the so called "cage effect" observed in condensed phases [59] is absent so that the primary quantum efficiencies in the gas phase are often higher[11]. This can be observed, e.g., in the photolysis of hydrogen peroxide, Eq. (2.30):

$$\text{HO–OH} + h\nu \rightarrow 2\ \text{OH} \tag{2.30}$$

Φ $(-H_2O_2)$ = 1.0 [60, 61]   Φ' (formation of OH) = 2.0

The corresponding values in water are 0.5 and 1.0.

The experimental work consists in measuring the quantum efficiencies and absorption spectra in the gas phase. If Φ is known, preferably at several wavelengths, k_{hv} can be calculated according to Eqs. (2.25) or (2.28) for various solar irradiations (spectral distribution and intensity). If the absorption in the gas phase is not known, a spectrum measured in a non-polar solvent can be used for molecules showing broadband spectra in the gas phase and in solution. Small molecules with a well-resolved vibrational/rotational structure of the electronic bands should not be treated in this way. In well-resolved spectra, the intervals $\Delta\lambda$ (2.28) have to be chosen according to the resolution of the spectra.

In the troposphere, the sink "direct photo-transformation" is relevant only for substances showing UV absorption at wavelengths $\lambda \geq 295$ nm. In the

---

[11] The cage effect hinders the rapid dissociation of the fragments in the case of a true photolysis and thus favours the recombination; the effective quantum efficiency is therefore smaller than in the gas phase.

stratosphere, however, which can be reached by extremely persistent compounds (e.g., chlorofluorohydrocarbons, perfluorohydrocarbons, SF_6 or persistent chlorohydrocarbons, e.g., CCl_4 [5, 28]), the sink can even be relevant for compounds absorbing only in the far-UV ($\lambda < 200$ nm).

2.2.2 Examples of Photochemical Reactions in the Gas Phase

There are many types of photochemical reactions contributing to the sink "direct photo-degradation", too many to enumerate here [50, 51, 59]. It should be noted, however, that some photochemical reactions cannot be regarded as real sinks, e.g., *cis–trans* isomerisations (*cis–trans* stilbene, etc.), especially if these reactions are reversible, either thermally or photochemically. Other photochemical processes lead to a splitting of the molecule (photolysis), often into radicals that may react further in dark reactions under atmospheric conditions and may finally bring about "mineralization", the ultimate goal of environmental degradation in all media [1]. Disappearance of the "mother molecule" is a first and necessary step in a degradation cascade and measuring this step may be considered as the "zero-th" solution to quantifying persistence. This is, of course, also true for the indirect photochemical reactions and any other sink.

The database of quantum efficiencies measured according to the OECD guideline [58] is meagre, to put it mildly. Some data on Φ and/or $k_{h\nu}$ are included in Chapter 3. They often refer to molecules studied for reasons of photochemical interest, the mechanistic and kinetic aspects, but only seldom to the potential behaviour of the molecules in the environment. The quantum efficiencies have been measured for some pesticides for registration, however, these data are not (yet) in the public domain. This environmental aspect was only considered in a few studies, but even in these it was often the degradation under specific climatic condition that was the focus of interest and no quantum efficiency was determined, or only with "white", i.e., natural or polychromatic UV radiation, allowing no unambiguous measurement of Φ. Knowing Φ, however, is the prerequisite for calculating rate constants and lifetimes for many different climatic conditions.

The situation is even worse for semi-volatile organic compounds (SOC) that are difficult to measure in the quartz cells typically used for studies of gases and volatile substances (e.g., solvents). For instance, it was not possible to find gas-phase studies for benzophenone, a compound which has been studied extensively in solution [50]. Klöpffer et al. [25] measured an approximate quantum efficiency in a smog chamber (irradiation only, no actual smog formation) of:

$$\Phi \text{ (–benzophenone)} \approx 5 \times 10^{-3}$$

In this experiment, propene and toluene were added as probes and showed the absence of any trace of OH and/or ozone during the photochemical reaction (see also Section 4.2).

The preferred objectives for investigations (organic molecules) are aldehydes and ketones. Acetone, the simplest ketone, has already been treated by Noyes and Leighton in 1941 [62]. The primary process, a true photolysis, is a so-called Norrish-I splitting (2.31a):

$$CH_3-(C=O)-CH_3 + h\nu \rightarrow CH_3-C=O + CH_3 \quad (2.31a)$$

$$CH_3-C=O \rightarrow CH_3 + CO \quad (2.31b)$$

This reaction mechanism can be concluded from the reaction products of radical recombination: ethane ($CH_3\cdot + CH_3\cdot$), biacetyl ($2 \times CH_3-CO$) and CO formed in the secondary reaction (Eq. 2.31b).

The quantum efficiency of CO formation at 313 nm under exclusion of oxygen (to prevent oxidative chain reactions) was measured at 298 K and a partial pressure between 67 and 200 hPa:

$$\Phi'(CO) = 0.11-0.12$$

The quantum efficiency of the disappearance of acetone, at the same partial pressure range and temperature was observed to be:

$$\Phi(-\text{acetone}) \approx 0.3$$

More work on the atmospheric chemistry of carbonyl compounds can be found in a review article by Carlier et al. [63] and, for acetone in particular, in Gardner et al. [64]. In contrast to older work (exclusion of oxygen), specific *tropospheric conditions* were simulated. One reason for these studies was the question: how can the relatively high mixing ratios of acetone (about 500 pptv) in remote areas be explained [65]?

According to these investigations, the primary process of the acetone reaction is still the Norrish-I splitting (2.31a) proceeding via T_1, the first excited triplet state[12]. In the presence of oxygen, the radicals formed in the photolytic step rapidly react (Eq. 2.32) to give peroxy radicals (M is an inert collision partner):

$$CH_3 + O_2 + M \rightarrow CH_3OO + M \quad (2.32a)$$

$$CH_3-CO + O_2 + M \rightarrow CH_3-(CO)-OO + M \quad (2.32b)$$

$$2\ CH_3-(CO)-OO \rightarrow 2\ CH_3-COO + O_2 \quad (2.32c)$$

$$CH_3-(CO)-OO + CH_3OO \rightarrow CH_3-COO + CH_3O + O_2 \quad (2.32d)$$

---

[12] The absorption in the relevant spectral region (near-UV) is due to a (spin-allowed) singlet–singlet transition, a relatively weak n–π^* transition; the (excited) S_1-state thus created is transformed by "inter-system crossing" (a radiationless process) into the lowest excited triplet state T_1, the reactive state. Competing with the photochemical process are two photophysical processes: phosphorescence (unlikely at normal temperature and in the presence of O_2) and inter-system crossing back to the ground state S_0. Quenching of T_1 by O_2 (collision followed by energy transfer) leads to O_2 ($^1\Delta_g$), a reactive excited state of the oxygen molecule, the so-called "singlet oxygen", see also Section 2.1.1.1, Eq. (2.1).

For reactions under tropospheric conditions it is important to know that CH_3–COO breaks down to yield carbon dioxide and methyl radicals, Eq. (2.33):

$$CH_3\text{–}COO \rightarrow CH_3 + CO_2 \qquad (2.33)$$

The key to this finding was the observation that Φ (–acetone) was equal to Φ' (CO_2) under all tropospheric conditions investigated:

$$\Phi/\Phi' (CO_2) = 1.02 \pm 0.03 \ (n = 40)$$

The quantum efficiency of the disappearance of acetone has been measured to be [64]:

$$\Phi \text{ (–acetone)} = 0.077$$

This quantum efficiency can be used for the lower troposphere (boundary layer) for the calculation of k_{hv} and the direct photochemical lifetime or $t_{1/2}$ ($h\nu$). According to Gardner, at a zenith angle (of the sun) of 40°,

$$k_{hv} = 7.85 \times 10^{-7} \ s^{-1}$$

resulting in a calculated direct photochemical lifetime of

$$\tau_{hv} = 1/k_{hv} = 14.7 \ d \approx 2 \text{ weeks}$$

A recalculation of τ_{hv} for Central Europe has been performed, using the more realistic solar irradiance data by Frank and Klöpffer [57]. The UV-absorption gas phase spectrum necessary for the calculation was taken from Finlayson-Pitts and Pitts [14] and Meyrahn et al. [66]. The data are presented in Table 2.

Table 2 Data used for the calculation of (direct) photochemical lifetimes of acetone in the troposphere (extract).

Wavelength λ [nm]	Abs. cross section σ [cm^2 molecule^{-1}][a]	Solar irrad. I (January) [photons s^{-1} cm^{-2} nm^{-1}] [57]	Solar irrad. I (July) [photons s^{-1} cm^{-2} nm^{-1}] [57]
295	4.3×10^{-20}	5.6×10^1	2.2×10^9
300	$2.95 \ (2.78) \times 10^{-20}$	1.4×10^7	1.9×10^{11}
310	$1.59 \ (1.44) \times 10^{-20}$	1.5×10^{11}	6.3×10^{12}
320	$0.55 \ (0.48) \times 10^{-20}$	2.2×10^{12}	2.0×10^{13}
330	$0.22 \ (0.08) \times 10^{-20}$	7.0×10^{12}	3.9×10^{13}

a) Left-hand values taken from [14], values in parentheses from [66].

As can be seen from Table 2, there is a dramatic difference between winter and summer in the European solar irradiance data, especially in the short-wavelength region near the cut-off at 300 nm. As many compounds show a weak tailing of the UV absorption instead of strong absorption bands in this spectral region, this difference may be of utmost importance for many substances showing a direct photochemical sink! Acetone has a relatively weak (n–π*) transition in the overlap region (Table 2) with a maximum at 270 nm [14].

Taking the simplified equation (2.28) and a constant $\Phi = 0.077$, summation from 295 to 330 nm and an interval of $\Delta\lambda = 2.5$ nm, the following results are obtained:

1. *Summer (July):*

$$k_{h\nu} = 2.1 \times 10^{-7} \text{ s}^{-1} \qquad \tau_{h\nu} \approx 55 \text{ d}$$

This result is in the same order of magnitude as that obtained by Gardner et al. [64], a (summer-) lifetime of about two weeks.

2. *Winter (January):*

$$k_{h\nu} = 2.1 \times 10^{-8} \text{ s}^{-1} \qquad \tau_{h\nu} \approx 550 \text{ d}$$

Of course, January has only 31 days, so the result is to be understood as "if the whole year over this low irradiance would prevail". This caveat is valid for all estimates of chemical lifetimes. These results show the great influence of season and region on (direct) photochemical reaction rates and lifetimes. The winter results also show that any calculation of $\tau_{h\nu}$ based on solar irradiances measured in California or Texas are misleading if used for moderate climatic conditions.

According to Meyrahn et al. [66], the quantum efficiency of acetone disappearance in air is *not* constant but increases from 330 to 290 nm from $\Phi = 0.033$ to 0.3 ($<\Phi> \approx 0.11$). The calculation of the photolytic lifetime by these workers at 40° N gives the following results:

	$\tau_{h\nu}$ [d]
January	420
July	64
Average (year)	120

Quantum efficiencies have also been determined for pyruvic acid [67], several chlorobenzenes [68] and organic nitrates [69, 70]. The few data available are compiled in Chapter 3.

3
Heterogeneous Degradation

3.1
Degradation on Solid Surfaces

3.1.1 Introduction

Particles in the troposphere can originate from human activities (soot, ashes, etc., about 15% of the total solid aerosol [71]) or from natural sources (sea salt aerosol [72], desert sand [71], particles from gas reactions [73]). These particles contain different fractions of organic materials. The average content of organic substances that can be extracted with ether has been reported to be about 1 µg m^{-3} under standard conditions of pressure and temperature [73]. Substance groups identified include organic acids, bases, aliphatic and aromatic hydrocarbons. As the composition turned out to be relatively independent of the collection site, it was concluded that the organic matter is predominantly of natural origin [71].

Anthropogenic compounds can occur in or at the solid phase, depending on their physical–chemical properties. As shown in [1], in the case of solid, dry surfaces, it is the vapour pressure of the substance that determines to which of the two following groups a substance belongs:

1. $p(20–25\ °C) < 10^{-7}$ Pa: This is the region of complete adsorption; there should be no free molecules in the gas phase, if there are any particles in the atmosphere at all.

2. 10^{-7} Pa $< p(20–25\ °C) < 5 \times 10^{-3}$ Pa: This is the region of the semi-volatile organic compounds (SOCs). These occur partly adsorbed, depending on the surface of aerosol per volume of air, temperature and specific interactions, which will not be considered for the moment. Bidleman [234] provided a useful review on the particle/air partitioning of semi-volatile substances.

According to the "Junge formula" [74, 75], adsorption starts at an upper limit of about 1 mPa, i.e., at a lower vapour pressure than given above. It should be taken into account, however, that vapour pressures are mainly given for room temperature and decrease exponentially to lower temperatures. Furthermore, locally, there may be higher concentrations of aerosol particles and thus higher surface areas per volume. Finally, the Junge formula is a very convenient approximation for estimates and modelling, but it is not a law of nature (see below).

How can substances with a very low vapour pressure reach the troposphere?

1. Emission of hot gases, e.g., at caloric power plants (especially in the initial phase), waste incineration plants, cars and diesel engines of all types (e.g., in ships) etc.

2. Emission together with carrier-particles, such as soot or ashes, followed by secondary adsorption during the cooling phase [76].

3. Volatilisation from water due to very low water solubility $C_{s,w}$ [1]; the Henry coefficient (H), which determines the volatility may be high, even for compounds with low vapour pressure p, Eq. (3.1):

$$H(T) = \frac{p(T)}{RT \times C_{s,w}(T)} \qquad (3.1)$$

where R = gas constant (8.314 510 J K^{-1} mol^{-1}); $C_{s,w}$ = water solubility [mol m^{-3}]; p = vapour pressure [Pa]; $H(T)$ = dimensionless Henry coefficient or air–water distribution coefficient (C_a/C_w [–]) as a function of T [K].

This is true for pure water; in the case of surface waters with a high content of suspended matter a competition occurs between volatilisation and adsorption to the particles with transport to the sediment.

4. The substance may be formed in the atmosphere by transformation from volatile precursors, e.g., nitrophenols from BTX aromatics [77, 132].

5. Formation in the droplet phase from water-soluble precursors, followed by evaporation of the droplets; a solid nucleus remains after the evaporation, which may carry substances with low vapour pressure.

SOCs are likely to be adsorbed from the gas phase of the troposphere into which they may enter during or after use in addition to the pathways 1–5 shown above. The adsorption to solid surfaces may be much more complicated and substance-specific, as predicted by the simple model proposed by Junge [74, 75]. This has been shown in work by Goss [78, 79, 235], according to which only structurally similar compounds can be related in a quantitative way. Furthermore, the interface between particle and air has to be known, e.g., whether there is a water layer or not. This depends on the chemical composition of the aerosol particles and on the relative humidity.

The problem is further aggravated by the fact that not all surfaces are inorganic, as in the case of sand or sea salt. Many particles contain organic matter (see above). In the case of prevalent organic particles, adsorption is perhaps not the dominant mode of uptake from the gas phase. It might rather be distribution between an organic phase (particle) and the gas phase. In order to describe this effect, an octanol/air distribution coefficient K_{oa} has been proposed instead of vapour pressure [80] (n-octanol representing, as in K_{ow}, the organic phase). Xiao and Wania [81] showed, however, that it is not possible to conclude which parameter is better – even in the restricted range of organic phase particles. The problem of consistent physicochemical data sets has been treated by Beyer et al. [82]. Data availability and data quality seem to be at the heart of the quantitative treatment of chemicals and especially of SOCs [83–85, 236].

In the following discussions a distinction will be made between experimental work on the degradation of substances adsorbed on soot and fly ash (Section 3.1.2) and artificial aerosols (see Section 3.1.3).

3.1.2 Degradation on Fly Ash and Soot

It has been known for a long time that polycyclic aromatic hydrocarbons (PAHs) are transported with the particle-phase of the atmosphere. Björseth et al. [86] found, in South Sweden and South Norway by analysis (GC/MS) of 30 single compounds, that

$$\sum \text{PAH (ads.)} \approx 0.5 - 100 \text{ ng m}^{-3}$$

The most important compounds per mass are (vapour pressures according to Rippen [95]):

- Fluoranthene (p_{25} = 1.25 mPa)
- Pyrene (p_{25} = 0.82 mPa)
- Phenanthrene (p_{25} = 22 mPa)
- Benzo[a]pyrene (BaP) (p_{25} = 0.0007 mPa)

BaP, a borderline case between SOC and adsorptive compound, contributes 5%. The highest concentrations were measured during SE–S–SW winds, bringing air masses from Western and Central Europe to Skandinavia, and during winter time.

Experiments conducted to measure the photo-degradation of adsorbed PAHs in the laboratory, using natural and artificial particles, produced contradictory results [87–91, 123]. The experiments showed that the PAHs were degraded by radiation that simulated solar radiation, but the time needed to degrade 50% varied considerably from ca. 1 hour to a day and longer. In one case [89], the PAHs (applied to coal fly ash) were described as "highly resistant"! These differences were discussed in terms of different (acidic or basic) surface groups. A re-evaluation in terms of particle size and UV absorption [1] showed that simple UV absorption in some of the larger particles may be one reason for the discrepancies observed. This would point to direct photochemical transformation as the main degradation mechanism; however, PAH molecules "buried" inside larger particles may be protected not only from UV radiation, but also from reactive radicals (OH, NO_3). It is interesting to note that Behymer and Hites [92] also concluded from their photo-degradation experiments (on fly ash) that the "colour" of the particles plays a decisive role in the reactivity of the organic molecules present on/in the particles. PAHs, readily reacting in the gas phase and on/in white particles, are protected from the incident radiation in black particles. In the case of strong absorption of UV and/or VIS radiation, a particle size of about 1 μm may be sufficient to effectively protect the molecules in the interior from the radiation [1].

3.1.3 Degradation on Artificial Aerosols

In the investigations discussed in Section 3.1.2, with few exceptions, relatively large particles were used to study the degradation of PAHs. In the natural troposphere, however, small particles (in the order of 0.1–1 μm) are more important due to their slow sinking velocity, stability and high surface to volume ratio. The last is even higher for the very small particles (< 0.1 μm), but these particles sink faster

Table 3 Approximate deposition behaviour of atmospheric particles.[a]

Diameter [µm]	Sinking velocity [cm s^{-1}]	Time needed for 100-m sinking
100	25	0.1 h
10	0.3	9 h
1	0.01	2 weeks
0.1	0.01	2 weeks
0.01	0.1	30 h

a) Data approximated based on the work by Warneck [15] (only dry deposition).

so that the average time spent in the atmosphere is shorter (see Table 3). The very small particles are also unstable and tend to aggregate.

The smallest deposition velocities are measured in the range 0.08–2 µm, the minimum being close to 0.1 µm, corresponding to a maximum of the residence time. In this range of diameters there is also 50% of the total mass of the solid aerosol[13], but only about 5% of the number of particles. The very small particles have the highest number, but they are unstable, coagulate and attract water (thus potentially acting as condensation nuclei).

This brief consideration of the basic facts concerning aerosols leads to the conclusion that to stimulate a natural aerosol an average particle diameter of 0.1–1 µm should be optimal. A large aerosol chamber that has been constructed for such experiments [177, 180] is described in Section 4.5. Three types of particles have been used as carriers for the substances (surface covering: monolayer or below):

1. Inert particles (e.g., SiO_2, Al_2O_3); no absorption at $\lambda \geq 300$ nm.
2. Potentially photochemical active particles (TiO_2, ZnO, Fe_2O_3); absorption at $\lambda \geq 300$ nm is a necessary but not sufficient condition.
3. Salt particles (especially NaCl).

The chemical that has been most thoroughly investigated is di(2-ethylhexyl)-phthalate (DEHP). DEHP is a typical SOC [$p(293\ K) = 18.1 \times 10^{-6}$ Pa; $p(298\ K) = 30 \times 10^{-6}$ Pa [95]] and tends to evaporate from some carriers [93, 94]. This seems to be a major problem, as a loss of substance could be misinterpreted as degradation. SOCs can occur by definition in/on the particles and in the gas phase. Because evaporation from the surface of the particles has also been observed for other substances, a test aerosol chamber (glass cylinder of 1 m diameter, 4 m high, about 3140 L) has been constructed, which can be cooled down to about –10 °C, and for which the first results are available [96]. DEHP was adsorbed in amounts

---

13) The "aerosol" is the sum of solid phase + gas (air), not the solid phase alone; this term designates a relatively (at least temporary) stable mixture of the two phases in the same way as "emulsion" designates a colloidal mixture of two liquids.

smaller than a monolayer onto SiO$_2$ flakes (Aerosil 200 from Degussa, specific surface 200 m^2 g^{-1}, average diameter 12 µm) mimicking the adsorption on desert sand aerosol particles.

Results on SiO$_2$ (Aerosil 200 from Degussa) with the original aerosol chamber were reported by Behnke et al. [94]. During this and similar experiments [96], a photosmog formed during irradiation, characterized by an OH concentration of [OH] = 5 × 10^6 radicals cm^{-3}. This corresponds to about ten times the average global day/night OH concentration <[OH]>$_{global}$. According to the observed pseudo first-order kinetics (the degradation rate depends linearly on [OH]), a chemical OH lifetime of

$$\tau_{OH} \text{ (DEHP)}_{ads} \approx 1.5 \text{ d}$$

is calculated for <[OH]>$_{global}$. The degradation of DEHP has not been measured in the gas phase due to its low vapour pressure (SOC). The OH lifetime can only be compared with τ_{OH} measured using an indirect method, the measurement in 1,2,2-trichlorotrifluoroethane (Freon 113) [97, 98, 114, 176]. In this method, the substance is dissolved in the completely OH-resistant solvent, saturated with the OH-source H$_2$O$_2$ and irradiated with UV (λ > 300 nm); k_{OH} is determined relative to a known reference substance. The relative values thus measured in the solvent correspond to the values measured in the gas phase. This method therefore gives estimates for the gas-phase bimolecular k_{OH}, which can be used to calculate the OH lifetime. In the case of DEHP:

$$\tau_{OH} \text{ (DEHP)}_g \approx 3 \text{ d}$$

More recent experiments with DEHP adsorbed on SiO$_2$ particles [96] showed a second-order rate constant of $k_{OH,ads}$ = (19 ± 3) × 10^{-12} cm^3 molecule^{-1} s^{-1} and, using the above average OH concentration

$$\tau_{OH} \text{ (DEHP)}_{ads} \approx 1.2 \text{ d}$$

It remains to be shown whether the use of second-order kinetics in heterogeneous systems of this type is correct and, hence, whether a comparison with gas phase reaction rates is allowed. It seems, however, that the order of magnitude is given correctly. The problem of accessibility in this type of heterogeneous photochemistry has been discussed further by Balmer et al. [237] for soil surfaces, where similar problems are encountered.

The degradation of DEHP on Fe$_2$O$_3$ particles gives similar results when compared with those obtained with the artificial SiO$_2$ aerosol, i.e., reaction with OH-radicals ($k_{OH,ads}$ = 12.8 ± 3 × 10^{-12} cm^3 molecule^{-1} s^{-1} [94]). Additional disappearance channels discussed by the researchers involve hydrolysis, evaporation and also possibly direct photochemical reactions due to the red iron oxide particles.

DEHP adsorbed on TiO$_2$ particles exhibits a different behaviour. The white pigment TiO$_2$ is well known for its photo-catalytic activity [238, 239]. The pigment has a strong UV absorption below 400 nm, which is due to the excitation of an electron from the "valence-band" to the "conduction-band"[14]. Under UV radiation, positive "holes" are formed in the valence band of the crystal, in addition to free electrons in the conduction band and trapped electrons (at lattice imperfections, at the surface and at chemical impurities). In the presence of NO_x, an increased OH concentration is observed in the gas phase surrounding the particles. The following reaction mechanism has been proposed (Eq. 3.2):

$$O_2 \text{ (g)} + e^- \text{ (from TiO}_2\text{*)} \rightarrow O_2^- \text{ (ads)} \tag{3.2a}$$

$$H^+ \text{ (aq)} + O_2^- \rightarrow HOO \text{ (ads)} \tag{3.2b}$$

$$OH^- \text{ (aq)} - e^- \text{ (to TiO}_2\text{*)} \rightarrow OH \text{ (ads)} \tag{3.2c}$$

The role of the nitrogen oxides can be explained by Eq. (3.3)

$$NO \text{ (g)} + OOH \text{ (ads)} \rightarrow OH \text{ (g)} + NO_2 \text{ (ads)} \tag{3.3}$$

These results are interesting, as they show the interaction with a photochemically active solid. The degradation also seems finally to be due to OH radicals. The relevance of the TiO_2 aerosol in the environment is questionable (close to pigment production sites?).

NaCl particles, on the other hand, are very relevant to the atmospheric environment, as NaCl is the main constituent of the sea salt aerosol. Pure NaCl particles can be considered as a model for sea salt particles. An increased degradation (relative to the inert SiO_2) was observed, which in this case is *not* due to OH. Zetzsch postulated the liberation of Cl atoms [93], which are extremely reactive towards organic molecules, e.g., vinyl chloride (VC) or ethane [100], both substances being only moderately reactive towards OH:

$$k_{Cl} \text{ (VC)} = 1.4 \times 10^{-10} \text{ cm}^3 \text{ molecule}^{-1} \text{ s}^{-1}$$

$$k_{Cl} \text{ (ethane)} = 6.2 \times 10^{-11} \text{ cm}^3 \text{ molecule}^{-1} \text{ s}^{-1}$$

The concentration of Cl atoms in the troposphere in general seems to be much lower than that of other reactive species, but Cl may play a role near the surface of the sea, so that a certain influence of chlorine atoms on the chemistry of the boundary layer seems possible [2, 100, 101]. Singh and Kasting [102] proposed that Cl atoms are formed according to Eq. (3.4), HCl being set free in the maritime NaCl aerosol:

---

[14] In contrast to molecular solids, in which the individual molecules retain their "individuality", i.e., they are the same entities as in the gas phase, in solution or in the melt, inorganic crystals show a collective behaviour of the groups of atoms forming the chemical structure. The energy levels of the "valence electrons" form a broad band (in terms of energy) which replaces the narrow individual energy levels. This band is filled and cannot contribute to the conductivity, unless an electron is removed, e.g., by excitation to the lowest empty band (conduction band).

$$\text{HCl} + \text{OH} \rightarrow \text{Cl} + \text{H}_2\text{O} \tag{3.4}$$

According to the same workers [102], the average maritime concentration of Cl amounts to:

$$<[\text{Cl}]>_{\text{mar}} \approx 1000 \text{ atoms cm}^{-3}$$

Wang et al. [201] conclude that, due to the very high rate constants of some H-abstraction reactions with Cl atoms (about 100-times faster than OH), degradation by Cl may play a role in the marine boundary layer and in coastal regions. They reported $k_{\text{Cl}} > 10^{-10}$ cm³ molecule⁻¹ s⁻¹, i.e., at the diffusion limit, for a series of methylnaphthalenes, whereas the mother compound reacts slowly:

$$k_{\text{Cl}} \text{ (naphthalene)} \leq 10^{-12} \text{ cm}^3 \text{ molecule}^{-1} \text{ s}^{-1}$$

Over the continents, the average concentration of Cl atoms is assumed to be an order of magnitude lower. Molecular chlorine (Cl_2) is not stable during daylight due to rapid photolysis to Cl. Chlorine atoms should be considered as a possible sink (especially in the maritime boundary layer) for molecules that are only moderately reactive towards OH, NO_3 and ozone, the "classical" reactive species [11b].

According to these results, it seems that genuine aerosol solid/gas reactions with environmental chemicals are rare. There are instances in which reactive species (OH, Cl) are formed at or near the particle/gas interface and others in which the particles only act as unreactive carriers. The latter is certainly true for SiO_2, which can be regarded as a model for desert sand. It seems not to be true for NaCl, a model for sea salt aerosol, but in this instance an increased reactivity due to the Cl atoms has to be taken into consideration. It should be noted, however, that there are no studies available dealing with natural aerosol particles, including water layers surrounding the solid nucleus if the humidity is high enough to allow such behaviour. The artificial aerosols have been studied in order to develop reproducible *testing methods*, not for reasons of basic research. This should be considered, if the missing knowledge concerning the "natural behaviour" of SOCs is deplored.

Polycyclic aromatic hydrocarbons (PAHs) have been given less attention in studies with artificial aerosol particles. These substances occur mostly associated with carbonaceous particles. Esteve et al. [103] studied 11 semi-volatile and non-volatile PAHs adsorbed on calibrated 1–2-μm graphite particles, in the form of sub-monolayers (calculated). The artificial gaseous environment created allowed the study of the reactions of OH, NO_2 and NO with the PAHs adsorbed. The observed first-order rate constants with OH are in the range of $k^{\text{I}}_{\text{OH}} \approx 0.1–0.2$ s⁻¹, irrespective of the chemical structure, ranging from three (phenanthrene) to five (benzo[a]pyrene) annealed aromatic rings. Using the mean OH concentration given (3.4×10^{10} radicals cm⁻³), second-order rate constants in the order of $k_{\text{OH}} \approx 3–6 \times 10^{-12}$ cm³ molecule⁻¹ s⁻¹ can, at least formally, be calculated. This is a low value for a PAH {see, for comparison, the gas-phase rate constants of k_{OH} (phenanthrene) = 2.7×10^{-11}; k_{OH} (fluoranthene) = 1.1×10^{-11}; k_{OH}

(anthracene) = 1.9×10^{-10} cm^3 molecule^{-1} s^{-1} [104]}. It should be noted, however, that [OH] was only estimated in this work, so that an actual lower concentration would yield higher k_{OH} values. The observation that k_{OH}^I does not depend on the chemical structure of the PAH is more puzzling and requires an in-depth analysis of the experimental conditions (variable and measured [OH] should be used) and on the correct use of second-order kinetics in case of heterogeneous reactions.

3.2
Degradation in Droplets

3.2.1
Direct Photochemical Transformation

Direct photochemical reactions can occur in the droplet phase, as in surface waters, if

- the substance absorbs UV (and VIS) radiation to a significant extent in the spectral region above 295 nm,
- the quantum efficiency in this region is Φ > 0, and
- the irradiation intensity is high enough.

Compared with the situation in surface water [1], there are additional conditions to be considered in the droplet phase of the atmosphere:

1. The irradiation intensity may be higher due to the albedo of the earth and due to the light scattering within the clouds. Whereas at the surface of the earth (horizontal water surfaces) the irradiation intensity can be quantified, e.g., in photons per area (taking into consideration the zenith angle of the sun), an irradiation density (e.g., in photons per volume) required for scattered light is difficult to measure. Calculations showed that at noon, within clouds there is a higher UV-radiation density compared with outside the cloud, all other conditions being equal [105].

2. In polluted air, the pH can deviate greatly from that of the pure troposphere (pH = 5.8), mainly towards the acid region. Because in the droplet phase polar molecules are likely to accumulate, acidic and basic groups should occur frequently, so that the pK-values have to be considered in order to determine the species (anion, cation, neutral molecule) that will possibly prevail.

3. The "penetration depth" in droplets is high compared with surface waters, i.e., the irradiation is more homogeneous.

4. There may be interface effects due to the high surface to volume ratio, especially in the case of very small droplets; surface films have been postulated [106] but not studied with regard to their potential role in photochemical transformations.

There are few, if any measured data on the direct photochemical degradation of organic substances in the droplet phase. As a substitute, quantum efficiencies and/or first-order rate constants measured in bulk water can be used to estimate the photochemical lifetime in the droplet phase, taking into consideration the items mentioned above. The equations are similar to those discussed for the gas phase (Section 2.2.1). Owing to the short lifetime of tropospheric droplets, it is to be expected that only molecules with high Φ and/or strong UV/VIS absorption can be degraded by this sink. Some quantum efficiencies of organic compounds in aqueous solution are given in *Environmental Organic Chemistry* by Schwarzenbach et al. [146].

3.2.2 Reactive Trace Compounds in Cloud, Fog and Rainwater

Reactive trace compounds (radicals, molecules) can be taken up from the surrounding gas phase (equilibrium according to Henry's law) or be formed in the aqueous phase from a precursor substance. The applicability of Henry's law to droplets has been questioned by Goss [240] for reasons of surface forces, which possibly invalidate the law; even so, H may be an approximate measure of the distribution. The relative importance of the different pathways does not always seem to be clear. Most scientific papers refer to the system sulfur dioxide and its oxidation to the sulfate ion [2]. A comprehensive review on rate constants and reactive particles in fog and cloud droplets was prepared in 2000 by Herrmann et al. 2000 [107]. The data have also been used to calculate concentrations of reactive particles. The short discussion below highlights the role of some reactive species in the droplet phase, especially of hydrogen peroxide [117, 118].

Hydrogen Peroxide

Measured concentrations of H_2O_2 in cloud, fog and rainwater are in the range of [1]:

$$[H_2O_2] < 0.3–250 \ [\mu mol \ L^{-1}]$$

Hydrogen peroxide is an extremely polar, hydrophilic substance and has a dimensionless Henry coefficient [35]:

$$H \ (293 \ K) = [H_2O_2]_g / [H_2O_2]_w = 2.9 \times 10^{-7} \ [–]$$

As the concentration of liquid water in air is low (in relatively "wet" clouds it corresponds to only about 1 mL m^{-3}), even this low Henry coefficient means that in equilibrium about one third of the hydrogen peroxide is outside the droplets. However, equilibrium estimates are questionable in reactive systems such as droplets in the troposphere, especially during the day. As found by Goss [240], larger molecules may deviate significantly from equilibrium distribution due to adsorption onto the surface. This could explain some strange effects with respect to pesticides in fog, reported by Glotfelty et al. [123]. These workers explained the discrepancies between measured concentrations both in the gas and droplet phase and calculations using H through the possible existence of surface layers, e.g., by surfactants [1]. A complete kinetic analysis of the uptake process also requires the knowledge of the "accommodation coefficients", which are known for a few small molecules, but not for organic molecules [1].

H_2O_2 is photolysed in the aqueous phase (as in air) to give OH-radicals (Eq. 3.5):

$$HO\text{--}OH + h\nu \rightarrow 2\ OH \tag{3.5}$$

OH is highly reactive in water (as in the gas phase) and reacts with a multitude of molecules, but in pure water there are also reactions leading to (much less reactive) OOH-radicals. OOH is a weak acid so that in water, in contrast to the gas phase, O_2^- ions are formed in a pH-dependent equilibrium (Eq. 3.6)

$$OOH \leftrightarrow O_2^- + H^+ \tag{3.6}$$

$pK_a = 4.8$ [109]

Owing to this dissociation, all reaction rate constants that depend on OOH are pH-dependent. In addition, O_2^- reacts with ozone to form OH (in contrast to the neutral form OOH) [105]:

$$O_2^- + O_3 \rightarrow OH + 2\ O_2 \tag{3.7}$$

$k\ (3.7) = 1.0 \times 10^9$ L mol^{-1} s^{-1} [15)] [105]

OOH radicals react (back) to H_2O_2, Eq. (3.8)

$$OOH + OOH \rightarrow H_2O_2 + O_2 \tag{3.8}$$

$k\ (3.8) = (8.3 \pm 0.7) \times 10^5$ L mol^{-1} s^{-1} [109]

The "mixed" reaction (3.9) is much faster, whereas two O_2^- ions do not react to hydrogen peroxide.

$$OOH + O_2^- + H_2O \rightarrow H_2O_2 + O_2 + OH^- \tag{3.9}$$

$k\ (3.9) = (9.7 \pm 0.6)10^7$ L mol^{-1} s^{-1} [109]

The maximum of the overall reaction (3.8 + 3.9) is pH = pK_a = 4.8 [109].

The total reaction scheme for H_2O_2 (excluding catalytic or photo-catalytic reactions involving metal ions) is shown in Fig. 1. The main reactions can be summarized as follows:

---

15) The unit of the second-order reaction rate constants in liquid solution is usually based on [mol L^{-1}] as the concentration unit, hence [L mol^{-1} s^{-1}].

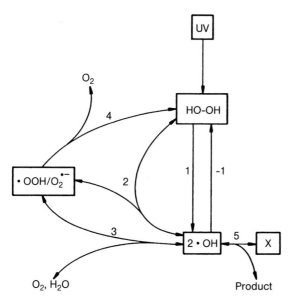

Fig. 1 Scheme for photochemical hydrogen peroxide reaction and re-formation in water.

$$HO–OH + h\nu \rightarrow 2\,OH \quad (1)$$

$$OH + OH \rightarrow H_2O_2 \quad (-1)$$

$$OH + H_2O_2 \rightarrow OOH + H_2O \quad (2)$$

$$OH + OOH/O_2^- \rightarrow O_2 + H_2O \quad (3)$$

$$OOH + OOH/O_2^- \rightarrow O_2 + H_2O_2 \quad (4)$$

$$OH + X \rightarrow products \quad (5)$$

Reaction (1) (= Eq. 3.5) is the photolysis of hydrogen peroxide and shows the following quantum efficiencies in water:

$$\Phi(-H_2O_2) = 0.5; \quad \Phi'(+OH) = 1.0\ [110, 111]$$

The UV absorption of H_2O_2 is weak ($\varepsilon \approx 1$ L mol^{-1} cm^{-1} at 300 nm [1]) and shows no vibrational structure, pointing to dissociation (1) as the primary photochemical reaction. Reaction (–1), although very fast (near to the diffusion limit $\approx 10^{10}$ L mol^{-1} s^{-1} in water) plays no practical role due to the extremely low concentration of OH-radicals. Reaction (2) is the most important source of OOH/O$_2^-$ at high concentrations of H_2O_2.

$$k(2) = 2.7 \times 10^7\ \text{L mol}^{-1}\ \text{s}^{-1}\ [112]$$

In atmospheric droplets, however, OOH formed in the gas phase and taken up by the droplets may be the primary source of H_2O_2 [105]. In this instance, H_2O_2 is formed via reaction (4). In the *absence of other sinks*, reaction (3) is the main elimination reaction for the reactive radicals. It also consists of two fast reactions, both near to the diffusion limit:

$$k(OH + OOH) = 7.1 \times 10^9 \text{ L mol}^{-1} \text{ s}^{-1}$$

$$k(OH + O_2^-) = 1.0 \times 10^{10} \text{ L mol}^{-1} \text{ s}^{-1}$$

The reaction designated as "X" in Fig. 1 (5) stands for any reaction of the reactive OH-radical with impurities present in the droplets, e.g., organic compounds of anthropogenic or natural origin. As in the gas phase of the atmosphere, most organic molecules also react with OH in water. A double-logarithmic plot of k_{OH}-values measured in the two media is linear and allows a rough estimate of k_{OH}(water) if k_{OH}(air) is known and *vice versa* [113–115]. Since the 1960s reaction rate constants have been measured and compiled for a large number of organic substances in an aqueous phase[16] [116]. A relative rate constant method using H_2O_2 as the OH source has been developed [117].

These reaction rate constants can, in principle, be used for predicting (OH) lifetimes in the droplet phase. Although it is not (yet) possible to recommend an average concentration of OH-radicals in droplets to be used in such calculations, it is clear that atmospheric droplets are a fairly reactive milieu, especially due to the OH-radicals.

If it is true that OOH is primarily taken up from the troposphere [105] (the same seems to be true for OH [107]), it is interesting to estimate the equilibrium concentration in the aqueous phase. The dimensionless Henry-coefficient amounts to

$$H(298 \text{ K}) = [OOH]_g/[OOH]_w = 3.4 \times 10^{-5} \text{ [105]}$$

Using the average daytime concentration in the gas phase $<[OOH]_g>_{day} \approx 10^8 \text{ cm}^{-3}$, the average concentration in the aqueous phase is calculated to be

$$[OOH]_w \approx 3 \times 10^{12} \text{ radicals cm}^{-3} (\approx 5 \text{ nmol L}^{-1})^{17)}$$

A calculated value for [OH], including the main reactions (sources and sinks) of OH has been reported by McElroy [105] to be

---

16) These extensive studies have most likely been performed in connection with reactor safety studies and/or studies aimed at the mechanism of radiation induced damage on living beings; this is indicated by the multitude of physiological molecules investigated. There is hardly any new literature in this field.

17) To convert the two types of expressions for the concentration: 1 mole contains $N_A = 6.02214 \times 10^{23}$ molecules, radicals, atoms or other "molecular" entities such as ions. The unit of volume in the "physical" definition of concentration [molecule cm^{-3}] is cm^3 ($\equiv$ mL), whereas in the "chemical" definition [mol L^{-1}], which is based on SI units, the volume has the unit litre (L $\equiv 10^{-3}$ m^3).

$[OH]_w \approx 2 \times 10^{-13}$ mol L^{-1} (≈ 0.2 pmol L^{-1})

More recently, Herrmann et al. [107] calculated maximal concentrations for $[OH]_w$ (the peaks at noon) under the following conditions:

Urban: 1.4 pmol L^{-1}
Remote continental: 1.7 pmol L^{-1}
Maritime: 1.9 pmol L^{-1}

From the 24-h $[OH]_w = f(t)$ curves shown in Ref. [107], an order-of-magnitude estimate for the average day and day/night concentration of $[OH]_w$ can be made:

$<[OH]>_{w,day/night} \approx 5 \times 10^{-13}$ mol L^{-1} (0.5 pmol L^{-1} $\approx 3 \times 10^8$ radicals cm^{-3})

$<[OH]>_{w,day} \approx 10 \times 10^{-13}$ mol L^{-1} (1.0 pmol L^{-1} $\approx 6 \times 10^8$ radicals cm^{-3})

The estimate of the day/night average is close to the older estimate by McElroy [105].

An additional source of OH, not considered by McElroy [105], is the photolysis of Fe^{III}–hydroxycomplexes (3.10) [119]:

$$[Fe^{III}OH]^{++} + h\nu \rightarrow OH + Fe^{++} \qquad (3.10)$$

$\Phi = 0.14 \pm 0.04$ at 313 nm; $\Phi = 0.017 \pm 0.003$ at 360 nm and 293 K

The species [Fe^{III}OH]$^{++}$ dominates in moderately acidic aqueous conditions between pH 2.5 and 5 [119], which are likely to occur in the troposphere ("acid rain"). Traces of Fe (and Mn) can be introduced into the droplets via aerosol particles.

Ozone

In contrast to H_2O_2, ozone is very hydrophobic, as shown by the dimensionless Henry coefficient:

$H = [O_3]_g/[O_3]_w = 4.3$ [35]

Using the estimated average tropospheric concentration of ozone (in the gas phase, see Section 2.1.3.1)

$<[O_3]>_{av,g} \approx 7 \times 10^{11}$ molecule cm^{-3}

the average concentration in the droplet phase is calculated to be:

$<[O_3]>_w = <[O_3]>_{av,g}/4.3 = 1.63 \times 10^{11}$ molecule cm^{-3} $\approx 2.7 \times 10^{-10}$ mol L^{-1}

Finlayson-Pitts and Pitts Jr. estimated [14], based on somewhat different input data,

$$<[O_3]>_w \approx 5 \times 10^{-10} \text{ mol L}^{-1} \text{ (0.5 nmol L}^{-1})$$

Ozone reacts in water both photochemically and in dark reactions. The UV/VIS absorption is similar to the gas-phase spectrum [14] {ε = 301 (300 nm), 25 (320 nm), 2.0 (340 nm), 5.1 (590 nm), 1.6 (675 nm) [L mol^{-1} cm^{-1}]}.

The reaction at short wavelengths ($\lambda < 320$ nm) leads, as in the gas phase, to excited oxygen species and OH (3.11):

$$O_3 + h\nu \ (\lambda \leq 320) \rightarrow O\ (^1D) + O_2\ (^1\Delta_g) \quad (3.11a)$$

$$O\ (^1D) + H_2O \rightarrow 2\ OH \rightarrow H_2O_2 \quad (3.11b)$$

Ozone is therefore connected to the H_2O_2 cycle. The quantum efficiency of reaction (3.11a) at 310 nm amounts to:

$$\Phi\ (-O_3) = 0.23\ [14]$$

Further connections of ozone to the H_2O_2 cycle are reactions (3.7) and (3.12).

$$O_3 + OH^- \rightarrow OOH + O_2^- \quad (3.12)$$

$$k\ (3.12) = 70 \pm 7 \text{ L mol}^{-1}\text{ s}^{-1}$$

Taking into account these reactions, the chemical lifetime of ozone in water at pH 7 is 1.6 days. Under acidic conditions, the reaction is slower and the lifetime correspondingly longer (165 days at pH 5).

Another important sink for ozone is H_2SO_3 [121], the efficiency of which increases with increasing pH.

Transition Metal Ions

Transition metal ions play an important role in oxidizing SO_2 and can produce OH radicals in the dark (Fenton [114, 122]) according to reaction (3.13):

$$Fe^{++} + H_2O_2 \xrightarrow{H^+} Fe^{+++} + OH^- + OH \quad (3.13)$$

Iron(III) oxalate is another effective source for H_2O_2 [124] and at the same time an important sink for oxalic acid, a common polar organic trace compound in the troposphere of natural origin. According to Jans and Hoigné [140], ubiquitous traces of Cu (ca. 1 nmol L^{-1}) are sufficient to catalyse the reaction of ozone to OH.

Nitrogen Species

- HONO (nitrous acid),
- HONO$_2$ (nitric acid),
- peroxyacetylnitrate (PAN).

The two acids are photolysed to give OH and the corresponding nitrogen oxide, Eqs. (3.14) and (3.15):

$$HO - NO + h\nu \rightarrow OH + NO \qquad (3.14)$$

$$HO - NO_2 + h\nu \rightarrow OH + NO_2 \qquad (3.15)$$

HONO absorbs UV radiation in the spectral range from 300 to 400 nm [14, 124], the absorption of HONO$_2$ starts below 320 nm. The quantum efficiencies of both reactions are about

$$\Phi \text{ (Eqs. 3.14, 3.15)} \approx 0.1 \text{ [14]}$$

The acids dissociate in aqueous solution, depending on pH, to give the photoactive anions nitrate (NO_3^-) and nitrite (NO_2^-), both yielding OH-radicals and oxygen atoms upon irradiation with UV [14].

3.2.3 Reactions of Organic Molecules

Degradation reactions of organic molecules in aqueous droplets can in principle be initiated by OH, O$_3$ and OOH/O$_2^-$. In this brief discussion, only reactions with OH are considered, because collections of reaction rate constants are available for this radical [114, 116, 125–127]. Furthermore, it is likely to be the most reactive oxygen-species present in atmospheric droplets.

As a prerequisite for reacting with OH or other reactive species in the droplet phase, this phase has to be reached by the organic molecules. This seems to be likely in the following instances:

- very polar/water-soluble compounds characterised by a dimensionless Henry coefficient (air–water partition coefficient) in the relevant temperature interval (approximately 260–300 K) – $H \leq 10^{-5}$;
- substances entering the troposphere using aerosol particles as carriers (e.g., PAH);
- polar substances formed in the troposphere from non-polar precursor molecules [77, 106, 128–130, 132], e.g., nitrophenols and chloroacetic acids.

For those organic compounds reaching the droplet phase despite these restrictive conditions, the droplet phase is a *highly reactive environment*. Unfortunately, there are no direct measurements of [OH] in the droplet phase. As a proxy, we therefore use the estimated average values, derived by Herrmann et al. [107], for day and day/night respectively (see above).

Slow reactions that are of interest in the gas phase should not play a great role in the droplet phase due to the short lifetime of the droplets, always oscillating between growing (ultimately forming rain drops under favourable conditions for growth) and evaporation (leaving aerosol particles as possible nuclei of renewed droplet formation). This oscillation is especially true for the average-sized droplets.

For a series of organic oxygenated compounds, likely to occur in droplets, the OH reaction rate constant amounts to [114, 115]:

$$\log k_{OH} \text{ (water)} \approx 8.7\text{--}9.3 \text{ L mol}^{-1} \text{ s}^{-1}$$

Members of this class of compounds are, e.g., formaldehyde, acetaldehyde and glycol.

Using $k_{OH} \approx 10^9$ L mol^{-1} s^{-1} and $<[OH]>_{w,day} \approx 10^{-12}$ mol L^{-1}, the chemical (OH)$_w$ lifetime of these compounds is estimated to be

$$\tau_{OH}(\text{droplet}) \approx \frac{1}{k_{OH} \times <[OH]>_{day}} = 1000 \text{ s} \approx 17 \text{ min}$$

As many organic substances in aqueous solution have k_{OH} values around 10^9 L mol^{-1} s^{-1}, a chemical (OH)$_w$ lifetime in the order of less than 1 h seems to be a good estimate for these substances of relative high OH reactivity. The OH reaction with phenols is even faster by an order of magnitude ($k_{OH} \approx 10^{10}$ L mol^{-1} s^{-1}), near the diffusion limit. Examples are phenol, p-cresol and o-nitrophenol[18] [114]. For these substances, again using the estimated day-average $<[OH]>_{w,day}$

$$\tau_{OH} \text{ (droplet)} \approx 100 \text{ s}$$

is calculated. It should be remembered, however, that [OH] is a calculated value and that the estimate may be not be particularly conservative.

At the other end of the scale, there is urea (H$_2$N–CO–NH$_2$), a very slowly reacting chemical used mainly as a fertilizer and which is thus likely to enter the environment. With $p_{20} = 0.86$ mPa [95], urea belongs to the SOCs. Using the estimated day/night average $<[OH]>_{w,day/night} = 0.5$ pmol L^{-1} for this persistent compound and

$$k_{OH} \text{ (water)} \approx 2 \times 10^6 \text{ L mol}^{-1} \text{ s}^{-1} \text{ [114]}$$

one obtains the following OH lifetime:

$$\tau_{OH}(\text{droplet}) \approx \frac{1}{k_{OH} \times <[OH]>_{w,day/night}} = 10^6 \text{ s} \approx 12 \text{ d}$$

---

[18] 2-Nitrophenol reacts only slowly with OH in the gas phase, $k_{OH} = 9.0 \times 10^{-13}$ cm^3 molecule^{-1} s^{-1} [132] and therefore does not fit into the linear log k_{OH}(water) versus log k_{OH}(air) plot; the same is true for the parent compound benzene [114].

It is not known whether or not this substance, which is easily bio-degraded in natural waters and in moist soils, actually occurs in the atmosphere. If it does, it could show long-range transfer due to the lack of any efficient sink in the troposphere. Washout is possible, however, due to the high water solubility ($S = 900$ g L^{-1}; $H_{20} = 2.6 \times 10^{-11}$ [95]).

Dominant organic compounds in the droplet phase are [106]:

Formic acid: <[HCOOH]> (rain in remote areas) ≈ 0.7 mg L^{-1}
k_{OH} (water) = 2.8×10^9 L mol^{-1} s^{-1} [126]
τ_{OH} (droplet, day) = 6 min

Acetic acid: <[CH$_3$COOH]> (rain in remote areas) ≈ 0.3 mg L^{-1}
k_{OH} (water) = 8.5×10^7 L mol^{-1} s^{-1} [126]
τ_{OH} (droplet, day) = 3.3 h

Oxalic acid is present in cloud water and fog in concentrations of 0.2–20 µmol L^{-1} [140]. This substance seems to be an important degradation product of high-molecular natural substances. The OH-reaction rate constant depends on the pH, the undissociated form can not really exist in natural clouds:

Oxalic acid (C$_2$H$_2$O$_4$) (water, pH 0.5) = 1.4×10^6 L mol^{-1} s^{-1} [141]

Oxalate (C$_2$HO$_4$)$^{1-}$ (water, pH 3.0) = 4.7×10^7 L mol^{-1} s^{-1} [141]
Oxalate (C$_2$O$_4$)$^{2-}$ (water, pH 6.0) = 7.7×10^6 L mol^{-1} s^{-1} [141]

τ_{OH} (droplet, pH 3, day) = 5.9 h
τ_{OH} (droplet, pH 6, day) = 36 h

Pure oxalic acid and its anions are degradable by OH in the droplet phase. As mentioned in the paragraph on metal ions, catalytic degradation appears to be the main sink [124].

3.2.4 Summary

The "abiotic degradation in the droplet phase" sink is probably an efficient system for those organic substances able to get there. The selective situation, which allows only a small fraction of the organic molecules to enter, is due to the stringent conditions for access. Only very polar substances (very low Henry coefficient) can enter from the gas phase under equilibrium conditions. Even this may be hindered by adsorption at the surface of the droplets [240]. Others may be "smuggled in" via aerosol particles, e.g., acting as condensation nuclei.

The only pollutant for which the droplet phase is known to be the decisive sink is the inorganic compound SO$_2$. A special case is the absorption of N$_2$O$_5$/NO$_3$ [131] causing the so-called "acid dew".

4
Experimental

4.1
Indirect Photochemical Degradation

4.1.1 Bimolecular Reaction with OH

4.1.1.1 Direct Methods for Measuring k_{OH}

The experimental procedure consists in measuring the second-order reaction rate constant k_{OH} of a gas-phase reaction, Eqs. (4.1) and (4.2) [2, 58]:

$$A + OH \rightarrow products \qquad (4.1)$$

$$-d[A]/dt = k_{OH}[A][OH] \qquad (4.2)$$

When using a direct method [2], the concentration of test substance A ([A]) is kept constant during the reaction and [OH] is measured as a function of time (t). The variants of the method differ according to the way [OH](t) is measured. Furthermore, various methods are used to create the reactive radical OH, e.g., by photolysis of H_2O, H_2O_2 or other precursors, such as HONO [2]. As the concentration of the test substance is constant and much higher than the concentration of OH

$$[A] = const. \gg [OH]$$

the reaction is first order with respect to OH (Eq. 4.3):

$$[OH](t) = [OH]_0 \exp(-k_{OH}[A]t) \qquad (4.3)$$

In a (linear) plot of ln[OH] versus t, the slope $-k_{OH}[A]$ can be used to determine k_{OH}.

The standard method used is known as the Resonance Fluorescence (RF) Method and dates back to the pioneering work of Stuhl and Niki [133]. The concentration of the hydroxyl radical is measured by the fluorescence of OH* as a function of time [19, 136]. As the electronically excited hydroxyl radicals are deactivated by oxygen[19], the measurement has to be performed under an inert gas (He, Ar, N_2) and preferably at reduced pressure.

The most important components are shown in Fig. 2.

1. N_2-flash lamp (formation of OH).
2. OH-resonance lamp (H_2O/microwave discharge at 2 hPa in Ar).
3. photomultiplier (detection of the OH-resonance fluorescence).

---

[19] "Quenching" of the fluorescence.

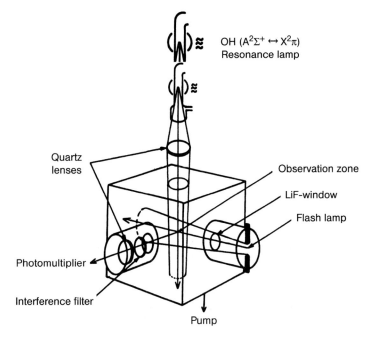

Fig. 2 Schematic representation of the flashphotolysis/resonance fluorescence system according to Zetzsch [136].

In another variant, the N_2 lamp is replaced by a pulsed excimer laser, giving more flexibility with respect to the OH source (see Zellner in [136]) and avoidance of the photolysis of photosensitive test substances.

The advantages of the direct method(s) are:

1. preciseness, absolute rate constant,
2. measurement of small k_{OH},
3. measurement of k_{OH} as a function of temperature and pressure,
4. can be automated.

There are also a few disadvantages:

1. artificial experimental conditions (exclusion of O_2, usually low pressure),
2. the test substance has to be very pure (especially in the case of slowly reacting substances)[20],
3. interferences due to photolysis of the test substance are possible.

There are other, more sophisticated OH-detection methods allowing the direct measurement at normal pressure in air [136]:

---

[20] In this instance, a small amount of a highly reactive impurity can produce wrong (too high) reaction rate constants.

- OH absorption using a long light path according to Wahner and Zetzsch [137],
- OH laser-induced fluorescence as described by Schmidt et al. [138].

"Discharge-flow" systems also belong to the direct methods. In these methods adsorption effects can disturb the measurement [139], especially at low temperature. Such effects can, however, never be completely avoided and pose severe problems in the case of SOCs.

4.1.1.2 Indirect Methods for the Measurement of k_{OH}

The indirect determination of k_{OH} in the gas phase, the "smog-chamber method", consists in measuring the degradation of the test substance *relative* to one or several reference substances in the presence of OH radicals [2, 19, 25, 58, 136, 142]. The absolute OH-reaction rate constants of reference compounds are known (e.g., "R" in the table of rate constants and quantum efficiencies in Chapter 3 of this book or the recommended values by Atkinson [20, 27] and Atkinson et al. [24, 147,148]). In order to become recommended, the rate constants should have been measured by at least two teams as a function of temperature in the relevant temperature range. Mechanistic details should have been discussed, at least for instances of pressure dependence or odd temperature behaviour, as are often encountered in OH addition reactions. Furthermore, a suitable reference compound should react with OH specifically (e.g., toluene), as ozone and OOH are always formed in OH-containing reaction mixtures. These less reactive species can, due to their lower reactivity, reach much higher concentrations compared with OH and in this way disturb the measurement of k_{OH}. NO_3, the "OH of the night" is not a competitor in relative k_{OH} rate constant measurements due to its very short photolytic lifetime.

The experiments are carried out in large vessels made of Pyrex (borosilicate) glass or "Teflon" (i.e., transparent Tedlar-foils)[21], so-called "smog-chambers" [11, 25, 58, 134, 136, 142]. Highly volatile and rapidly degrading chemicals can also be measured in small vessels [136], so that the range of volumes of smog chambers that can be used amounts to about 10–10 000 L (10 m^3).

A convenient version of a smog chamber (see Fig. 3) consists of a Pyrex-glass cylinder surrounded by special low-pressure Hg lamps showing maximum radiation intensity in the spectral range 300–400 nm. The pressure within the chamber is usually kept slightly above 1 bar (0.1 MPa) and, if it is anticipated that no temperature control will be required, slightly above room temperature (ca. 300 K). Teflon chambers are operated at the pressure of the surrounding air.

The intensity of the radiation used to produce the OH-radicals is conveniently measured by k_1 of reaction (4.4):

$$NO_2 + h\nu \xrightarrow{k_1} NO + O \tag{4.4}$$

---

[21] Pure Teflon $(C_2F_4)_n$ is not suitable for thin foils; Tedlar® (DuPont) is poly(vinyl fluoride) or a copolymer.

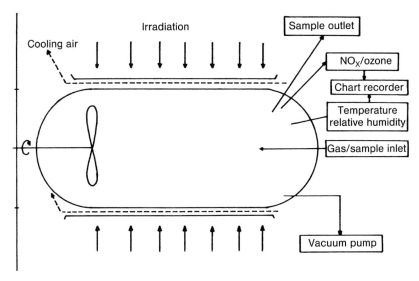

Fig. 3 Schematic representation of a smog chamber for measuring relative k_{OH} reaction rate constants at room temperature [136].

The values reported for this first-order rate constant are in the range:

$$0.2 \text{ min}^{-1} < k_1 < 1 \text{ min}^{-1}$$

The corresponding value for the short-wavelength fraction of the solar radiation (at the surface of the earth) amounts, under favourable conditions, to $k_1 \approx 0.5 \text{ min}^{-1}$ [143–145]. High radiation intensity is believed to shorten the time required for an experiment. In contrast to direct photochemical transformation ("photolysis"), however, there is no direct correlation between the radiation intensity and the time needed to produce a measurable decay of the test substance. More important is the concentration of OH-radicals [OH], which depends on k_1, the OH precursor (UV absorption, quantum efficiency), and secondary reactions of OH with other components in the reaction mixture.

Readily degradeable compounds (k_{OH} near to the diffusion limit), e.g., aniline, most unsaturated terpenes and the somewhat less reactive olefins, can be studied with a "weak" OH source, such as photochemical smog. In order to create such a smog in the chamber, it is sufficient to UV-irradiate ordinary polluted laboratory air. The hydrocarbons (VOC) and NO_x always present in the air form the well-known photochemical smog in which OH radicals play a central role. Even with this simple method of generating OH it is advisable to use synthetic air + NO + olefine (1 ppmv each) to begin the smog reaction. Ozone accumulates as soon as the NO is spent [2, 14, 35] and may reach high concentrations – as in the lower troposphere under smog conditions. During the measurement of k_{OH}, high [O_3] can seriously disturb the measurement if the test substance also reacts with O_3 (e.g., propene).

In the troposphere, OH is formed by photolysis of ozone and reaction of electronically excited oxygen atoms formed in reactions (2.1 + 2.2). As reaction (2.1) requires UV radiation at wavelengths shorter than 320 nm, whether the reaction will take place or not depends on the material used for the smog chamber. It certainly will if Tedlar foil is used, as this material does not absorb UV radiation in the spectral region 290–320 nm. Pyrex glass, however, in the thickness used, will absorb most of the radiation in this region. In Pyrex chambers, OH is most likely formed according to reaction (2.3), the photolysis of HONO into OH and NO.

The average [OH] formed in this way, which stays constant over several hours, amounts to [14, 25, 142, 149]:

$$\langle[OH]\rangle_{photosmog} \approx (1\text{–}5) \times 10^6 \text{ radicals cm}^{-3}$$

The average [OH] determined by Akimoto et al. [149] ($n = 24$) is

$$\langle[OH]\rangle_{av} = (2.5 \pm 1.4) \times 10^6 \text{ radicals cm}^{-3}$$

Much higher initial concentrations of [OH] up to 10^8 radicals cm^{-3} can be achieved with the help of externally generated HONO as the OH source [25] leading to a corresponding shortening of the time necessary to perform the experiment. However, the *average* [OH] achievable is lower, compared with the initial values [25]:

$$\langle[OH]\rangle_{av}(HONO) \approx (1\text{–}5) \times 10^7 \text{ radicals cm}^{-3}$$

Using this method Cox and Derwent [150] achieved an average ($n = 19$)

$$\langle[OH]\rangle_{av}(HONO) = (3.9 \pm 1.3) \times 10^7 \text{ radicals cm}^{-3}$$

Similar results were reported by Klöpffer and coworkers [25, 142] and by Kerr and Sheppard [151]. As HONO is rapidly spent ($\Phi = 1$; absorption in the near-UV), it has to be supplied every 15 min in the form of gas [25]. With this method, relatively low k_{OH} values in the order of several 10^{-13} cm^3 molecule^{-1} s^{-1} can be measured using the relative method.

Another suitable source of OH is the photolysis of alkyl nitrites [152]. This reaction primarily yields alkoxy radicals RO (4.5a), which react with oxygen to give hydroperoxyl radicals (4.5b) and are finally reduced to OH by NO (4.5c):

$$RO\text{–}NO + h\nu \rightarrow RO + NO \tag{4.5a}$$

$$RO + O_2 \rightarrow OOH + \text{carbonyl} \tag{4.5b}$$

$$NO + OOH \rightarrow NO_2 + OH \tag{4.5c}$$

The OH sources considered so far have the disadvantage of inevitably inducing smog formation due to the additional introduction of NO_x and eventually leading to undesirable side reactions under long irradiations (as required in the case of recalcitrant molecules). Looking for a "clean" OH source, photolysis of hydrogen peroxide seems to be a good choice [61]. H_2O_2 is photolysed to OH (Eq. 4.6a), similar to in water (Section 3.2), but with doubled quantum yields:

$\Phi(-H_2O_2)_{gas} = 1.0$ [24]

$$HO-OH + h\nu \rightarrow 2\,OH \tag{4.6a}$$

the primary quantum yield related to OH formation $\Phi'(OH) = 2.0$.

The follow-up reactions are very similar to those occurring in water (Fig. 1 and subsequent reactions) except for the fact that the hydroperoxyl radical (OOH) cannot dissociate in the gas phase and, thus, any complications due to O_2^- do not occur. The basic scheme was proposed by Volman in his pioneering work 1949 [154]; the key reaction is the OOH formation with loss of OH (Eq. 4.6b):

$$OH + HOOH \rightarrow HOH + OOH \tag{4.6b}$$

$k\,(4.6b) = 1.7 \times 10^{-12}$ cm^3 molecule^{-1} s^{-1} [155]

In analogy to the corresponding reaction in water, hydrogen peroxide is partly regenerated in reaction (4.6c):

$$OOH + OOH\,(+M) \rightarrow HOOH + O_2 \tag{4.6c}$$

$k\,(4.6c)\,(p = 0.1\text{ MPa}) = 2.5 \times 10^{-12}$ cm^3 molecule^{-1} s^{-1} [60]

A complete reaction scheme is given in Ref. [61]. Owing to the high quantum yield, the weak UV absorption of hydrogen peroxide above 295 nm is sufficient to produce OH. Some experiments using H_2O_2 as the OH source resulted in disappointingly low [OH] ≈ 2×10^6 OH-radicals cm^{-3}; much higher concentrations could be expected on the basis of known reaction-rate constants for practically all partial processes [61]. Very similar results were obtained by Krüger et al. (2005) [96], higher stationary [OH] being reported for methyl nitrite as the OH source or the dark-OH source hydrazine/ozone.

Very high OH concentrations up to [OH] = 5.3–12 $\times 10^{10}$ radicals cm^{-3} have been reported by Chen et al. (2003) [170] in relative rate measurements. The OH source is photolysis of O_3 in the presence of water vapour, the same reaction that produces most of the OH in the troposphere. The high concentrations were not obtained in conventional smog chambers in the stationary mode, but rather with a flow-through system and steady-state kinetics. Chen et al. [170] reported that they were able to measure (relative) k_{OH} values down to 10^{-16} cm^3 molecule^{-1} s^{-1}.

Some test substances are very sensitive towards UV radiation (photolysis). In these instances it is advisable to use one of the following dark sources of OH:

1. $H_2O_2 + NO_2 + CO$ (Eq. 4.7) [157]

   Source:

   $$H_2O_2 + NO_2 \rightarrow OH + HNO_3 \quad (4.7a)$$

   $$OH + CO \rightarrow CO_2 + H \quad (4.7b)$$

   $$H + NO_2 \rightarrow OH + NO \quad (4.7c)$$

   Sink:

   $$OH + NO_2 \, (+M) \rightarrow HNO_3 \quad (4.7d)$$

   M = inert collision partner (e.g., Ar)

2. Peroxynitric acid + NO (Eq. 4.8) [158]

   $$HO_2NO_2 \, (+M) \leftrightarrow OOH + NO_2 \quad (4.8a)$$

   $$OOH + NO \rightarrow OH + NO_2 \quad (4.8b)$$

   k (4.8b) $= 1.1 \times 10^{-11}$ cm^3 molecule^{-1} s^{-1}

3. Hydrazine + ozone (Eq. 4.9) [159]

   $$N_2H_4 + O_3 \rightarrow N_2H_3 + OH + O_2 \quad (4.9)$$
   + follow-up reactions

   $\langle[OH]\rangle \approx 2 \times 10^7$ radicals cm^{-3}

The actual measurement in all indirect or relative methods consists in following up the concentrations of test and reference substance(s). As [OH] is constant over time spans comparable to the time needed for the experiment, first-order kinetics are frequently observed (see Section 5). The unknown k_{OH} can simply be calculated from the relative slopes of ln[A] vs. t and ln[R] vs. t (A = test substance; R = reference) and the known $k_{OH}(R)$. Alternatively, [OH] can be calculated first, Eq. (4.10)

$$\ln[R]/[R]_0 = -k_{OH}(R)[OH]t \quad (4.10)$$

Using the now known concentration of [OH], the unknown $k_{OH}(A)$ of the test substance is calculated according to Eq. (4.11):

$$\ln[A]/[A]_0 = -k_{OH}(A)[OH]t \quad (4.11)$$

There is a general evaluation scheme that can also be used if [OH] is *not* constant during the experiment. It is based on the following argumentation by Cox and Sheppard (1980) [160], Eqs. (4.12) and (4.13):

$$d[X]/dt = -k_{OH,X}\,[X][OH] \tag{4.12}$$

where [X] = concentration of either A (test substance) or R (reference substance) in the reaction chamber; $k_{OH,X}$ = second-order reaction rate constant of either A or R.

It is clear from Eq. (4.12) that the velocity of the disappearance of X varies over time if the concentration of the OH radicals changes during the experiment.

Integration of Eq. (4.12) yields (Eq. 4.13):

$$\ln\frac{[A]_0}{[A]_t} = \frac{k_{OH,A}}{k_{OH,R}} \times \ln\frac{[R]_0}{[R]_t} \tag{4.13}$$

A plot of all pairs of corresponding concentration data (measured at equal times t and divided by the initial concentrations of A and R at $t = 0$) has to have its origin at 0 and the slope = ratio of the OH reaction rate constants. Such a plot is shown schematically in Fig. 4.

This "universal plot", from Cox and Sheppard, should always be used if [OH] changes during the experiment, a case which can easily be recognized if the decay of [A] and [R] does not follow first-order kinetics and, thus, an unambiguous calculation of [OH] according to Eq. (4.10) is not possible. Note that no absolute concentrations of R and A are necessary in this evaluation. The relative concentration of A has to vary during the experiment significantly, so that the quotient $[A]_0/[A]_t$ can be determined accurately and t has to be chosen accordingly. There is a practical limit, however, in the case of slowly degrading

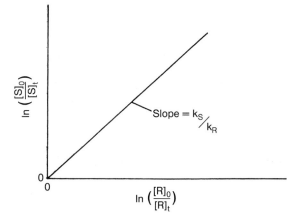

Fig. 4 Evaluation of relative reaction rate constants after Cox and Sheppard (reproduced from [58]); "S" corresponds A in Eq. (4.13).

substances, as the experiments cannot be prolonged indefinitely. There is also the prerequisite that OH is actually the *only* sink for A. Ozone and direct photochemical transformation ("photolysis") are obvious candidates for additional sinks if UV is used to create OH radicals. If dark OH sources are used, ozone (e.g., if used together with hydrazine) and NO_3 are potential competitors.

The advantages of the smog chamber methods are:
1. The reaction conditions are close to the natural ones with regard to pressure, temperature and chemical composition of the gas.
2. The method can be used for the more volatile end of the SOCs, in favourable cases down to $p(A) \approx 1$ cPa.[22]
3. The test substance can contain a certain amount of impurities, as is often the case in technical grade chemicals, if substance-specific analytical methods are used.
4. Several substances can be measured in one experiment.

There are also some disadvantages:
1. Pressure (P) and temperature (T) cannot be varied significantly (with the exception of some "luxury chambers", which are not used very often, probably for reasons of cost).
2. The method can in general only be used for $k_{OH} > 10^{-13}$ cm^3 molecule^{-1} s^{-1} (reason: time of measurement).
3. The analytical determinations may be fairly difficult, especially in the case of SOCs (low p = low concentration = low absolute amount in the chamber).
4. Long experimentation times are required for multiple measurements; this disadvantage is partly off-set by advantage 4.

4.1.2 Bimolecular Reaction with NO_3

4.1.2.1 Introduction

The measurement of the second-order reaction rate constant of the general reaction (4.14a) and the corresponding kinetics (4.14b)

$$NO_3 + A \rightarrow products \tag{4.14a}$$

$$-d[A]/dt = k_{NO_3}[NO_3][A] \tag{4.14b}$$

can, as in the case of the corresponding reactions for OH, be performed with indirect (relative) or direct (absolute) methods. The advantages and disadvantages of the two methods are similar to those discussed for OH. The main difference in the indirect method consists in the fact that NO_3 is always in equilibrium with NO_2 and N_2O_5. This equilibrium depends on the temperature, Eqs. (2.12) and (2.13) and Table 2.1, and can be used to create a suitable concentration $[NO_3]$ to

---

[22] A general pressure limit cannot be given, as specific interactions of compounds with the wall appear to exist; this was the experience with phenols in a Pyrex glass chamber.

conduct the measurement of k_{NO_3} in the dark. Remember that the radical NO_3 is photolytically unstable, so that the measurement has to be carried out in the dark. Exclusion of UV alone is not sufficient, as the long-wavelength absorption of the radical is situated in the visible region of the electromagnetic spectrum.

4.1.2.2 Absolute Measurement

The absolute measurement methods set up by the groups working with Schindler [161, 162], Wayne [40, 163] and Le Bras [164] use a flow tube for temporal resolution. In this method, NO_3 is created mainly by reacting fluorine atoms with gaseous nitric acid (4.15):

$$F + HNO_3 \rightarrow HF + NO_3 \quad (4.15)$$

$$k\ (4.15) = (2.1 \pm 1) \times 10^{-11}\ cm^3\ molecule^{-1}\ s^{-1}\ [161]$$

The F-atoms are formed in a microwave-discharge of F_2 in He carrier gas. The flow tube has a diameter of 2.5 cm [161] or 3.8 cm [163] and length about 1 m. The reaction can also be carried out, in principle, with a surplus of NO_3 or test substance in order to give first-order kinetics, as discussed in Section 4.1.1 for OH. As a rule of thumb, one component should be in at least a ten-fold excess in order to guarantee pseudo-first order kinetics [29].

The method of detection is different in the two systems: the British group uses the absorption of NO_3 at 662 ± 2 nm [163], which requires a surplus of the test substance (otherwise [NO_3] would be constant). The German group uses detection by a quadrupole mass spectrometer, allowing both NO_3 and substance excess. Furthermore, this variant of the direct method allows for the identification of the reaction products (4.14a).

A detailed description of the properties of the nitrate radical, including its creation and detection, photochemistry and absorption properties, has been given by Wayne et al. [29]. In this context, several kinetic NO_3 measurement methods are shown in detail. Detailed descriptions of experimental procedures can also be found in a more recent book by Finlayson-Pitts and Pitts Jr. (2000) [2].

Berndt et al. (1996) [165] also used a tube, but created NO_3 by heating N_2O_5. The source is stored externally at ca. 220 K, the heating occurs at 450 K in a silicon oil bath. The detection system consists of GC-MS/FID. The concentration of the organic molecules (cyclic, conjugated dienes) was in the range (2–12) × 10^{12} molecule cm^{-3}, the conversion achieved was 30–90% at a flow velocity of ca. 1–3 m s^{-1}. It should be taken into account that this type of "zero-persistent" compounds (terpenes and related substances) react very rapidly with NO_3 (as with OH and also O_3).

4.1.2.3 Relative Measurements

Many NO_3 reaction rate constants were determined using relative methods [36, 37, 41, 43]. A smog chamber was used to take these measurements, as described for OH, but without a light source. The NO_3 is produced by thermal decomposition of N_2O_5 (see Table 1). If the temperature can be kept at precisely the same temperature during the experiment, this method could even be used for absolute measurements, without a reference compound [166]. In general, however, reference compounds with recommended k_{NO_3} values should be used, measured according to an absolute method and preferably as a function of temperature. They should not react with N_2O_5, as some PAHs do [2]. In contrast to NO_3, N_2O_5 is a nitrating agent and reacts with PAH to give nitroarenes.

A reference substance recommended by Atkinson [41] is *trans*-butene:

$$k_{NO_3} (296 \pm 2 \text{ K}) = (3.87 \pm 0.45) \times 10^{-13} \text{ cm}^3 \text{ molecule}^{-1} \text{ s}^{-1}$$

For other reference compounds see substances marked "R" (recommended) in the table of reaction rate constants and quantum efficiencies in Chapter 3. More detailed information on the experimental conditions can be obtained in the reviews by Wayne et al. [29], Atkinson [27, 43] and Atkinson et al. [147, 148].

4.1.3 Bimolecular Reaction with Ozone

The measurement of the second-order reaction rate constant (test substance + $O_3 \rightarrow$ products) is almost exclusively carried out in smog chambers, using either the test substance (A) or ozone in excess in order to obtain pseudo-first order kinetics [25, 47, 58, 167]. It is recommended, however, to use test substance (A) in relative excess (OECD 1992 [58]) in order to avoid interferences due to high concentrations of ozone, e.g., spurious formation of OH radicals:

$$[A] = \text{const.} \gg [O_3]$$

Any reaction not due to ozone has to be avoided. This is particularly important because this species is less reactive by orders of magnitude compared with OH, and less pronounced relative to NO_3. Thus, very inefficient side reactions of ozone may contribute to the decay of A and simulate too high a value for k_{O_3}!

With [A] constant and much higher than $[O_3]$, a pseudo-first order kinetics with regard to ozone applies (linear $\ln[O_3]$ vs. t; slope k^I):

$$k^I = k_{O_3} \times [A] \tag{4.16}$$

This determination of k_{O_3} is absolute and a substance-specific analytical measurement is not necessary unless the concentration of [A] is being checked. The actual experiment consists in the measurement of $[O_3]$ as a function of time. This seems to be too simple to be true, and indeed there are at least two restrictions:

- The test substance, which has to be applied in excess, has to be rather volatile; the lower limit with respect to the vapour pressure is given by OECD [58] $p > 10$ Pa.
- It has to be very pure in order to avoid any decay of ozone not due to A. This is especially critical in very slowly reacting test substances (i.e., most of them), as a very low concentration of a highly O_3-reactive impurity would give a completely wrong k_{O_3}.

Ozone also reacts frequently with the wall of the chamber {recommended by the OECD [58]: "Teflon" (Tedlar®)}. This effect has to be quantified separately and, if measurable, be taken into account as a correction of the ozone decay.

In the second method (not recommended by the OECD [58], but what happens to the low volatile test compounds or SOCs?), $[O_3]$ is kept constant and well above the concentration of the test substance:

$$[O_3] = \text{const.} \gg [A]$$

As ozone is a gas, technically this is always possible. The decrease in the test substance $\{[A] = f(t)\}$ is measured using a substance-specific analytical procedure. The method is, if $[O_3]$ is known, also absolute. A reference substance increases, however, the reliability of the determination of k_{O_3}. The method is, in principle, also independent of the purity of the test substance. Occasionally deviations from the simple pseudo-first order kinetics have been observed, most probably due to catalytic chain reactions. Interferences due to traces of OH, which can be formed in dark reactions, are best excluded through an excess of a volatile organic substance that reacts with OH but not at all with O_3, e.g., n-octane [48].

Reference compounds marked "R" (recommended) should be used. More details can be obtained in reviews by Atkinson and Carter [47] and Atkinson [27].

4.2
Direct Photo-transformation

4.2.1 Determination of the Quantum Efficiency in the Gas Phase

4.2.1.1 **Gas Cuvette and Monochromatic Radiation**

The photochemistry in the gas phase was presented for the first time in the classic monograph by Noyes and Leighton (1941) [62]. In this book and in another classic, Calvert and Pitts [51], the basic experimental techniques of gas-phase photochemistry were described. They have not changed much since then, at least not if compared with the progress in ultra-short kinetic and spectroscopic photochemical research. Lasers are not necessary to perform a reasonable determination of a quantum yield. Modern analytical equipment (e.g., GC/MS) may help, if additional information about the reaction products is required.

The key information to be obtained is the quantum efficiency $\Phi(\lambda)$, which can be used to calculate the direct photochemical lifetime $\tau_{h\nu}$ for any known spectral irradiance $I(\lambda)$ to be encountered in the natural environment (see Section 2.2.1). The only additional information required is the UV/VIS absorption spectrum of the test substance at wavelengths $\lambda \geq 290$ nm (the short wavelength cut-off for solar radiation in the troposphere). Ideally, this spectrum should be measured in the gas phase, but in the (very frequent) case of low resolution of the vibrational fine structure of the spectra, an absorption spectrum recorded in a non-polar solvent is also sufficient and suitable for calculating the lifetime.

The OECD Environment Monograph No. 61 [58] recommends a method originally developed by Mill et al. for the US-EPA [241]. It essentially consists of a gas cuvette, as used in gas-phase photochemistry, and an actinometric method for measurement of the irradiation intensity.

In order to establish whether a substance can undergo a direct photo-transformation in the atmospheric environment at all, the test substance should be pre-selected using Eq. (4.17):

$$k_{hv} \leq \sum \sigma(\lambda)\, I(\lambda)\, \Delta\lambda \quad [s^{-1}] \tag{4.17}$$

Equation (4.17) is derived from Eq. (2.28) using the approximation $\Phi = 1$, the highest possible quantum efficiency of disappearance of a compound. This calculation gives an upper limit of the first-order rate constant k_{hv} and, thus, a *lower limit of the photochemical lifetime*, as

$$\tau_{hv} = 1/k_{hv}$$

If in this estimate the lifetime turns out to be many years, a measurement of Φ is not meaningful.

To measure the quantum efficiency, a cylindrical reaction vessel with flat quartz windows is proposed (see Fig. 5) with an optional water jacket for thermostatic control. The vessel should fit into a UV/VIS absorption spectrophotometer in order to follow the reaction directly (in favourable cases, i.e., no disturbance by reaction products). If other analytical techniques have to be used (GC, GC/MS, HPLC), frequently the whole content of the cuvette has to be sacrificed in order to provide sufficient substance for an analysis. In this instance, one experiment only yields one point on the degradation curve.

The whole experimental set-up should preferably be mounted on an optical bench. If a strong source is used (e.g., 1000 W Hg high pressure) a water filter should be placed between the lamp and the cuvette to eliminate the IR radiation. The radiation should illuminate the gas volume as completely as possible, but not the wall. The cuvette is connected to a gas-handling system, to be used for filling the reaction cell with either the test substance or the actinometer gas. The

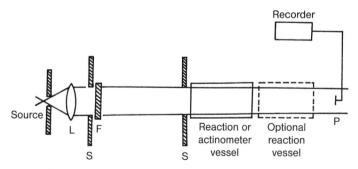

Fig. 5 Cylindrical reaction vessel with flat quartz windows.
L = less; S = slits; F = filter; P = probe (detector) (Reproduced from [58]).

wavelength used for irradiation is selected either by means of an interference filter for the appropriate wavelength or with a small monochromator. Alternatively, a gas laser or a dye laser can be used as a monochromatic light source.

The actual measurement consists in analysing the test substance (A) as a function of irradiation time. A semi-logarithmic plot of [A] vs. t should give a straight line, indicating first-order kinetics. In this instance, the (negative) slope is the first-order rate constant of direct photochemical degradation ($k_{h\nu}$). In general, this is the situation for short periods of irradiation, because at longer times reaction products may interfere and complicate the reaction path, either by slowing down or by accelerating the decay of the test substance. These additional reactions may be very interesting from a scientific point of view, but in testing the ability of a substance to degrade it is rather a nuisance.

In order to secure simple reaction conditions, the absorbance ($E = \log I_0/I$) of the test substance at the irradiation wavelength (λ_{irr}), E', should be small:

$$E' \leq 0.02$$

In order to determine a possible (frequently observed) dependence of $k_{h\nu}$ and Φ on λ_{irr}, the experiment should be conducted with at least two irradiation wavelengths. This is not necessary if, as is often the case, the test substance shows only an absorption tail in the spectral region $\lambda \geq 290$ nm. In the case of well developed absorption bands in the near-UV or even in the visible range, two or more λ_{irr} should be used.

The quantum efficiency Φ_λ is calculated for each λ_{irr} according to Eq. (4.18)

$$\Phi_\lambda = k_{h\nu}/\sigma_\lambda I_\lambda \tag{4.18}$$

Integration of Eq. (4.18) over the whole UV/VIS range yields Eq. (2.28), but $k_{h\nu}$ and I_λ are related to the experimental set-up, not to the environment. $k_{h\nu}$ is the experimental first-order rate constant at λ_{irr} and the specific conditions of the irradiation experiment. I_λ is the spectral irradiance at λ_{irr} and depends on the lamp, the monochromator or filter and other experimental details. It has to be measured with an independent experiment using an actinometer gas of known quantum efficiency [51] at λ_{irr}. OECD [58] recommends NO_2 as the actinometer gas at wavelengths $\lambda_{irr} < 390$ nm ($\Phi = 1.9 \pm 0.1$)[23] and NOCl for the region $380 < \lambda_{irr} < 600$ nm ($\Phi = 2.0$). The calculation of I_λ can also be made using Eq. (4.18), $k_{h\nu}$ being the measured first-order rate constant of the actinometer gas at λ_{irr}, σ_λ the absorption cross section of the actinometer gas at the same wavelength and Φ the quantum efficiency of the actinometer gas.

Mill's method as adopted by OECD [58] is restricted to substances that occur in the gas phase to at least 80%. This condition limits the method to volatile substances:

---

[23] In this case Φ is greater than 1, as a secondary reaction removes a second molecule of NO_2 or NOCl.

$p_{25} \geq 10$ Pa

It is therefore not suitable for measuring SOCs.

4.2.1.2 Smog-chamber Method

The potential usefulness of smog chambers for measuring Φ of less volatile compounds has been pointed out by Klöpffer et al. [25]. In this method, the compound is tested in a large vessel, as used for measuring relative reaction rates of OH and O_3. The irradiation can occur through the Pyrex-glass cylinder mantle or through (in this instance) the flat top of the vessel. In the latter, the radiation can be conducted parallel to the long axis of the cylinder, giving a better defined irradiance (W cm^{-2} or photons cm^{-2}).

The experiment consists in measuring the concentration of the test substance as a function of the irradiation time [A](t). If first-order kinetics are observed, k_{hv} can easily be obtained from the slope of ln[A] vs. t provided that other reaction channels can be excluded. If, furthermore, the UV/VIS absorption spectrum of the substance and the spectral irradiance within the cylinder are known, Φ can be calculated under the assumption that the quantum efficiency does not depend on the wavelength.

In this experiment, strict exclusion of NO_x is mandatory, as even traces of this gas induce the formation of photochemical smog and, thus, OH radicals and ozone! In order to secure the absence of these reactive species, non-UV-absorbing substances, such as propene (high k_{OH} and k_{O_3}), should be added and measured together with the test substance. If the concentration of the probe remains constant over the irradiation time, no smog has formed then the decay of the test substance should indeed be due to direct photo-transformation.

It should be noted that the use of large outdoor chambers, such as the EUPHORE facility in Valencia, Spain [169] is related to the suggested smog-chamber method. Under such "semi-natural" conditions OH reactions also occur (same order of magnitude in the case of aldehydes), thus the radicals formed in the photolytic reaction (Norrish-I splitting) have to be measured and the yield is compared with a theoretical value (Φ = 1). In this way "effective quantum efficiencies" have been determined. The spectral irradiance during the experiment has to be measured and the absorption spectra of the substances must be known.

4.2.2 Outlook

Although methods for measuring the photochemical parameters required for estimating the strength of this sink in the atmospheric environment do exist, there are few papers dealing explicitly with the testing and assessment aspects. Aldehydes, ketones and a few other classes, known to be photochemically active, have been investigated for scientific reasons. In environmental fate assessment, direct photo-transformation is largely ignored or, worse, the quantum efficiency is assumed to be unity. This is clearly nonsense, as pointed out by Moortgat [169]. Most quantum efficiencies are much lower, even in the known classes mentioned.

Compounds showing strong absorption in the near-UV or in the visible range (coloured compounds) can degrade in the atmospheric environment, even if Φ is very low. Absorption in these regions often requires large molecules belonging to the SOCs or even non-volatile compounds. These compounds are difficult or impossible to measure in the gas phase. If this is so, why not investigate the adsorbed state specifically with respect to direct photo-transformation? Inert solvents may also be used as "proxies" for the gas phase, as shown by Palm et al. and Zetzsch et al. for brominated diphenyl ethers [171a, b].

Pesticides are among the most carefully watched (and regulated) chemical products, the active ingredients belonging predominantly to the class of SOCs. In a review dated 1999 [172], the first-order rate constant k_{hv} is given for only six such compounds, mostly approximate values.

4.3
Degradation in the Adsorbed State

4.3.1 Introduction

Semi-volatile organic compounds (SOCs) offer a special challenge with regard to the measurement of abiotic degradation. They can occur in the gas phase [relatively high vapour pressure (p) and/or low n-octanol/air partition coefficient] or in the adsorbed phase depending on the abundance of aerosol particles and their total surface per volume of air, temperature and other, more specific factors [75, 78–81]. As many pollutants, e.g., most pesticides and all POPs identified for a global ban belong to the group of SOCs, it is of utmost importance to quantify their abiotic degradation.

The *gas-phase degradation* of SOCs (most studies are restricted to OH radicals) can be measured or estimated in several ways:

- Measurement in large smog chambers: this is possible for SOCs at the upper limit of the vapour pressure range, which defines the SOCs (see Section 3.1.1). A clear limit cannot be given, however, due to specific substance–wall interactions [25, 142].

- Measurement at elevated temperatures and extrapolation to relevant temperatures by means of an Arrhenius plot [104, 173–175]. This requires the measurement at three or more temperatures in the range of negligible adsorption effects. Furthermore, it has to be assumed that, in the temperature range below the measuring range and down to room temperature, there is Arrhenius behaviour (single activation energy). This condition is not granted in the case of OH- and/or NO_3-addition to unsaturated compounds or complicated kinetics.

- Measurement in another medium and extrapolation to the gas phase by means of "log–log plots": whereas the correlation of k_{OH} values measured in water and in the gas phase shows several serious outliers [113, 115, 140], measurements in the non-polar, OH non-reactive solvent 1,1,2-trichlorotrifluoroethane ("Freon 113") show a very good correlation with relative gas-phase reaction rate constants [97, 98, 176].

4.3.2 Aerosol Chambers

Aerosol chambers suitable for measuring the abiotic *degradation in the adsorbed state* require considerable effort in the construction, maintenance and carrying out of the experiments. In order to obtain the maximum of time during which the particles (artificial aerosol particles, mostly SiO_2) can be kept in the gas phase, a diameter near to the optimum of 0.1–1 µm is chosen. Furthermore, a small temperature gradient is applied in order to prevent the sedimentation of the solid phase of the aerosol. An advanced aerosol chamber of 2.4 m^3 volume is shown in Fig. 6 [177]. A more recent chamber, based on the same principle but designed to work at low temperature (6.5 °C, 3140 L, to prevent SOCs from evaporation) is discussed by Zetzsch and his group [96].

The chamber (Fig. 6) consists of a glass cylinder of 1-m diameter and 4-m high. Irradiation is through the upper flat Pyrex-glass cover and a 3-cm water filter, so that IR and UV ($\lambda < 310$ nm) radiation is removed. Together the lamps have an installed power of 8 $kW_{el.}$. The UV irradiation strength in the interior of the chamber, measured as k_1 of the NO_2-photolysis, amounts to:

$$k_1 = 1.15 \text{ min}^{-1}$$

The particles loaded with test and reference substances are generated by spraying an aqueous emulsion into dry, very pure air. The aerosol water droplets are dried and then introduced into the smog chamber. An exact thermostatic control is necessary to prevent convective flows leading to a deposition of the particles on the walls of the chamber. The degradation experiments are performed with about one monolayer of test substance adsorbed at the aerosol particles. The OH-radicals are produced by irradiation of OH sources (e.g., methyl nitrite or hydrazine/ozone). The decay of the substance is measured by analytical methods.

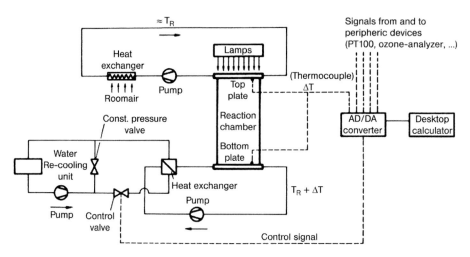

Fig. 6 Aerosol chamber designed for studying the degradation of SOCs adsorbed on artificial aerosol particles (reproduced from [136]).

As already discussed in Section 3.1.3, the results depend partly on the chemical composition of the particles. Aerosil (SiO_2 from Degussa) simulates a sand aerosol and is chemically neutral. The degradation is brought about by OH-radicals, the influence of which can be increased by adding an OH source into the chamber [96]. Semi-volatile pesticides have been successfully measured in the adsorbed state (OH, O_3) by Palm et al. [178, 179]:

$$k_{OH,ads} \text{ (terbuthylazine, 300 K)} = (1.1 \pm 0.2) \times 10^{-11} \text{ cm}^3 \text{ molecule}^{-1} \text{ s}^{-1} \text{ [178]}$$

$$k_{OH,ads} \text{ (pyrifenox, 299 K)} = (1.8 \pm 0.4) \times 10^{-11} \text{ cm}^3 \text{ molecule}^{-1} \text{ s}^{-1} \text{ [179]}$$

$$k_{O_3,ads} \text{ (pyrifenox, 299 K)} = (2 \pm 1) \times 10^{-19} \text{ cm}^3 \text{ molecule}^{-1} \text{ s}^{-1} \text{ [179]}$$

For a review of the earlier work, see Zetzsch (1991) [180]. An overview of recent techniques and detailed measurement results is available as a report to Umweltbundesamt, Berlin 2005 [96].

4.3.3 Alternative Measurements of $k_{OH,ads}$

The main disadvantage of the aerosol-chamber method is the great experimental expenditure, which hinders the measurement of a large number of substances on a reasonable time-scale. It was therefore timely to try a simplified method for measuring, above all, $k_{OH,ads}$. Basically, Palm et al. [181] coated aerosol particles (Aerosil 200 m^2 g^{-1}) with sub-monolayers of test substances, which were collected on filters and exposed to OH-radicals produced in a medium-sized chamber. The OH-radicals are created through conventional photochemical OH sources at 298 K, yielding [OH] of about $(0.7–8.0) \times 10^6$ cm^{-3}.

Under the experimental conditions given, first-order decay of the adsorbed SOCs was observed. In some instances it was possible to calculate the decay using $k_{OH,ads}$ data measured in the aerosol chamber. These comparisons showed that the filter method seems to underestimate the second-order reaction rate constant, i.e., the substances appear more persistent in the adsorbed state than they really are. The most plausible reason put forward by these workers (see also [96]) is the uneven distribution of the OH-radicals. The [OH] concentration may be smaller at the outer surface of the filter probes because of fast reaction, and the inner surfaces arc sheltered from OH-radical reactions. At high OH concentrations (see also [96]), the reaction at the outer surface may be faster than the diffusion of the test substance from the inner parts to the outer surface with the effect that the k_{OH} rate constant decreased with higher OH-radical concentrations. This hypothesis is supported by the findings of Balmer et al. [237], who established proof that such migration processes for photo-irradiation processes in probes of 2–3 mm thickness are actually occurring.

In another research project at the Umweltbundesamt Berlin, Rühl [182] reported a sophisticated method that served the same purpose: to measure $k_{OH,ads}$ in a small and compact apparatus (ca. 1500 cm^3) that would perhaps, if standardized, be suitable for measuring a greater number of test substances. This remains to be seen, however. The principle of this method is the observation of a single microparticle suspended in an "electrodynamic trap". This method has been developed by Paul [130] and Davis et al. [99]. A monolayer coating is used and the decay of the test substance is measured by Raman spectroscopy. In one instance (lindane = γ-hexachlorocyclohexane), it was possible to compare the measured $k_{OH,ads}$ (lindane) = (3.45 ± 3) × 10^{-13} cm^3 molecule^{-1} s^{-1} (room temperature), with good agreement, to $k_{OH,gas}$ = (1.4–2.5) × 10^{-13} cm^3 molecule^{-1} s^{-1} extrapolated from higher temperatures to room temperature by Brubaker and Hites [173]. According to this result it seems that the OH reactivities of molecules adsorbed on neutral (i.e., non-interacting) particles do not differ by much from those measured in the gas phase at the same temperature.

5
Additional Information Necessary for Calculating Lifetimes

5.1
Atmospheric Lifetimes

The measurement of reaction rate constants, quantum efficiencies, etc., is interesting in itself, as the results give insight into the reactivity as a function of chemical structure and may help in establishing quantitative structure reactivity relationships (QSRR). Although direct measurements are preferable to theoretical estimates, it is a fact that the sheer number of chemicals subject to the chemical legislation (not to mention the total number of known molecules), ca. 30 000 chemicals existing in Europe, necessitates the assistance of theory. A large number of carefully measured reaction rate constants is the best (or only) basis for scientifically meaningful QSRR to be used in the practice of chemical assessment and evaluation.

Reaction rate constants can be used in calculating the "lifetime" of a substance introduced into the environment. As we are dealing here with the atmosphere, more precisely with the lower 10–15 km of the atmosphere, the troposphere[24], only tropospheric lifetimes are discussed. Globally averaged or multi-media lifetimes are core-elements in the discussion and quantification of "persistence", the much debated central criterion of environmental hazard assessment of chemicals [183–189]. The elaboration of lifetimes restricted to the troposphere should be considered as a contribution to a solution of the persistence problem (including the related topic "long-range transport"), but only a partial one: the whole figure is more complex (see also Chapter 1).

---

24) The troposphere is separated from the stratosphere by a thermocline called the tropopause (10 km above the sea level near the poles, ca. 15 km near the equator); the warmer stratosphere "is sitting" on the very cold upper part of the troposphere.

The general procedure follows the path:

Reaction rate constant (general) $\xrightarrow{\text{(Environmental conditions-specific or average)}}$ Characteristic time

The reaction rate constants (or quantum efficiencies) can be used to calculate "characteristic times" for different regions, or, using global averages, global lifetimes. The specificity is introduced by the "environmental conditions", which determine for instance average OH concentrations, irradiation intensities, seasons, geographical regions, climate and temperature. This procedure is *superior to the older site-specific testing procedures* (e.g., *in situ* irradiation tests), which may yield reasonable results for a specific region, but cannot be used for extrapolations to other regions and climates. These latter testing methods, still much used in environmental assessment (especially of pesticides), should be replaced by the above reaction rate-based concept as far as possible (i.e., if a theory exists which allows the connection between "universal" measurement results and potential local or regional exposure).

If the environmental conditions are not known exactly, they can be approximated using average values or scenarios, especially if global boundary conditions have to be applied. In order to illustrate the general approach, a bimolecular reaction of environmental relevance (sink) is considered (Eq. 5.1):

$$A + B \rightarrow \text{products} \tag{5.1}$$

where A is the substance to be assessed and B is a reactive species present in the troposphere (OH, NO_3, O_3). The decay of A due to reaction (5.1) is described by the differential equation (5.2):

$$-d[A]/dt = k^{II}[A][B] \tag{5.2}$$

It is assumed that the substance A is only present in trace amounts, so that the average concentration of the reactive species B is not altered to a significant extent by A. Using this assumption (Eq. 5.3), the kinetics simplify to first order (Eq. 5.4), i.e., the reaction velocity depends only on a constant[25], the first-order reaction rate constant k^I and on the actual concentration of A at time t:

$$k^I = k^{II}[B] \tag{5.3}$$
(if [B] ≈ const.)

$$-d[A]/dt = k^I[A] \tag{5.4}$$

---

[25] Because reaction rate "constants" in most instances depend on the temperature and in certain instances on the pressure, it is often said that the k values should be called "reaction rate coefficients". We use the more usual term.

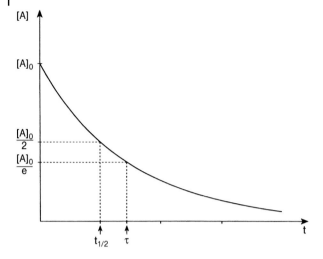

Fig. 7 Characteristic times in first-order kinetics.

Integration of (5.4) yields the exponential decay characteristic for a first-order reaction (5.5a) and Fig. 7.

$$A(t) = A_0 \exp(-k^I t) \qquad (5.5a)$$

The chemical lifetime τ is defined as the reciprocal of the first-order reaction rate constant k^I (Eq. 5.5b):

$$\tau = 1/k^I \qquad (5.5b)$$

$A_0/e \approx 0.37\ A_0$

The time after which A_0 decreased to $A_0/2$ (half-life, $t_{1/2}$) is, of course, shorter than the chemical lifetime τ (Eq. 5.5c). The term "half-life" is often used due to the better compatibility with more accepted language.

$$t_{1/2} = \tau \ln 2 \approx 0.69\ \tau \qquad (5.5c)$$

In Fig. 7, the two most frequently used characteristic times τ and $t_{1/2}$ are indicated. A third characteristic time is $t_{0.9}$, i.e., the time for 90% decrease of $[A]_0$, but this time is not used much for the atmosphere compartment (in contrast to soil science and ecotoxicity).

For use in chemicals assessment, Fig. 7 can be interpreted as follows.

It is assumed that there is a homogeneous medium (in the atmosphere, this is evidently a better assumption than in soil), that there are constant environmental conditions (T, p, ...) and that other degradation processes are absent.

After rapid introduction of A into the compartment being considered (a "pulse"), a concentration $[A]_0$ is created. This compartment contains the reactive partner at a constant concentration. The concentration $[A]_0$ decreases to 50% after $t_{1/2}$ and to ca. 37% (1/e) after the time τ. The chemical lifetime – hence the name – is also the *average lifetime of an individual molecule A* present in the compartment considered, irrespective of how it entered the compartment (continuously or pulsed). It should also not be forgotten that these lifetimes are "partial" lifetimes, *related to one specific sink*, which should always be designated (e.g., τ_{OH}, the chemical lifetime due to reaction with OH radicals).

If there are several reactions of the type (Eq. 5.1) involving reactive species B, C, ..., the (total) chemical lifetime τ_{tot} is given by Eq. (5.6):

$$\tau_{tot} = 1/\sum k^I = 1/\{k_B^{II}[B] + k_C^{II}[C] + ...\} \qquad (5.6)$$

Any additional reaction channel shortens the total chemical lifetime. This is also true for a direct photochemical transformation ("photolysis"), which can be added into the braces as first-order rate constant $k_{h\nu}^I$ (Eq. 5.7):

$$\tau_{tot} = 1/\sum k^I = 1/\{k_B^{II}[B] + k_C^{II}[C] + ... + k_{h\nu}^I\} \qquad (5.7)$$

The *chemical lifetime* has to be strictly distinguished from the *residence time* of a molecule in a compartment of the environment. The average residence time of A in a compartment is in general shorter than the chemical lifetime, as a consequence of one or several of the physical transfer processes between the compartments, such as volatilisation, dry and wet deposition, etc. [1, 146]. Including these transfer processes to and from adjacent compartments, the concentration of A in a compartment x $[A]_x$ can be described according to Eq. (5.8) [184]:

$$d[A]_x/dt = I_x + \sum_i k_i[A]_i - [A]_x \sum_j k_j^I - [A]_x \sum_s k_s^I \qquad (5.8)$$

where I_x = direct input of A from the technosphere into compartment x (per time and volume); k_i = first-order rate constant of the transfer to compartment x (from an adjoining compartment i); $[A]_i$ = concentration of A in compartment i; k_j = first-order rate constant of the transfer from compartment x to an adjoining compartment j; k_s = first-order reaction rate constants of A in compartment x according to Eqs. (5.3) and (5.4).

In many instances a few sinks and transfer processes dominate, hence the overall behaviour does not always have to be as complex, as suggested by Eq. (5.8). Multimedia models allow the calculation of complex situations, provided the rate constants of the transfer processes and sinks are known.

5.2
Indirect Photochemical Degradation

5.2.1 Average OH Concentration in the Troposphere

The reaction with OH-radicals is the most important sink for organic substances in the troposphere. For this reason, OH has been called "the detergent of the atmosphere" in order to emphasize the "cleaning" function against which only a few classes of substances are totally resistant. In order to translate the k_{OH} values into chemical lifetimes (τ_{OH}) or half-lifes ($t_{1/2,OH}$) (see Section 5.1), an average concentration of the OH radicals <[OH]> is needed. Among the many possible definitions of <[OH]>, the global yearly (day/night) average is the most important one, as this average is decisive in questions of persistence and long-range transport to remote areas, entry of substances into the stratosphere, inter-hemispheric (N/S) mixing, etc. Local questions, such as the degradation during smog events, can be answered using measured [OH] or experiences gained in the laboratory (smog chambers, see Section 4.1.1.2).

Estimations of <[OH]>$_{global}$ were published in the 1970–1980s:

<[OH]>$_{global}$ [radicals cm^{-3}]	Authors
3×10^5	Crutzen and Fishman, 1977 [190]
$(3-4) \times 10^5$	Singh et al., 1979 [191]
5×10^5	Wagner and Zellner, 1979 [198]
9.7×10^5	Bottenheim and Strausz, 1980 [45]
6.5×10^5	Volz et al., 1981 [192]
6×10^5	Derwent, 1982 [193]
7.7×10^5	Prinn et al. 1987 [196]
$7 \pm 3 \times 10^5$	Class and Ballschmiter 1987 [156]: northern hemisphere
$17 \pm 7 \times 10^5$	Class and Ballschmiter 1987 [156]: southern hemisphere

Based on these and similar estimates, BUA [17] recommended the following global average:

$$<[OH]>_{global} = (6 \pm 2) \times 10^5 \text{ radicals cm}^{-3}$$

The European Chemicals Bureau, Ispra, recommended in the Technical Guidance Document (TGD) of 1996 [16], in accordance with the textbooks of the time [1, 14, 15] as a conservative value for risk assessment:

$$<[OH]>_{global} = 5 \times 10^5 \text{ radicals cm}^{-3} \text{ [16, 194]}$$

This recommendation remained *unchanged* in the second, revised edition of the TGD, Ispra, in 2003 [194]. The value 5×10^5 has been widely used in chemicals assessment work, although more recent work, discussed below, showed that a value of about 10^6 OH per cubic centimetre might be nearer to the truth. Actually, the Stratospheric Ozone Assessment Panel of the WMO recommended in its 1998 Report (published 1999) [6] 10^6 cm^{-3} and so did Finlayson-Pitts and Pitts Jr. in 2000 [2] and Schwarzenbach et al. in 2002[26] [146]. What happened?

In 1995, Prinn et al. [195], following up earlier work on methyl chloroform (MCF) and [OH] [196], published results obtained by a mass-balance and (OH) lifetime study using analytical measurements of 1,1,1-trichloroethane (methylchloroform, MCF) at five globally distributed stations (4–12 measurements daily in the period July 1978 to June 1994). These results were combined with industrial emission inventories and k_{OH} reaction rate constants of MCF. One result of this long-term study was that the concentration of MCF in the lower atmosphere reached a maximum in 1991 and has been decreasing since (MCF is a Montreal Protocol substance). Interestingly, the calculated weighted average OH concentration did *not* change in the period of observation with an uncertainty of only $0.0 \pm 0.2\%$. The density-averaged OH concentration is given as

$$<[OH]>_{global} = (9.7 \pm 0.6) \times 10^5 \text{ radicals cm}^{-3} \text{ [195]}$$

The uncertainty of the k_{OH} measurement is not included in the uncertainty reported [195]. According to Atkinson [27], this uncertainty amounts to $\pm 30\%$ for k_{OH} at 298 K.

Krol et al. (1998) [199] used another, independent model (the "MOGUNTIA model")[27] in order to evaluate the same experimental data as used by Prinn et al. [195] and challenged the result that the weighted, global average of the OH concentration was almost constant over the 16-year period studied. According to these workers, the experimental results can also be explained with a yearly increase of [OH] of 0.46% (the most likely value; possible range from −0.1 to +1.1%). This result is sensitive, however, to the MCF emission estimates. If these estimates are correct, the average [OH] increased by 7% over the time-span of the experiment, from $(1.0 + 0.09 - 0.15) \times 10^6$ in 1978 to $(1.07 + 0.09 - 0.17) \times 10^6$ [radicals cm^{-3}] in 1993. Although the question of whether [OH] increases globally or stays constant cannot be answered fully, the magnitude of

$$<[OH]>_{global} \approx 1.0 \times 10^6 \text{ radicals cm}^{-3} \text{ [199]}$$

has been confirmed, provided that the emission inventories of MCF are correct.

---

26) Second edition 2002; the first edition of this leading text book on "Environmental Organic Chemistry" (1993) did not deal with abiotic degradation in the atmosphere at all, in contrast to the aqueous phase.
27) Moguntia is the Latin name of the Roman city later to become Mainz; the model was developed at the MPI for Chemistry in Mainz by Zimmermann and Crutzen, *loc. cit.* [199].

In an update to the earlier work [195, 196], Prinn et al. (2001) [197] extended the analysis of the measuring results of the five stations to June 2000. At the end of the time period, the average concentration of MCF declined to values lower than those in 1978, when the measurements began (ca. 40 pptv at all measuring stations in 2000 vs. 60–90 pptv in 1978; the maximum was found in 1992, ca. 120–150 pptv). In contrast to Krol et al. [199], they reached the conclusion that there was a small *decrease* in $<[OH]>_{global}$ with a rate of –0.64 ± 0.60% per year. For the time-period analysed by Krol et al. [199], a similar increase in [OH] was then calculated by Prinn et al. [197]; the two models do not necessarily contradict each other, the differences seem to be due to the larger time-horizon available in the more recent paper.

The weighted, global average of the OH concentration did not vary, however, compared with the value 1978–1994 [195]:

$$<[OH]>_{global} = (9.4 \pm 1.3) \times 10^5 \text{ radicals cm}^{-3} \text{ [197]}$$

The uncertainty in the k_{OH} measurement has now been included in the total uncertainty [197]. A recalculation of the *previous* value [195] gives $(9.7 \pm 1.3) \times 10^5$ [197]. The average OH concentration in the southern hemisphere (SH) seems to be about 14% higher compared with the northern hemisphere (NH), but the uncertainties are significantly higher than for the global average, 1978–2000 [197]:

$$<[OH]>_{SH} = (9.93 \pm 2.02) \times 10^5 \text{ radicals cm}^{-3}$$

$$<[OH]>_{NH} = (8.98 \pm 2.02) \times 10^5 \text{ radicals cm}^{-3}$$

This result may be an artefact, however, as the reported MCF concentrations after the phase-out in 1996 (in developed countries) are too low [204]; actually, it seems that in Europe even in 2000 there was an emission of 20 000 metric tons of MCF (continuing MCF emissions have also been reported for the US [204]). This amount, if genuine, is not alarming with respect to the stratospheric ozone layer, but it is certainly an unpleasant surprise for the calculations based on the MCF emissions reported by industry. Krol and Lelieveld [203] cast further doubt on some results indicating a decrease in $<[OH]>_{global}$ during the 1990s; the calculated decrease vanishes under the assumption that 65 000 t MCF reported to be emitted in the early 1990s were actually delayed and emitted in the late 1990s. Direct local long-term measurements of [OH] in Southern Germany (1999–2003) also showed no trend in OH concentration, but a strong correlation with solar UV radiation [108]. Interestingly, no correlation was found with any of the atmospheric pollutants measured simultaneously; this shows that either all chemical effects cancel each other and/or that there are serious gaps in our understanding of the OH budget.

Bousquet et al. [200] concluded that any trends in OH concentrations derived from high-resolution models and the best analytical and kinetic data available at present depend on the reliability of the MCF emission inventories.

Summing up, the favoured globally averaged OH concentration in all recent scientific publications is <[OH]>$_{global}$ ≈ 1.0×10^6 radicals cm^{-3}. It has furthermore been claimed that the global tropospheric OH concentration has been fairly stable over the last 100 years [205], reductions over the tropical oceans being compensated by increases over the continents, driven by NO$_x$ emissions.

It should be kept in mind, however, that all recent global OH trends and absolute concentrations have been deduced using models of varying complexity, but all results are based on primary information obtained with MCF. Whereas the analytical measurements of [MCF] over a very long period of time seem to be unique (a critical review showed that MCF is indeed the best investigated chemical (Spivakovsky et al., [206]), it has been pointed out by most workers [195, 197, 202, 203, 205] that the results depend critically on the reliability of the MCF emission estimates. Alternative evaluations have either not been performed or did not provide conclusive results.

Unfortunately, direct measurements of site-specific [OH], as pioneered by Hübler et al. [207], Ortgies and Comes [208], Campbell et al. [209] and Altshuller [210], Dorn et al. [135], Brauers et al. [153] cannot replace the global estimates for reasons of sensitivity and for how representative they are (too many sites would have to be measured at different locations and seasons).

The absolute measurements of ground-level [OH], obtained under optimal conditions (noon, sunshine) with long-path laser absorption at 308 nm [120] showed

$$<[OH]>_{noon} = 1.6 \times 10^6 \text{ radicals cm}^{-3}$$

It is interesting to note that this value coincides well with the very first estimate made by Levy in 1971 [8] under similar conditions. Altshuller [210] reported the range of high, i.e., measurable, [OH] ≈ $(2-6) \times 10^6$ OH-radicals cm^{-3}. These values are less than an order of magnitude higher compared with the "new" global [OH], which are averages over day and night and periods of high and low radiation intensities. Summer noon peak hydroxyl concentration values in 2001 in the Mediterranean area (north-east coast of Crete) have been reported as [OH]$_{peak}$ = 1.5×10^7 radicals cm^{-3} (MINOS campaign, [213, 214]). The daytime average over the whole measuring campaign was

$$<[OH]>_{summer, day} = (8.2 \pm 1.6) \times 10^6 \text{ radicals cm}^{-3} \; [213]$$

The OH concentrations measured by Rohrer and Berresheim [108], every 5 min over nearly 5 years, are in the range between 0 and about 10^7 radicals cm^{-3}.

It has to be concluded that the "true" global average of the tropospheric OH concentration can only be approximated to be in the range $(5-10) \times 10^5$ cm^{-3}.

For assessment of the atmospheric persistence and of the long-range transport potential it seems it is appropriate to use the lower, "conservative" (i.e., not underestimating the persistence) value, as recommended by the TGD [194].

This value has recently (2005) also been derived from a trans-atlantic transport study (Li et al. [211]). It should be mentioned, however, that the use of a global [OH] average for the estimation is only meaningful for chemicals that are not easily degraded. Furthermore, to obtain the total chemical lifetime of a chemical, all relevant tropospheric sinks (OH, NO_3, O_3, $h\nu$) have to be considered, see Eq. (5.7). Measured reaction rate constants are preferable to calculated ones, especially if the database of structurally related compounds (used for establishing the QSRR-algorithms) is poor.

In particular circumstances, e.g., in assessing the fate of a substance during a smog period, higher [OH] values (about 10^7 cm^{-3}) can be used in order to estimate a realistic chemical lifetime or half-life. In winter time, on the other hand, much lower values of [OH] are appropriate, the extreme case being the arctic night.

5.2.2 Average NO_3 Concentration in the Troposphere

A globally averaged NO_3 concentration cannot be determined in the same way as discussed for OH. The reason is that no chemical reacting specifically with the nitrate radical (if there is such a substance) has been discovered and continuously measured over the years (such as MCF for OH). The same is true for ozone, a molecule associated with nitrate in night-time tropospheric chemistry.

Hence, for the above reason an average concentration $<[NO_3]>_{global}$ can only be roughly estimated from site-related measurements of $[NO_3]_{local}$. As discussed briefly in Section 2.1.2.1, the range of early (1980s) measurements of $[NO_3]$ during the night was about 10–100 pptv with maxima in the evening of up to several 100 pptv. Such measurements ($[NO_3]$ as a function of time throughout the night) were performed by Platt et al. using Differential Optical Absorption Spectroscopy (DOAS) [30, 212]. In the mid-1980s, an average 12-h night-time $[NO_3]$ of

$$<[NO_3]>_{night} \approx 10^9 \text{ radicals cm}^{-3}\ [28)]$$

was used by Atkinson et al. [36]. Wayne et al. [29] stated that the concentration of nitrate radicals shows a strong variability in time and space. Factors influencing the abundance of NO_3 are – besides light, the most important factor – water, aerosol particles and the presence of high concentrations of terpenes (rapid reaction with NO_3), e.g., over forests, and of highly reactive dimethyl sulphide (DMS, $k_{NO_3} = 7.0 \times 10^{-13}$ cm^3 molecule^{-1} s^{-1}). DMS, the most important compound of S^{II} in the troposphere, is produced by phytoplankton in the ocean and during the decay of biomass in general [29]. This variability makes the recommendation of an average $[NO_3]$ even more difficult. Wayne et al. [29] used the following value, which is said to be typical for remote *continental* $[NO_3]$, which corresponds to 10 pptv:

$$<[NO_3]>_{remote, cont.} \approx 2.5 \times 10^8 \text{ radicals cm}^{-3}$$

---

28) At 25 °C $\equiv$ 298.15 K, a (volume or molar) mixing ratio of 10 pptv corresponds to a concentration of 2.46×10^8 molecule cm^{-3} at sea level, 100 pptv corresponds to 2.46×10^9 molecule cm^{-3}.

The general caveat of high variability, mentioned again in the context of this value, also extends to the remote maritime troposphere. As the concentration of nitrogen oxides is very low ($[NO_x] \leq 25$ pptv), the concentration of NO_3 has to be much lower than over the continents [29]. The fact that two-thirds of the earth's surface is covered by oceans also has to be taken into account in proposing a global average.

In 2004 Vrekoussis et al. [213] reported on the results of the measuring campaign MINOS in Crete (28 July to 17 August 2001). NO_3 was measured daily during the nights at Finokalia (noth-eastern coast, 150 m above sea level) by means of DOAS with a total light path of 10.4 km. Typical trace-gas conditions at that time in the Mediterranean area were reported to be 0.5 ppbv NO_2 and 50 ppbv O_3 (the precursors of NO_3). The average night-time NO_3 level over the whole campaign amounted to

$$<[NO_3]>_{night} = (1.1 \pm 1.1) \times 10^8 \text{ radicals cm}^{-3} \text{ [213]}$$

This value is almost identical to a recent continental long-term observation at Lindenberg (near Berlin) [215] and at the lower end of a series of coastal measurements made in the 1990s, including Tenerife and three other stations (5–10 pptv; 1.35–2.7×10^8 cm^{-3}) reported Carslaw et al. in 1997 [216], Heintz et al. in 1996 [217], Allan et al. in 1999 [218] and Allan et al. in 2000 [219]. The last cited paper reported a mixing ratio average of 5 pptv, very close to 4.5 pptv (MINOS), for the "remote marine boundary layer", casting some doubt on the statement by Wayne et al. [29] that the remote marine area is supposed to be virtually free from NO_3.

In the light of these long-term observations, the night-time [NO_3] average used by Atkinson [36] seems to be roughly an order of magnitude too high. It might be that California was the right place to detect NO_3, but smog-related data are not representative.

A realistic tropospheric night-time average may be

$$<[NO_3]>_{global, night} \approx 1 \times 10^8 \text{ radicals cm}^{-3}$$

and the global 24-h average as 50% of this value:

$$<[NO_3]>_{global} \approx 5 \times 10^7 \text{ radicals cm}^{-3}$$

5.2.3 Average O_3 Concentration in the Troposphere

The estimation of globally averaged ozone concentrations encounters similar problems to those for the nitrate radical, but O_3 is less reactive and, thus, the fluctuations are less pronounced. There is a clear distinction between ozone in photochemical smog events (ground layer "Los Angeles" type) and the tropospheric background.

Atkinson and Carter proposed in 1984 [47] an average mixing ratio of 30 ppbv, corresponding to an average concentration at sea level of:

$$<[O_3]>_{average} = 7 \times 10^{11} \text{ molecule cm}^{-3}$$

Wayne et al. [29] proposed for the Northern Hemisphere the higher value (due to more NO_x pollution) of 40 ppbv or 9.8×10^{11} molecule cm^{-3}.

A 2004 review by Vingarzan [220] showed that the background mixing ratios in the Northern Hemisphere at mid-latitudes are in the range of 20–45 ppbv with median values in the range of 23–34 ppbv. These values are similar worldwide and increased in the last 100 years by a factor of two. The trend in the last 30 years was 0.5–2% a^{-1} with a steeper increase in the 1970s and 1980s compared with the 1990s. The medians given agree well with the average given above, recommended by Atkinson and Carter [47]. It is interesting to note that the annual means measured at the south pole (2835 m) between 1992 and 2001 were also in the same range (26–30 ppbv). The decrease observed since 1975 (–0.70% a^{-1}) is attributed to the yearly ozone hole and the increased photochemical activity during its opening [6].

The WMO ozone report of 1998 [6] gave trends for the various regions of the world, showing in general a levelling-off of the increases, except for the remote stations in Europe where ozone concentrations have been increasing. The average present-day mixing ratio of 15 remote stations[29)] (NH and SH) is 33.2 ± 12 ppbv, corresponding to a concentration (at sea level, 298 K) of

$$<[O_3]>_{average} = (8.1 \pm 2.9) \times 10^{11} \text{ molecule cm}^{-3}$$

This value is so close to the recommendation by Atkinson and Carter that it is not necessary to change it. It is proposed, therefore, to continue to use this value [47], corresponding to a mixing ratio of 30 ppbv as the global average:

$$<[O_3]>_{global} = 7 \times 10^{11} \text{ molecule cm}^{-3}$$

In special cases, especially in the assessment of chemicals during smog events, much higher ozone concentrations can be used. The highest concentrations are reached after complete oxidation of NO, hours after the start of smog formation (and often many kilometres away from the source of pollution). In these situations, the mixing ratio of ozone may surpass 250 ppbv (> 6×10^{12} molecule cm^{-3}, ca. 500 µg m^{-3})[30)]. It should be noted that this "mature" photo-smog, rich in NO_2 and O_3, is an excellent precursor for elevated concentrations of NO_3 during the following night (see Section 5.2.2).

---

29) Table 8-3 in Ref. [6].
30) Health-related limiting values are often around 200 µg m^{-3} (average values per hour or per 8 h, different definitions in the various laws); ozone is also a strong poison for plants, hence the discussion of a possible contribution to the forest die-back observed in Europe.

5.3
Direct Photochemical Degradation

5.3.1 Introduction

The primary information necessary for estimating chemical lifetimes ($\tau_{h\nu}$) or half-lives related to direct photo-transformation is the quantum efficiency Φ, which should be known as a function of the wavelength λ, see Section 4.4. The basic equation relating Φ, the first-order rate constant $k_{h\nu}$ and the chemical lifetime $\tau_{h\nu}$ is Eq. (5.9) (see also Eq. 2.25)

$$k_{h\nu} = 1/\tau_{h\nu} = \int \sigma(\lambda)\, \Phi(\lambda)\, I(\lambda)\, d\lambda \quad [\text{s}^{-1}] \tag{5.9}$$

In order to calculate the first-order reaction constant, two further pieces of information are needed:

- UV/VIS absorption spectrum $\sigma(\lambda)$, where σ is the absorption cross section [cm^2 molecule^{-1}] and λ is the wavelength [nm]
- spectral photon irradiance $I(\lambda)$ [photons s^{-1} cm^{-2} nm^{-1}].

5.3.2 Absorption Spectrum

The absorption spectrum is specific to the molecule and determined by its chemical structure and symmetry, and the combined electronic–vibronic transitions of the molecule [50, 51, 222]. The absorption spectra of organic molecules are similar in the condensed phase and in the gas phase, although the latter may be much better resolved, i.e., the vibrational structure of the electronic bands can be clearly distinguished in favourable cases. Frequently, however, spectra of large molecules look very similar irrespective of the surroundings (gas or solvent), especially if non-polar solvents are used in recording the spectra (aliphatic hydrocarbons, ethers, etc.).

The measurement of gas-phase spectra is described in OECD 1992 [58]. The gas-phase spectra of several volatile, photochemically active alkyliodides have been measured as a function of temperature in the range from 200 to 300 K and at wavelengths $\lambda > 200$ nm [227]. Gas-phase spectra of highly volatile substances can be found in the literature, especially if there is a scientific interest in their spectroscopic or photochemical, and in a few cases also atmospheric/climatic, behaviour. No systematic investigations have been performed for the sake of chemical assessments, despite the emphasis given to this much neglected sink[31] in the OECD monograph [58].

Semi-volatile organic compounds (SOCs) can not really be measured according to the OECD method and should be measured in solvents, using conventional UV-absorption spectrophotometric techniques [223–226]. Many UV spectra can also be found in spectra collections, e.g., [221, 228, 229]. For use in Eq. (5.9), only part of the σ ($\lambda > 290$ nm) is needed, as in the solar irradiance spectrum $I(\lambda)$ shorter

---

31) NO_3, on the other hand, is not treated at all in this monograph.

wavelengths are filtered out by stratospheric ozone. This is true for the estimation of *tropospheric* lifetimes. For estimations of *stratospheric* lifetimes, spectra are required over the full UV range (easily done with commercial spectrophotometers and transparent solvents down to 200 nm). Of course, in this instance, also $\Phi(\lambda)$ and $I(\lambda)$ have to be known down to the shorter wavelength UV. Care has to be taken with molecules able to dissociate (acid/base); such processes are less likely in the gas-phase compared with solutions.

Most UV/VIS absorption spectra of organic substances in solution are presented in the "chemical" unit $\varepsilon(\lambda)$, the "molar decadal absorption coefficient", which is based on the molar unit of concentration C and the decadal version of absorbance $E = \log_{10}(I_0/I)$ [32] (5.10):

$$\varepsilon(\lambda) = \frac{E(\lambda)}{C \times d} \ [\text{L mol}^{-1} \text{ cm}^{-1}] \tag{5.10}$$

where ε = molar decadal absorption coefficient [L mol^{-1} cm^{-1}]; $E(\lambda)$ = extinction (or absorbance); d = pathlength of the cell (cuvette) [cm]; and C = concentration of the test substance [mol L^{-1}].

Spectra are often presented as $\varepsilon(\nu')$ instead of $\varepsilon(\lambda)$, i.e., the molar decadal absorption coefficient as a function of wavenumber with the usual unit [cm^{-1}], Eq. (5.11) [225]:

$$\nu' \ [\text{cm}^{-1}] = 1/\lambda \ [10^7 \text{ nm}^{-1}] \tag{5.11}$$

where ν' = wavenumber.

This presentation has the advantage that the wavenumber is proportional to the photon energy $h\nu$ so that, for instance, vibrational structures become more regular because the spacings in the spectra are proportional to the energy differences ΔE (5.12)[33]:

$$h\nu = h \, c/\lambda = h \, c \, \nu' \ [\text{J}] \tag{5.12}$$

where h = Planck constant ($6.626\,075\,5 \times 10^{-34}$ [J s]); ν = frequency [s^{-1}]; c = velocity of light in vacuum (299 792 458 [m s^{-1}]).

The transformation into absorption cross sections $\sigma(\lambda)$ [cm^2 molecule^{-1}] is given in Section 2.2, Eqs. (2.26) and (2.27). It should be remembered that both forms are distinguished only by different units and are different expressions of one and the same principle: Lambert–Beer's law.

---

[32] E (extinction) [223], the preferred name is now absorbance [225].
[33] $\nu \lambda = c$; the velocity of light in vacuum [c] is the product of frequency [ν] and wavelength [λ] of the radiomagnetic radiation.

5.3.3 Spectral Photon Irradiance

The intensity and spectral distribution of solar radiation is the most important *environmental* factor in the direct photochemical transformation of organic molecules. Together with the (more or less) inherent properties quantum efficiency and UV/VIS absorption (Section 5.3.2), the spectral photon irradiance $I(\lambda)$ determines the lifetime of a molecule with respect to this specific sink.

The total intensity of the solar radiation outside the atmosphere of the earth is called the "solar constant" and amounts to 1368 W m^{-2} [14]. Only part of this power[34] (radiation energy arriving per time and area) outside the atmosphere arrives at or near the surface of the earth (maximum ca. 1000 W m^{-2}). In Eq. (5.9) the spectral photon irradiance $I(\lambda)$ is required, an energy density (per wavelength interval and area) defined as the number of photons (of wavelength λ) arriving at the unit surface per unit time. As the energy per photon depends on λ, Eq. (5.12), an equal number of photons arriving per unit time and area does not mean an equal amount of power. The spectral photon irradiance is the presentation of the solar radiation density needed for the calculation of Eq. (5.9) and photochemical, i.e., quantum, processes in general [230]. If the data are available in power [W] per area and wavelength interval (the other usual presentation), the following transformation of data can be used:

$$I \text{ [photons s}^{-1}\text{ cm}^{-2}\text{ nm}^{-1}] = I \text{ [W cm}^{-2}\text{ nm}^{-1}] \times \lambda \text{ [nm]} \times 5.035 \times 10^{15}$$

In Eq. (5.9), a flat surface is assumed to be the recipient of the solar radiation, as required for calculating chemical lifetimes in surface water [1, 54, 146, 232, 233]. The same assumption is valid for the troposphere near to the surface, if reflection of light from the surface is neglected. The definition of irradiance is more complicated within clouds, where the scattered radiation "comes from all directions". In this instance, the use of the "spectral actinic flux density" [photons s^{-1} cm^{-2} nm^{-1}] is more appropriate, as this flux density refers to the *surface of a sphere* rather than to a flat surface [230]. Here, we consider only the situation for a flat surface.

The actual irradiance to be used for the calculation depends on latitude, season, hour and cloudiness. It would be time-consuming and eventually not very instructive to take into account all of these variables, on the other hand one global average does not seem appropriate for this sink. There are two types of irradiance tables in the literature:

- tables prepared for clear-sky conditions and different latitudes and seasons [54, 58],
- tables taking into account average weather conditions [57].

The difference between these data may be important, as seen in Fig. 8. It is especially pronounced in the low wavelength region around 300 nm, where in winter time there is virtually no solar radiation intensity. *This region is important, however, as many environmental chemicals show the tail of the UV absorption spectrum*

---

34) The derived SI unit of power is the Watt [W]; 1 W ≡ 1 J s^{-1} ≡ 1 kg m^2 s^{-3} ≡ 1 V A.

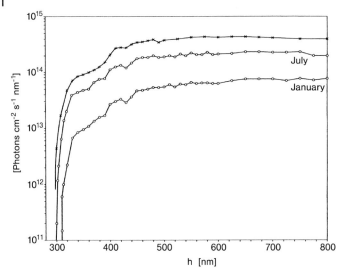

Fig. 8 Spectral photon irradiance of solar radiation at the surface of the earth (from [1]); ×–×–× 40° N, summer-noon [54]; o–o–o 52° N, daily average July (upper curve) and January (lower curve), Central Europe [57].

in this spectral region, yielding a small overlap integral with the edge of the spectral photon irradiance curve. The "summer-noon" curves, therefore, are *not conservative*, i.e., they imply a much too high a degradation constant and, thus a much too low a lifetime.

The tables presented in the OECD monograph [58] belong to the clear-sky type, but take into account all seasons and four latitudes: 40, 50, 60 and 70° N. The spectral photon irradiance in units of $[10^{19}$ photons d^{-1} cm$^{-2}]^{35)}$ is given numerically for the spectral range between 290 and 800 nm in 10-nm intervals. To take into account the seasons, the following data are given: Summer Solstice, Equinox, Winter Solstice, Fall or Winter Average and Spring or Summer Average.

Tables for Central Europe and the adjacent North Sea, taking into account average weather conditions (cloudiness), have been published by Frank and Klöpffer [57]. These tables contain the average spectral solar photon irradiance in units of [photons s^{-1} cm^{-2} nm^{-1}] for each month in wavelength intervals of 2.5 nm (292.5–330 nm), 10 nm (340–600 nm) and 25 nm (625–800 nm). The data refer to horizontal irradiance at sea level and have been calculated as average daytime values.

The choice of the appropriate table depends on the type of problem. In most instances it will be meaningful to calculate a range of lifetimes to see in which span the direct photochemical lifetime may be in the real environment. In approximating the integral (5.9) with Eq. (5.13), the intervals $\Delta\lambda$ chosen for $I(\lambda)_{av}$ and $\sigma(\lambda)_{av}$ (both averaged over $\Delta\lambda$) have to be equal.

---

35) 1 day [d] = 24 h = 86 400 s; day and hour are units used together with SI units without belonging to the SI.

$$k_{h\nu} = 1/\tau_{h\nu} \approx \sum \sigma(\lambda)\, \Phi(\lambda)\, I(\lambda)\, \Delta\lambda \quad [\text{s}^{-1}] \tag{5.13}$$

The same is true for Φ, if it depends on the wavelength. This has to be considered in data preparation.

The same equation can be used for calculating the lifetime of direct photo-transformation in the uppermost layer of surface waters, using the absorption spectrum and the quantum efficiency measured in water [1, 54, 146b, 231, 232]. This is also true for the droplet phase, if the more complicated geometry of radiation [230] is taken into account; for a rough estimate, however, the $I(\lambda)$ tables quoted may be sufficient. Direct photo-transformation in the adsorbed state (e.g., SOCs adsorbed at soil surfaces) should be considered as an emerging research topic.

5.3.4 Final Comments on Direct and Indirect Photochemical Transformation

It could be asked why is this sink so neglected in environmental research and assessment of chemicals? One reason may be that many small, volatile organic molecules do not show any significant absorption in the spectral region above 290 nm. These substances can be degraded by direct photochemical transformation only in the stratosphere, which cannot be reached if there are efficient sinks (indirect photochemical degradation) in the troposphere. A few small molecules contain chromophores[36] (e.g., carbonyl groups) which are photochemically active and often cause the splitting of the molecule after absorption of a photon. This process, e.g., a "Norrish I" splitting, is a truly photolytic process, the probability of which is given by the quantum efficiency Φ. Such molecules have frequently been studied for purely scientific reasons and also in a few instances for environmental purposes.

The great majority of chemicals that are sufficiently volatile to enter the atmosphere, or are formed there from volatile precursors, are not considered for direct photochemical transformation. It is true that many substances show only weak UV absorption above 290 nm (e.g., many molecules showing only one aromatic or hetero-aromatic ring and no bathochromic substituents[37]). Frequently, however, the absorption above 290 nm is weak, but not zero. In combination with a high Φ, this may lead to efficient degradation in the summer time. Larger chromophores, e.g., two or more condensed aromatic rings + bathochromic groups render a molecule less volatile and cause the problems discussed for the SOCs. The real reason for neglecting these sinks may therefore be connected with the difficulty in measuring photochemical transformations for the large majority of "candidate molecules", e.g., about 2000 existing industrial chemicals produced or

---

36) In a strict sense, the term "chromophore" refers to the ability of a group of atoms to render a molecule coloured, i.e., absorbing in the visible part of the spectrum (light in the narrow sense). In a broader definition, chromophores are also parts of a molecule absorbing electromagnetic radiation in the near-UV (commonly defined as the spectral region between 200 and 400 nm).

37) A bathochromic substituent is a shifting of the UV/VIS absorption towards longer wavelengths; if this occurs in the visible part of the spectrum, the colour becomes "deeper" (hence the name), going from yellow to red and blue.

imported within the European Community in amounts exceeding 1000 tonnes per year [168]. As a strong UV/VIS absorption may easily overcome a small $\Phi^{38)}$, the sink may be appreciable for many SOCs that might be suspected of being persistent but in reality are not.

The only solution to this problem is more research and method development for degradation studies of SOCs in the gas phase and in the adsorbed state. This is not only true for the sink direct photochemical degradation, but also for the indirect photochemical reactions of SOCs. As most POPs belong to this group, this is a real challenge for future work in the field of persistence and long-range transport research and assessment.

References

1 Klöpffer, W.: Verhalten und Abbau von Umweltchemikalien. Physikalisch-chemische Grundlagen. ecomed Publishers, Landsberg/Lech (Bavaria, Germany) 1996.
2 Finlayson-Pitts, B. J.; Pitts, Jr., J. N.: Chemistry of the Upper and Lower Atmosphere. Theory, Experiments, and Applications. Academic Press, San Diego 2000.
3 Mackay, D.: Multimedia Environmental Models. The Fugacity Approach. Second Edition. Lewis Publishers, Boca Raton, FL 2001.
4 Beyer, A.; Matthies, M.: Criteria for Atmospheric Long-range Transport Potential and Persistence of Pesticides and Industrial Chemicals. UBA Report 299 65 402. Berichte 7/02. Erich Schmidt Verlag, Berlin 2002.
5 Rowland, F. S.; Molina, M. J.: Chlorofluoromethanes in the Environment. Rev. Geophys. Space Phys. 13 (1975) 1–35.
6 World Meteorological Organization: Global Ozone Research and Monitoring Project – Report No.44. Scientific Assessment of Ozone Depletion: 1998. Geneva, February 1999.
7 Intergovernmental Panel on Climate Change(IPCC): Third Assessment Report 2001. Houghton, J. T. et al. (Eds.): Climate Change 2001: The Scientific Basis. Cambridge University Press, Cambridge 2001.
8 Levy II, H.: Normal Atmosphere: Large Radical and Formaldehyde Concentrations Predicted. Science 173 (1971) 141–143.
9 Leighton, P. A.: Photochemistry of Air Pollution. Academic Press, New York 1961.
10 Wayne, R. P.: Chemistry of Atmospheres. An Introduction to the Chemistry of the Atmospheres of Earth, the Planets, and their Satellites. Clarendon Press, Oxford 1985.
11 a) Atkinson, R.: Kinetics and Mechanisms of the Gas Phase Reactions of the Hydroxyl Radical With Organic Compounds Under Atmospheric Conditions. Chem. Rev. 86 (1986) 69–201; b) Atkinson, R.; Arey, J.: Atmospheric Degradation of Volatile Organic Compounds. Chem. Rev. 103 (2003) 4605–4638.
12 Bohn, B.; Zetzsch, C.: Formation of HO_2 from OH and C_2H_2 in the presence of O_2. J. Chem. Soc. Faraday Trans. 94 (1998) 1203–1210.
13 Frost, M. J.; Sharkey, P.; Smith, I. W. M.: Reaction Between OH (OD) Radicals and CO at Temperatures down to 80 K: Experiment and Theory. J. Phys. Chem. 97 (1993) 12254–12259.
14 Finlayson-Pitts, B. J.; Pitts, Jr., J. N.: Atmospheric Chemistry. Fundamentals and Experimental Techniques. Wiley-Interscience, New York 1986.
15 Warneck, P.: Chemistry of the Natural Atmosphere. International Geophysics Series, Vol. 41, Academic Press, San Diego 1988.

---

38) For example, early synthetic dyestuffs fading in sunlight.

16 European Chemicals Bureau: Technical Guidance Document (TGD) in Support of The Commissions Directive 93/67/EEC on Risk Assessment for the Notified Substances and the Commission Regulation (EC) 1488/94 on Risk Assessment for Existing Substances. European Commission, Joint Research Centre, Ispra/Italy 1996, 2nd edn (see also [194]).

17 Beratergremium für umweltrelevante Altstoffe (BUA) der Gesellschaft Deutscher Chemiker (Hrsg.): OH-Radikale in der Troposphäre. BUA-Stoffbericht 100. S. Hirzel, Stuttgart 1993.

18 Handbook of Chemistry and Physics. 58th edn. CRC Press, Boca Raton, FL (1977–1978).

19 Atkinson, R.; Darnall, K. R.; Lloyd, A. C.; Winer, A. M.; Pitts, Jr., J. N.: Kinetics and Mechanisms of the Reaction of the Hydroxyl Radical with Organic Compounds in the Gas Phase. Advances in Photochemistry, Vol. 11, Wiley, New York (1979) 375–488.

20 Atkinson, R.: Kinetics and Mechanisms of the Gas-Phase Reactions of the Hydroxyl Radical with Organic Compounds. J. Phys. Chem. Ref. Data Monograph 1, Am. Inst. Phys., New York 1989.

21 Hägele, J.; Lorenz, K.; Rhäsa, D.; Zellner, R.: Rate Constants and CH_3O Product Yield of the Reaction OH + CH_3OH Products. Ber. Bunsenges. Phys. Chem. 87 (1983) 1023–1026.

22 Lorenz, K.; Zellner, R.: Kinetics of the Reactions of OH-Radicals with Benzene, Benzene-d6 and Naphthalene. Ber. Bunsenges. Phys. Chem. 87 (1983) 629–636.

23 Witte, F.; Urbanik, E.; Zetzsch, C: Temperature Dependence of the Rate Constants for the Addition of OH to Benzene and to Some Monosubstituted Aromatics (Aniline, Bromobenzene, and Nitrobenzene) and the Unimolecular Decay of the Adducts. Kinetics into a Quasi-Equilibrium. 2. J. Phys. Chem. 90 (1986) 3251–3259.

24 Atkinson, R.; Baulch, D. L.; Cox, R. A.; Hampson, Jr., R. F.; Kerr, J. A.; Troe, J.: Evaluated Kinetic and Photochemical Data for Atmospheric Chemistry. Supplement III IUPAC Subcommittee on Gas Kinetic Data Evaluation for Atmospheric Chemistry. Kinetics. J. Phys. Chem. Ref. Data 18 (1989) 881–1097.

25 Klöpffer, W.; Haag, F.; Kohl, E.-G.; Frank, R.: Testing of the Abiotic Degradation of Chemicals in the Atmosphere: The Smog Chamber Approach. Ecotox. Environ. Safety 15 (1988) 298–319.

26 Wallington, T. J.; Neumann, D. M.; Kurylo, M. J.: Kinetics of the Gas Phase Reaction of Hydroxyl Radicals with Ethane, Benzene, and a Series of Halogenated Benzenes over the Temperature Range 234–438 K. Int. J. Chem. Kinetics 19 (1987) 725–739.

27 Atkinson, R.: Gas-phase Tropospheric Chemistry of Organic Compounds. J. Phys. Chem. Ref. Data Monograph 2, published by Am. Chem. Soc., Am. Inst. Phys., National Institute of Standards and Technology, New York 1994.

28 Hautecloque, S.: Photolyse de CH_3CCl_3 dans l'UV proche. J. Photochem. 12, (1980) 187–196.

29 Wayne, R. P.; Barnes, I.; Biggs, P.; Burrows, J. P.; Canosa-Mas, C. E.; Hjorth, J.; Le Bras, G.; Moortgat, G. K.; Perner, D.; Poulet, G.; Restelli, G.; Sidebottom, H.: The Nitrate Radical: Physics, Chemistry, and the Atmosphere. Atmos. Environ. 25A (1991) 1–203.

30 Platt, U.; Perner, D.; Winer, A. M.; Harris, G. W.; Pitts, Jr., J. N.: Detection of NO_3 in the Polluted Troposphere by Differential Optical Absorption. Geophys. Res. Lett. 7 (1980) 89–92.

31 Killus, J. P.; Whitten, G. Z.: Behavior of Trace NO_x Species in the Nighttime Urban Atmosphere. J. Geophys. Res. 90 (1985) 2430–2432.

32 Russel, A. G.; Cass, G. R.; Seinfeld, J. H.: On Some Aspects of Nighttime Atmospheric Chemistry. Environ. Sci. Technol. 20 (1986) 1167–1172.

33 a) Klöpffer, W.: Photochemical Processes Contributing to Abiotic Degradation in the Environment. In: Calamari, D. (Ed.): Chemical Exposure Predictions. Lewis Publishers, Boca Raton, FL (1993) 13–26; b) Klöpffer, W.: Environmental Hazard Assessment of Chemicals and Products. Part VI. Abiotic Degradation in the Troposphere. Chemosphere 33 (1996) 1083–1099.

34 Davis, D. D.; Machado, G.; Conaway, B.; Oh, Y.; Watson, R.: A Temperature Dependent Kinetics Study of the Reaction of OH with CH_3Cl, CH_2Cl_2, $CHCl_3$, and CH_3Br. J. Chem. Phys. 65 (1976) 1268–1274.

35 Seinfeld, J. H.: Atmospheric Chemistry and Physics of Air Pollution. Wiley, New York (1986).

36 Atkinson, R.; Aschmann, S. M.; Winer, A. M.; Pitts, Jr., J. N.: Kinetics and Atmospheric Implications of the Gas-Phase Reactions of NO_3 Radicals with a Series of Monoterpenes and Related Organics at 294 ± 2 K. Environ. Sci. Technol. 19 (1985) 159–163.

37 Atkinson, R.; Aschmann, S. M.; Goodman, M. A.: Kinetics of the Gas-Phase Reactions of NO_3 Radicals with a Series of Alkynes, Haloalkanes, and α,β-Unsaturated Aldehydes. Int. J. Chem. Kinet. 19 (1987) 299–307.

38 Phousongphouang, P. T.; Arey, J.: Rate Constants for the Gas-Phase Reactions of a Series of Alkylnaphthalenes with the Nitrate Radical. Environ. Sci. Technol. 37 (2003) 308–313.

39 Atkinson, R.; Arey, J.; Aschmann, S. M.: Gas-Phase Reactions of Azulene with OH and NO_3 Radicals and O_3 at 298 ± 2 K. Int. J. Chem. Kinet. 24 (1992) 467–480.

40 Canosa-Mas, C.; Smith, S. J.; Toby, S.; Wayne, R. P.: Reactivity of the Nitrate Radical towards Alkynes and some other Molecules. J. Chem. Soc., Faraday Trans. II. 84 (1988) 247–262.

41 Atkinson, R.; Aschmann, S. M.; Pitts, Jr., J. N.: Rate Constants for the Gas-Phase Reactions of the NO_3 Radical with a Series of Organic Compounds at 296 ± 2 K. J. Phys. Chem. 92 (1988) 3454–3457.

42 Tyndall, G. S.; Burrows, J. P.; Schneider, W.; Moortgat, G. K.: Rate Coefficient for the Reaction between NO_3 Radicals and Dimethyl Sulphide. Chem. Phys. Lett. 130 (1986) 463–466.

43 Atkinson, R.: Kinetics and Mechanisms of the Gas-Phase Reactions of the NO_3 Radical with Organic Compounds. J. Phys. Chem. Ref. Data. 20 (1991) 459–507.

44 Fishman, J.; Crutzen, P. J.: The Origin of Ozone in the Troposphere. Nature, 274 (1978) 855–858.

45 Bottenheim, J. W.; Strausz, O. P.: Gas-Phase Chemistry of Clean Air at 55° N Latitude. Environ. Sci. Technol. 14 (1980) 709–718.

46 Wayne, R. P.: The Photochemistry of Ozone. Atmos. Environ. 21 (1987) 1683–1694.

47 Atkinson, R.; Carter, W. P. L.: Kinetics and Mechanisms of the Gas Phase Reactions of Ozone with Organic Compounds under Atmospheric Conditions. Chem. Rev. 84 (1984) 437–470.

48 Paulson, S. E.; Flagan, R. C.; Seinfeld, J. H.: Atmospheric Photooxidation of Isoprene. Part I: The Hydroxyl Radical and Ground State Atomic Oxygen Reactions. Int. J. Chem. Kinet. 24 (1992) 79–101.

49 Paulson, S. E.; Flagan, R. C.; Seinfeld, J. H.: Atmospheric Photooxidation of Isoprene. Part II: The Ozone – Isoprene Reaction. Int. J. Chem. Kinet. 24 (1992) 103–125.

50 Turro, N. J.: Modern Molecular Photochemistry. Benjamin/Cummings, Menlo Park, CA (1978).

51 Calvert, J. G.; Pitts, Jr., J. N.: Photochemistry. Wiley, New York (1966).

52 Grotthuß, T. v.: Abhandlungen über Elektrizität und Licht. Oettingen, A. v. (Hrsg.): Ostwalds Klassiker der exakten Wissenschaften No 152, Leipzig 1906. Reprint after Jahresverhandlungen der kurländischen Gesellschaft für Literatur und Kunst. 1 (1819) 119–189.

53 IUPAC Commission on Photochemistry: Glossary of Terms used in Photochemistry, Part I. European Photochemistry Association Newsletter No. 25 (1985) 13–33; Part II in No. 26 (1986) 15–22; Part III in No. 27 (1987).

54 Zepp, R. G.; Cline, D. M.: Rates of Direct Photolysis in Aquatic Environment. Environ. Sci. Technol. 11 (1977) 359–366.

55 Umweltbundesamt, Berlin (Ed.): Draft Test Guideline on Phototransformation of Chemicals in Water. Prepared for OECD Workshop on November 26–27 (1987).

56 Koller, L. R.: Ultraviolet Radiation. Wiley, New York (1965).

57 Frank, R.; Klöpffer, W.: Spectral Solar Photon Irradiance in Central Europe and the Adjacent North Sea. Chemosphere 17 (1988) 985–994.
58 Organization for Economic Co-operation and Development (Ed.): The Rate of Photochemical Transformation of Gaseous Organic Compounds in Air under Tropospheric Conditions. OECD Environment Monographs No. 61 OCDE/GD (92)172. Paris 1992.
59 Bünau, G. v.; Wolff, T.: Photochemie. Grundlagen, Methoden, Anwendungen. VCH Verlagsges., Weinheim (1987).
60 Baulch, D. L.; Cox, R. A.; Crutzen, P. J.; Hampson, Jr., R. F.; Kerr, J. A.; Troe, J.; Watson, R. T.: Evaluated Kinetic and Photochemical Data for Atmospheric Chemistry. Supplement I CODATA Task Group on Chemical Kinetics J. Phys. Chem. Ref. Data 11 (1982) 327–496.
61 Klöpffer, W.: How to Improve the H_2O_2 Photolysis as an OH Source in Simulated Atmospheric Degradation Testing of Chemicals? European Photochemistry Association (EPA) Newsletter No. 31 (1987) 4–12.
62 Noyes, Jr., W. A.; Leighton, P. A.: The Photochemistry of Gases. Dover Publ., New York (1966). Reprint of the 1st edn, Reinhold (1941).
63 Carlier, P.; Hannachi, H.; Mouvier, G.: The Chemistry of Carbonyl Compounds in the Atmosphere. A Review. Atmos. Environ. 20 (1986) 2079–2099.
64 Gardner, E. P.; Wijayaratne, R. D.; Calvert, J. G.: Primary Quantum Yields of Photodecomposition of Acetone in Air under Tropospheric Conditions. J. Phys. Chem. 88 (1984) 5069–5083.
65 Penkett, S. A.: Non-methane Organics in the Remote Troposphere. In Goldberg, E. D. (Ed.): Atmospheric Chemistry, Dahlem Conference. Springer, Berlin (1982) 329–355.
66 Meyrahn, H.; Pauly, J.; Schneider, W.; Warneck, P.: Quantum Yields for the Photodissociation of Acetone in Air and an Estimate for the Life Time of Acetone in the Lower Troposphere. J. Atmos. Chem. 4 (1986) 277–291.
67 Berges, M. G. M.; Warneck, P.: Product Quantum Yields for the 350 nm Photodecomposition of Pyruvic Acid in Air. Ber. Bunsenges. Phys. Chem. 96 (1992) 413–416.
68 Bunce, N. J.; Landers, J. P.; Langshaw, J.-A; Nakal, J. S.: An Assessment of the Importance of Direct Solar Degradation of Some Simple Chlorinated Benzenes and Diphenyls in the Vapor Phase. Environ. Sci. Technol. 23 (1989) 213–218.
69 Luke, W. T.; Dickerson, R. R.; Nunnermacker, L. J.: Direct Measurements of the Photolysis Rate Coefficients and Henry's Law Constants of Several Alkyl Nitrates. J. Geophys. Res.-Atmos. 94 (D12) (1989) 14 905–14 921.
70 Roberts, J. M.; Fajer, R. W.: UV Absorption Cross Sections of Organic Nitrates of Potential Atmospheric Importance and Estimation of Atmospheric Lifetimes. Environ. Sci. Technol. 23 (1989) 945–951.
71 Jaenicke, R.: Staub – Ein wichtiger Bestandteil der Atmosphäre. Forsch. Magazin d. Johannes Gutenberg Universität Mainz, April (1988) 19–24.
72 Blanchard, D. C.; Woodcock, A. H.: Bubble Formation and Modification in the Sea and its Meteorological Significance. Tellus 9 (1957) 145–158.
73 Ketseridis, G.; Hahn, J.; Jaenicke, R.; Junge, C.: The Organic Constituents of Atmospheric Particulate Matter. Atmos. Environ. 10 (1976) 603–610.
74 Junge, C. E.: Transport Mechanism for Pesticides in the Atmosphere. Pure Appl. Chem. 42 (1975) 95–104.
75 Junge, C. E.: Basic Considerations about Trace Constituents in the Atmosphere as Related to the Fate of Global Pollutants. In Suffet, I. H. (Ed.) Fate of Pollutants in the Air and Water Environments. Part 1. Wiley, New York (1977) 7–25.
76 Miguel, A. H.; Korfmacher, E. H.; Wehry, E. L.; Mamantov, G.; Natusch, D. F. S.: Apparatus for Vapor-Phase Adsorption of Polycyclic Organic Matter onto Particulate Surfaces. Environ. Sci. Technol. 13 (1979) 1229–1232.
77 Rippen, G.; Zietz, E.; Frank, R.; Knacker, T.; Klöpffer, W.: Do Airborne Nitrophenols Contribute to Forest Decline? Environ. Technol. Lett. 8 (1987) 475–482.

78 Goss, K. U.; Schwarzenbach, R. H.: Gas/Solid and Gas/Liquid Partitioning of Organic Compounds: Critical Evaluation of the Interpretation of Equilibrium Constants. Environ. Sci. Technol. 32 (14) (1998) 2025–2032.

79 Goss, K. U.: The Role of Air/Surface Adsorption Equilibrium for the Environmental Partitioning of Organic Compounds. Habilitation Thesis. ETH Zürich, 2001.

80 Finzio, A.; Mackay, D.; Bidleman, T. F.; Harner, T.: Octanol-Air Partition Coefficient as a Predictor of Partitioning of Semi-volatile Organic Chemicals to Aerosols. Atmos. Environ. 31 (1997) 2289–2296.

81 Xiao, H.; Wania, F.: Is Vapor Pressure or the Octanol-Air Partition Coefficient a Better Descriptor of Partitioning Between Gas Phase and Organic Matter? Atmos. Environ. 37 (2003) 2867–2878.

82 Beyer, A.; Wania, F.; Gouin, T.; Mackay, D.; Mathies, M.: Selecting Internally Consistent Physicochemical Properties of Organic Compounds. Environ. Toxicol. Chem. 21 (2002) 941–953.

83 Klöpffer, W.: Bewertung von Literaturdaten. In: Fachgespräche über Persistenz und Ferntransport von POP-Stoffen. UBA-Texte 16/02, Berlin (2002) 58–61.

84 Klöpffer, W.: Physikalisch-chemische Kenngrößen von Stoffen zur Bewertung ihres atmosphärisch-chemischen Verhaltens: Datenqualität und Datenverfügbarkeit. GDCh Monographie Vol. 28, Frankfurt am Main 2004, 133–136.

85 Eganhouse, R. P.; Pontolillo, J.: (2002): Assessing the Reliability of Physico-Chemical Property Data (K_{ow}, S_w) for Hydrophobic Organic Compounds: DDT and DDE as a Case Study. SETAC Globe 3 (4) 34–35.

86 Björseth, A.; Lunde, G.; Lindskog, A.: Long-range Transport of Polycyclic Aromatic Hydrocarbons. Atmos. Environ. 13 (1979) 45–53.

87 Güsten, H.: Photocatalytic Degradation of Atmospheric Pollutants on the Surface of Metal Oxides. In: Jaeschke, W. (Ed.) Chemistry of Multiphase Atmospheric Systems. Springer, Berlin (1986) 567–592.

88 Dalsey, J. M.; Lewandowski, C. G.; Zorz, M.: A Photoreactor for Investigations of the Degradation of Particle-Bound Polycyclic Aromatic Hydrocarbons under Simulated Atmospheric Conditions. Environ. Sci. Technol. 16 (1982) 857–861.

89 Korfmacher, W. A.; Wehry, E. L.; Mamantov, G.; Natusch, D. F. S.: Resistance to Photochemical Decomposition of Polycyclic Aromatic Hydrocarbons Vapor-Adsorbed on Coal Fly Ash. Environ. Sci. Technol. 14 (1980) 1094–1099.

90 Kamens, R. M.; Guo, Z.; Fulcher, J. N.; Bell, D. A.: Influence of Humidity, Sunlight, and Temperature on the Decay of Polyaromatic Hydrocarbons on Atmospheric Soot Particles. Environ. Sci. Technol. 22 (1988) 103–108.

91 Fox, M. A.; Olive, S.: Photooxidation of Anthracene on Atmospheric Particulate Matter. Science 205 (1979) 582–583.

92 Behymer, T. D.; Hites, R. A.: Photolysis of Polycylic Aromatic Hydrocarbons Adsorbed on Fly Ash. Environ. Sci. Technol. 22 (1988) 1311–1319.

93 Zetzsch, C.: Simulation of Atmospheric Photochemistry in the Presence of Solid Airborne Particles. VCH-Verlagsgesellschaft, Weinheim, Dechema-Monographie, Vol. 104 (1987) 187–212.

94 Behnke, W.; Nolting, F.; Zetzsch, C.: The Atmospheric Fate of Di(2-ethylhexyl)phthalate, Adsorbed on Various Metal Oxide Model Aerosols and on Coal Fly Ash. J. Aerosol Sci. 18 (1987) 849–852.

95 Rippen, G.: Handbuch Umweltchemikalien. Edition 5/2003; CDrom; ecomed Publishers, Landshut/Lech, 2003.

96 Krüger, H.-U.; Gavrilov, R.; Liu, Q. L.; Zetzsch, C.: Entwicklung eines Persistenz-Messverfahrens für den troposphärischen Abbau von mittelflüchtigen Pflanzenschutzmitteln durch OH-Radikale. Research project UFOPLAN 201 67 424/02. Umweltbundesamtes (Ed.). Berlin, February 2005.

97 Klöpffer, W.; Kohl, E.-G.: Determination of the Rate Constant k_{OH}(air) Using Freon 113 as an Inert Solvent. In: Mansour, M. (Ed.): Proceedings of the 3rd Workshop on Study

and Prediction of Pesticides Behaviour in Soils, Plants and Aquatic Systems. Munich-Neuherberg, May 30–June 1, 1990. Munich (1991) 296–310.
98 Klöpffer, W.; Kohl, E.-G.: Bimolecular OH-Rate Constants of Organic Compounds in Solution. Part 2. Measurements in 1,2,2-Trichloro-trifluoroethane Using Hydrogen Peroxide as OH-Source. Ecotox. Environ. Safety 26 (1993) 346–356.
99 Davis, E. J.; Buehler, M. F.; Ward, T. L.: The Double Ring Electrodynamic Balance for Microparticle Characterization. Rev. Sci. Instrum. 61 (1990) 1281–1288.
100 Atkinson, R.; Aschmann, S. M.: Kinetics of the Gas-Phase Reactions of Cl Atoms with Chloroethenes at 298 ± 2 K and Atmospheric Pressure. Int. J. Chem. Kinet. 19 (1987) 1097–1105.
101 Hov, O.: The Effect of Chlorine on the Formation of Photochemical Oxidants in Southern Telemark, Norway. Atmos. Environ. 19 (1985) 471–485.
102 Singh, H. B.; Kasting, J. F.: Chlorine-Hydrocarbon Photochemistry in the Marine Troposphere and Lower Stratosphere. J. Atmos. Chem. 7 (1988) 261–285.
103 Esteve, W.; Budzinski, H.; Villenave, E.: Relative Rate Constants for the Heterogeneous Reactions of OH, NO_2 and NO Radicals with Polycyclic Aromatic Hydrocarbons Adsorbed on Carbonaceous Particles. Part 1: PAHs Adsorbed on 1–2 µm Calibrated Graphite Particles. Atmos. Environ. 38 (2004) 6063–6072.
104 Brubaker, W. W.; Hites, R. A.: OH Reaction Kinetics of Polycyclic Aromatic Hydrocarbons and Polychlorinated Dibenzo-p-dioxins and Dibenzofurans. J. Phys. Chem. A 102 (1998) 915–921.
105 McElroy, W. J.: Sources of Hydrogen Peroxide in Cloudwater. Atmos. Environ. 20 (1986) 427–438.
106 Ehhalt, D. H.; Rudolph, J.; Schmidt, U.: On the Importance of Light Hydrocarbons in Multiphase Atmospheric Systems. In: Jaeschke, W. (Ed.) Chemistry of Multiphase Atmospheric Systems. Proceedings of the NATO Advanced Study Institute Symposium, Corfu, Sept. 26–Oct. 8, 1983. Springer, Berlin (1986) 321–350.
107 Herrmann, H.; Ervens, B.; Jacobi, H.-W.; Wolke, R.; Nowacki, P.; Zellner, R.: CAPRAM2.3: A Chemical Aqueous Phase Radical Mechanism for Tropospheric Chemistry. J. Atmos. Chem. 36 (2000) 231–284.
108 Rohrer, F.; Berresheim, H.: Strong Correlation Between Levels of Tropospheric Hydroxyl Radicals and Solar Ultraviolet Radiation. Nature 442 (2006) 184–187.
109 Bielski, B. H. J.; Cabelli, D. E.; Arudi, R. L.; Ross, A. B.: Reactivity of HO_2/O_2^- Radicals in Aqueous Solution. J. Phys. Chem. Reference Data 14 (1985) 1041–1100.
110 Weeks, J. L.; Matheson, M. S.: The Primary Quantum Yield of Hydrogen Peroxide Decomposition. J. Am. Chem. Soc. 78 (1956) 1273–1278.
111 Baxendale, J. H.; Wilson, J. A.: The Photolysis of Hydrogen Peroxide at High Light Intensities. Trans. Faraday Soc. 53 (1956) 344–356.
112 Christensen, H.; Sehested, K.; Corfitzen, H.: Reactions of Hydroxyl Radicals with Hydrogen Peroxide at Ambient and Elevated Temperatures. J. Phys. Chem. 86 (1982) 1580–1590.
113 Güsten, H.; Klasinc, L.; Maric, D.: Prediction of Abiotic Degradability of Organic Compounds in the Troposphere. J. Atmos. Chem. 2 (1984) 83.
114 Klöpffer, W.; Kaufmann, G.; Frank, R.: Phototransformation of Air Pollutants: Rapid Test for the Determination of k_{OH}. Z. Naturforsch. 40a (1985) 686–692.
115 Güsten, H.: Predicting the Abiotic Degradation of Organic Pollutants in the Troposphere. Chemosphere 38 (1999) 1361–1370.
116 Fahartaziz; Ross, A. B.: Selected Specific Rates of Reactions of Transients from Water in Aqueous Solution. III. Hydroxyl Radical and Perhydroxyl Radical and Their Radical Ions. Report, U.S. Department of Commerce, NTIS PB-263 198 (1977).
117 Klöpffer, W.; Kohl, E.-G.: Bimolecular OH-Rate Constants of Organic Compounds in Solution. Part 1. Measurements in Water Using Hydrogen Peroxide as OH-Source. Ecotox. Environ. Safety 22 (1991) 67–78.
118 Luňák, S.; Sedlák, P.: Photoinitiated Reactions of Hydrogen Peroxide in the Liquid Phase. J. Photochem. Photobiol. A: Chemistry 68 (1992) 1–34.

119 Faust, B. C.; Hoigné, J.: Photolysis of Fe(III)-Hydroxy Complexes as Sources of OH Radicals in Clouds, Fog and Rain. Atmos. Environ. 24A (1990) 79–89.

120 Jaeschke, W. (Ed.): Chemistry of Multiphase Atmospheric Systems. Proceedings of the NATO Advanced Study Institute Symposium, Corfu, Sept. 26–Oct. 8, 1983. Springer, Berlin (1986).

121 Hoigné, J.; Bader, H.; Haag, W. R.; Staehelin, J.: Rate Constants of Reactions of Ozone with Organic and Inorganic Compounds in Water-III. Inorganic Compounds and Radicals. Water Res. 19 (1985) 993–1004.

122 Walling, C.: Fenton's Reagent Revisited. Acc. Chem. Res. 8 (1975) 125–131.

123 Mill, T.; Mabey, W. R.; Bomberger, D. C.; Choi, T. W.; Hendry, D. G.: Laboratory protocols for evaluating the fate of organic chemics in air and war. U.S. Environmental Protection Agency Report EPA-600/3-82-022. 338 pp. (1982).

124 Graedel, T. E.; Weschler, C. J.: Chemistry Within Aqueous Atmospheric Aerosol and Raindrops. Rev. Geophys. Space Phys. 19 No.4 (1981) 509–539.

125 Anbar, M.; Neta, P.: A Compilation of Specific Bimolecular Rate Constants for the Reactions of Hydrated Electrons, Hydrogen Atoms and Hydroxyl Radicals with Inorganic and Organic Compounds in Aqueous Solution. Int. J. Appl. Radiat. Isotop. 18 (1967) 493–523.

126 Dorfman, L. M.; Adams, G. E.: Reactivity of the Hydroxyl Radical in Aqueous Solution. Natl. Stand. Ref. Data Series, Natl. Bur. Stand. (U.S.) 46 (1973).

127 Buxton, G. V.; Greenstock, C. L.; Helman, W. P.; Ross, A. B.: Critical Review of Rate Constants for Reactions of Hydrated Electrons, Hydrogen Atoms and Hydroxyl Radicals ($\cdot$OH/O$^-\cdot$) in Aqueous Solution. J. Phys. Chem. Ref. Data. 17 (1988) 513–886.

128 Klöpffer, W.: Photochemical Degradation of Pesticides and other Chemicals in the Environment: A Critical Assessment of the State of the Art. Sci. Total Environ. 123/124 (1992) 145–159.

129 Leuenberger, C.; Czuczwa, J.; Tremp, J.; Giger, W.: Nitrated Phenols in Rain: Atmospheric Occurrence of Phytotoxic Pollutants. Chemosphere 17 (1988) 511–515.

130 Paul, W.: Electrodynamic Traps for Charged and Neutral Particles. Rev. Mod. Phys. 62 (1990) 531–540.

131 Chang, T. Y.; Kuntasal, G.; Pierson, W. R.: Night-Time N_2O_5/NO_3 Chemistry and Nitrate in Dew Water. Atmos. Environ. 21 (1987) 1345–1351.

132 Harrison, M. A. J.; Barra, S.; Borghesi, D.; Vione, D.; Arsene, C.; Olariu, R. I.: Nitrated Phenols in the Atmosphere: A Review. Atmos. Environ. 39 (2005) 231–248.

133 Stuhl, F.; Niki, H.: Flash Photochemical Study of the Reaction OH + NO + M Using Resonance Fluorescent Detection of OH. J. Chem. Phys. 57 (1972) 3677–3679.

134 Akimoto, H.; Hoshino. M.; Inoue, G.; Samaki, F.; Washida, N.; Okuda, M: Design and Characterization of the Evacuable and Bakable Photochemical Smog Chamber. Environ. Sci. Technol. 13 (1979) 471–475.

135 Dorn, H. P.; Brandenburger, U.; Brauers, T.; Hausmann, M.; Ehhalt, D. H.: In-situ Detection to Troposheric OH Radicals by Folded Long-path Laser Absorption. Results from the POPCORN Field Campaign in August 1994. Geophys. Res. Lett. 23 (1996) 2537–2540.

136 Becker, K. H.; Biehl, H. M.; Bruckmann, P.; Fink, E. H.; Führ, F.; Klöpffer, W.; Zellner, R.; Zetzsch, C. (Eds.): Methods of the Ecotoxicological Evaluation of Chemicals. Photochemical Degradation in the Gas Phase. Vol. 6. OH Reaction Rate Constants and Tropospheric Lifetimes of Selected Environmental Chemicals. Report 1980–1983. Kernforschungsanlage Jülich GmbH, Projektträgerschaft Umweltchemikalien. July–Sept. 279, 1984.

137 Wahner, A.; Zetzsch, C.: The Reaction of OH with C_2H_2 at Atmospheric Conditions Investigated by CW UV Laser Longpath Absorption of OH. Ber. Bunsenges. Phys. Chem. 89 (1985) 323–325.

138 Schmidt, V.; Zhu, G.-Y.; Becker, K. H.; Fink, E. H.: Study of OH Reactions at High Pressures by Excimer Laser Photolysis Dye Laser Fluorescence. Ber. Bunsenges. Phys. Chem. 89 (1985) 321–322.

139 Wayne, R. P.; Canosa-Mas, C. E.; Heard, A. C.; Parr, A. D.: On Discrepancies Between Different Laboratory Measurements of Kinetic Parameters for the Reaction of the Hydroxyl Radical with Halocarbons. Atmos. Environ. 26A (1992) 2371–2379.

140 Jans, U.; Hoigné, J.: Atmospheric Water: Transformation of Ozone into OH-Radicals by Sensitized Photoreactions or Black Carbon. Atmos. Environ. 34 (2000) 1069–1085.

141 Getoff, N.; Schwoerer, F.; Markovic, V. M.; Sehested, K.; Nielsen, S. O.: Pulse Radiolysis of Oxalic Acid and Oxalates. J. Phys. Chem. 75 (1971) 749–755.

142 Klöpffer, W; Frank, R; Kohl, E.-G.; Haag, F.: Quantitative Erfassung der photochemischen Transformationsprozesse in der Troposphäre. Chem.-Ztg. 110 (1986) 57–62.

143 Holmes, J. R.; O'Brien, R. J.; Crabtree, J. H.; Hecht, T. A.; Seinfeld, J. H.: Measurement of Ultraviolet Radiation Intensity in Photochemical Smog Studies. Environ. Sci. Technol. 7 (1973) 519–523.

144 Heicklen, J.: Atmospheric Chemistry. Academic Press, New York (1976).

145 Bahe, F. C.; Schurath, U.; Becker, K. H.: The Frequency of Nitrogen Dioxide Photolysis at Ground Level, as Recorded by a Continuous Actinometer. Atmos. Environ. 14 (1980) 711–718.

146 a) Schwarzenbach, R. P.; Gschwend, P. M.; Imboden, D. M.: Environmental Organic Chemistry. John Wiley & Sons, New York 1993; b) Schwarzenbach, R. P.; Gschwend, P. M.; Imboden, D. M.: Environmental Organic Chemistry. Wiley-Interscience, 2nd edn, Hoboken, New Jersey 2002.

147 Atkinson, R.; Baulch, D.; Cox, R.; Crowley, J.; Hampson, R.; Hynes, R.; Jenkin, M.; Rossi, M.; Troe, J.: Evaluated Kinetic and Photochemical Data for Atmospheric Chemistry: Part 1 – Gas Phase Reactions of O_x, HO_x, NO_x and SO_x Species. Atmos. Chem. Phys. Discuss. 3 (2003) 6179–6699.

148 Atkinson, R.; Baulch, D.; Cox, R.; Crowley, J.; Hampson, R.; Hynes, R.; Jenkin, M.; Rossi, M.; Troe, J.: Evaluated Kinetic and Photochemical Data for Atmospheric Chemistry: Volume I – Gas Phase Reactions of O_x, HO_x, NO_x and SO_x Species. Atmos. Chem. Phys. 4 (2004) 1461–1738.

149 Akimoto, H.; Sakamaki, F.; Inoue, G.; Okuda, M.: Estimation of OH Radical Concentration in a Propylene-NO_x-Dry Air System. Environ. Sci. Technol. 14 (1980) 93–97.

150 Cox, R. A.; Derwent, R. G.: Atmospheric Oxidation Reactions, Rates, Reactivity, and Mechanism for Reaction of Organic Compounds with Hydroxyl Radicals. Environ. Sci. Technol. 14 (1980) 57–61.

151 Kerr, J. A.; Sheppard, D. W.: Kinetics of the Reaction of Hydroxyl Radicals with Aldehydes Studied under Atmospheric Conditions. Environ. Sci. Technol. 15 (1981) 960–963.

152 Niki, H.; Maker, P. D.; Savage, C. M.; Breitenbach, L. P.: An FTIR Study of Mechanisms for the HO Radical Initiated Oxidation of C_2H_4 in the Presence of NO: Detection of Glycolaldehyde. Chem. Phys. Lett. 80 (1981) 499–503.

153 Brauers, T.; Aschmutat, U.; Brandenburger, U.; Dorn, H. P.; Hausmann, M.; Heßling, M.; Hofzumahaus, A.; Holland, F.; Plass-Düllmer, C.; Ehhalt, D. H.: Intercomparison of Tropospheric OH Radical Measurements by Multiple Folded Long-path Laser Absorption and Laser Induced Fluorescence. Geophys. Res. Lett. 23 (1996) 2545–2548.

154 Volman, D. H.: The Vapor-Phase Photo Decomposition of Hydrogen Peroxide. J. Chem. Phys. 17 (1949) 947–950.

155 Kurylo, M. J.; Murphy, J. L.; Haller, G. S.; Cornett, K. D.: A Flash Photolysis Resonance Fluorescence Investigation of the Reaction OH + H_2O_2 → HO_2 + H_2O. Int. J. Chem. Kinet. 14 (1982) 1149.

156 Class, T.; Ballschmiter K.: Global Baseline Pollution Studies. Part X. Atmospheric Halocarbons: Global Budget Estimation for Tetrachloroethane, 1,2-Dichloroethane, 1,1,1,2-Tetrachloroethane, Hexachloroethane and Hexachlorobutadiene. Estimation of the Hydroxyl Radical Concentrations in the Troposphere of the Northern and Southern Hemisphere. Fresenius Z. Anal. Chem. 327 (1987) 198–204.

157 Audley, G. J.; Baulch, D. L.; Campbell, I. M.: Gas-phase Reactions of Hydroxyl Radicals with Aldehydes in Flowing $H_2O_2 + NO_2 + CO$ Mixtures. J. Chem. Soc., Faraday Trans. I. 77 (1981) 2541–2549.

158 Barnes, I.; Bastian, V.; Becker, K. H.; Fink, E. H.; Zabel, F.: Reactivity Studies of Organic Substances toward Hydroxyl Radicals under Atmospheric Conditions. Atmos. Environ. 16 (1982) 545–550.

159 Tuazon, E. C.; Carter, W. P. L.; Atkinson, R.; Pitts, Jr., J. N.: The Gas-Phase Reaction of Hydrazine and Ozone: A Nonphotolytic Source of OH Radicals for Measurements of Relative OH Radical Rate Constants. Int. J. Chem. Kinet. 15 (1983) 619–629.

160 Cox, R. A.; Sheppard, D.: Reactions of OH Radicals with Gaseous Sulphur Compounds. Nature 284 (1980) 330–331.

161 Rahmann, M. M.; Becker, E.; Benter, Th.; Schindler, R. N.: A Gasphase Kinetic Investigation of the System $F + HNO_3$ and the Determination of Absolute Rate Constants for the Reaction of the NO_3 Radical with CH_3SH, 2-Methylpropene, 1,3-Butadiene and 2,3-Dimethyl-2-Butene. Ber. Bunsenges. Phys. Chem. 92 (1988) 91–100.

162 Benter, Th.; Schindler, R. N.: Absolute Rate Coefficients for the Reaction of NO_3 Radicals with Simple Dienes. Chem. Phys. Lett. 145 (1988) 67–70.

163 Canosa-Mas, C.; Smith, S. J.; Toby, S.; Wayne, R. P.: Temperature Dependences of the Reactions of the Nitrate Radical with some Alkynes and with Ethylene. J. Chem. Soc., Faraday Trans. II. 84 (1988) 263–272.

164 Lançar, I. T.; Daele, V.; Le Bras, G.; Poulet, G.: Étude de la réactivité des radicaux NO_3 avec le diméthyl-2,3 butène-2, le butadiène-1,3 et le diméthyl-2,3 butadiène-1,3. J. Chim. Phys. 88 (1991) 1777 1792.

165 Berndt, T.; Böge, O.; Kind, I.; Rolle, W.: Reaction of NO_3 Radicals with 1,3-Cyclohexadiene, α-Terpinene, and α-Phellandrene: Kinetics and Products. Ber. Bunsenges. Phys. Chem. 100 (1996) 462–469.

166 Anderson, L. G.; Hailu, M. A.; Chapman, R. W.; Nelson, J. S.: A Technique for Studying the Kinetics of NO_3 and N_2O_5 Reactions with Organics. In: American Chemical Society (ACS), Division of Environmental Chemistry: Preprints of Papers Presented at the 196th ACS Natl. Meeting, Los Angeles, CA, Sept. 25–30, 1988. Vol. 28 No. 2 (1988) 108–111.

167 Nolting, F.; Behnke, W.; Zetzsch, C.: A Smog Chamber for Studies of the Reaction of Terpenes and Alkanes with Ozone and OH. J. Atmos. Chem. 6 (1988) 47–59.

168 Council Regulation (EEC) No. 793/93 of 23 March 1993 on the Evaluation and Control of the Risk of Existing Substances. OJ No L84 (5 April 1993), pp. 1–68.

169 Moortgat, G. K.: Important Photochemical Processes in the Atmosphere. Pure Appl. Chem. 73 (2001) 487–490.

170 Chen, L.; Kutsuna, S.; Tokuhashi, K.; Sekiya, A.: New Technique for Generating High Concentrations of Gaseous OH Radicals in Relative Rate Measurements. Int. J. Chem. Kinet. 35 (2003) 317–325.

171 a) Zetzsch, C.; Palm, W.-U.; Krüger, H.-U.: Photochemistry of 2,2′,4,4′,5,5′-HexaBDE (BDE-153) in THF and Adsorbed on SiO_2: First Observation of OH Reactivity of BDEs on Aerosol. Organohalog. Comp. 66 (2004) 2256–2262; b) Palm, W.-U.; Kopetzky, R.; Sossinka, W.; Ruck, W.; Zetzsch, C.: Photochemical Reactions of Brominated Diphenylethers in Organic Solvents and Adsorbed on Silicon Dioxide in Aqueous Suspension. Organohalog. Comp. 66 (2004) 2269–2274.

172 Atkinson, R.; Guicherit, R.; Hites, R. A.; Palm, W.-U.; Seiber, J. N.; de Voogt, P.: Transformations of Pesticides in the Atmosphere: A State of the Art. Water, Air, Soil Pollut. 115 (1999) 219–143.

173 Brubaker, W. W.; Hites, R. A.: OH Reaction Kinetics of Gas-phase α- and γ-Hexachlorocyclohexane and Hexachlorobenzene. Environ. Sci. Technol. 32 (1998) 766–769.

174 Anderson, P. N.; Hites, R. A.: OH Radical Reactions: The Major Removal Pathway for Polychlorinated Biphenyls from the Atmosphere. Environ. Sci. Technol. 30 (1996) 1756–1763.

175 Anderson, P. N.; Hites, R. A.: System to Measure Relative Rate Constants of Semi-volatile Organic Compounds with Hydroxyl Radicals. Environ. Sci. Technol. 30 (1996) 301–306.

176 Dilling, W. L.; Gonsior, S. J.; Boggs, G. V.; Mendoza, C. G.: Organic Photochemistry. 20. Relative Rate Measurements for Reactions of Organic Compounds with Hydroxyl Radicals in 1,1,2-Trichlorotrifluoroethane Solution – A New Method for Estimating Gas Phase Rate Constants for Reactions of Hydroxyl Radicals with Organic Compounds Difficult to Study in the Gas Phase. Environ. Sci. Technol. 22 (1988) 1447–1453.
177 Behnke, W.; Holländer, W.; Koch, W.; Nolting, F.; Zetzsch, C.: A Smog Chamber for Studies of the Photochemical Degradation of Chemicals in the Presence of Aerosols. Atmos. Environ. 22 (1988) 1113–1120.
178 Palm, W.-U.; Elend, M.; Krüger, H.-U.; Zetzsch, C.: OH Radical Reactivity of Airborne Terbuthylazine Adsorbed on Inert Aerosol. Environ. Sci. Technol. 31 (1997) 3389–3396.
179 Palm, W.-U.; Elend, M.; Krüger, H.-U.; Zetzsch, C.: Atmospheric Degradation of a Semi-volatile Aerosol-Borne Pesticide: Reaction of OH with Pyrifenox (an Oxime-ether), Adsorbed on SiO_2. Chemosphere 38 (1999) 1241–1252.
180 Zetzsch, C.: Photochemischer Abbau in Aerosolphasen. UWSF-Z. Umweltchem. Ökotox. 3 (1991) 59–64.
181 Palm, W.-U.; Millet, M.; Zetzsch, C.: OH Radical Reactivity of Pesticides Adsorbed on Aerosol Materials: First Results of Experiments with Filter Samples. Exotoxicol. Environ. Safety 41 (1998) 36–41.
182 Rühl, E.: Messung von Reaktionsgeschwindigkeitskonstanten zum Abbau von langlebigen, partikelgebundenen Substanzen durch indirekte Photooxidation. Research project UFOPLAN 202 67 434, Umweltbundesamt (Ed.), Berlin 2004.
183 Frische, R.; Esser, G.; Schönborn, W.; Klöpffer, W.: Criteria for Assessing the Environmental Behavior of Chemicals: Selection and Preliminary Quantification. Ecotox. Environ. Safety 6 (1982) 283–293.
184 Klöpffer, W.; Rippen, G.; Frische, R.: Physicochemical Properties as Useful Tools for Predicting the Environmental Fate of Organic Chemicals. Ecotox. Environ. Safety 6 (1982) 294–301.
185 Klöpffer, W.: Persistenz und Abbaubarkeit in der Beurteilung des Umweltverhaltens anthropogener Chemikalien. UWSF-Z. Umweltchem. Ökotox. 1 (1989) 43–51.
186 Klöpffer, W.: Environmental Hazard Assessment of Chemicals and Products. Part II. Persistence and Degradability. ESPR – Environ. Sci. Pollut. Res. 1 (1994) 108–116.
187 Scheringer, M.: Persistence and Spatial Range of Environmental Chemicals. New Ethical and Scientific Concepts for Risk Assessment. Wiley-VCH, Weinheim 2002.
188 Müller-Herold, U.: A Simple General Limiting Law for the Overall Decay of Organic Compounds with Global Pollution Potential. Environ. Sci. Technol. 30 (1996) 586–591.
189 Scheringer, M.: Characterization of the Environmental Distribution Behavior of Organic Chemicals by Means of Persistence and Spatial Range. Environ. Sci. Technol. 31 (1997) 2891–2897.
190 Crutzen, P. J.; Fishman, J.: Average Concentrations of OH in the Troposphere, and the Budgets of CH_4, CO, H_2 and CH_3CCl_3. Geophys. Res. Lett. 4 (1977) 321–324.
191 Singh, B.; Salas, L. J.; Shigeishi, H.; Scribner, E.: Atmospheric Halocarbons, Hydrocarbons, and Sulfur Hexafluoride: Global Distributions, Sources, and Sinks. Science 203 (1979) 899–903.
192 Volz, A.; Ehhalt, D. H.; Derwent, R. G.: Seasonal and Latitudinal Variation of ^{14}CO and the Tropospheric Concentration of OH Radicals. J. Geophys. Res. 86 (1981) 5163–5171.
193 Derwent, R. G.: On the Comparison of Global, Hemispheric, One-dimensional and Two-dimensional Model Formulations of Halocarbon Oxidation by OH Radicals in the Troposphere. Atmos. Environ. 16 (1982) 551–561.
194 European Chemicals Bureau: Technical Guidance Document on Risk Assessment (TGD). Part II, revised edition 2003 (1st edn 1996). European Commission, Joint Research Centre, Ispra, Italy. http://ecb.jrc.it.
195 Prinn, R. G.; Weiss, R. F.; Miller, B. R.; Huang, J.; Alyea, F. N.; Cunnold, D. M.; Fraser, P. J.; Hartley, D. E.; Simmonds, P. G.: Atmospheric Trends and Lifetime of CH_3CCl_3 and Global OH Concentrations. Science 269 (1995) 187–192.

196 Prinn, R.; Cunnold, D.; Rasmussen, R.; Simmonds, P.; Aleya, F.; Crawford, A.; Fraser, P.; Rosen, R.: Atmospheric Trends in Methylchloroform and the Global Average for the Hydroxyl Radical. Science 238 (1987) 945–950.

197 Prinn, R. G.; Huang, J.; Weiss, R. F.; Cunnold, D. M.; Fraser, P. J.; Simmonds, P. G.; McCulloch, A.; Harth, C.; Salameh, P.; O'Doherty, S.; Wang, R. H. J.; Porter, L.; Miller, R. B.: Evidence for Substantial Variations of Atmospheric Hydoxyl Radicals in the Past Two Decades. Science 292 (2001) 1882–1888.

198 Wagner, H. G.; Zellner, R.: Die Geschwindigkeit des reaktiven Abbaus anthropogener Emissionen in der Atmosphäre. Angew. Chem. 91 (1979) 707–718.

199 Krol, M.; van Leeuwen, P. J.; Lelieveld, J.: Global OH Trend Inferred from Methylchloroform Measurements. J. Geophys. Res. 103 (D9) (1998) 10 697–10 711.

200 Bousquet, P.; Hauglustaine, D. A.; Peylin, P.; Carouge, C.; Ciais, P.: Two Decades of OH Variability as Inferred by an Inversion of Atmospheric Transport and Chemistry of Methyl Chloroform. Atmos. Chem. Phys. Discuss. 5 (2005) 1679–1731.

201 Wang, L.; Arey, J.; Atkinson, R.: Reactions of Chlorine Atoms with a Series of Aromatic Hydrocarbons. Environ. Sci. Technol. 39 (2005) 5302–5310.

202 Krol, M. C.; Lelieveld, J.; Oram, D. E.; Sturrock, G. A.; Penkett, S. A.; Brenninkmeijer, C. A. M.; Gros, V.; Williams, J.; Scheeren, N. A.: Continuing Emissions of Methyl Chloroform from Europe. Nature 421 (2003) 131–135.

203 Krol, M.; Lelieveld, J.: Can the Variability in Tropospheric OH be deduced from Measurements of 1,1,1-Trichloroethane (Methyl Chloroform)? J. Geophys. Res. 108 (2003) D3, 4125, doi: 10.1029/2002JD002423.

204 Millet, D. B.; Goldstein, A. H.: Evidence of Continuing Methylchloroform Emissions from the United States. Geophys. Res. Lett. 31 (2004) 4026, doi: 10.1029/2004GL020166.

205 Lelieveld, J.; Peters, W.; Dentener, F. J.; Krol, M. C.: Stability of Trospheric Hydroxyl Chemistry. J. Geophys. Res. 107 D23, 4715 (2002), doi: 10.1029/2002JD002272.

206 Spivakovsky, C. M. et al.: Three-dimensional Climatological Distribution of Tropospheric OH: Update and Evaluation. J. Geophys. Res. 105 (2000) 8931–8980.

207 Hübler, G.; Perner, D.; Platt, U.; Tönnissen, A.; Ehhalt, D. H.: Groundlevel OH Radical Concentration: New Measurements by Optical Absorption. J. Geophys. Res. 89 No. D1 (1984) 1309–1319.

208 Ortgies, G.; Comes, F. J.: A Laser Optical Method for the Determination of Tropospheric OH Concentrations. Appl. Phys. B 33 (1984) 103–113.

209 Campbell, M. J.; Farmer, J. C.; Fitzner, C. A.; Henry, M. N.; Sheppard, J. C.; Hardy, R. J.; Hopper, J. F.: Radiocarbon Tracer Measurements of Atmospheric Hydroxyl Radical Concentrations. J. Atmos. Chem. 4 (1986) 413–427.

210 Altshuller, A. P.: Ambient Air Hydroxyl Radical Concentrations: Measurements and Model Predictions. J. Air Pollut. Control Assoc. (JAPCA) 39 (1989) 704–708.

211 Li, Y.; Campana, M.; Reimann, S.; Schaub, D.; Stemmler, K.; Staehelin, J.; Peter, T.: Hydrocarbon Concentrations at the Alpine Mountain Sites Jungfraujoch and Arosa. Atmos. Environ. 39 (2005) 1113–1127.

212 Platt, U.; Winer, A. M.; Biermann, H. W.; Atkinson, R.; Pitts, J. N.: Measurements of Nitrate Radical Concentrations in Continental Air. Environ. Sci. Technol. 18 (1984) 365–369.

213 Vrekoussis, M.; Kanakidou, M.; Mihalopoulos, N.; Crutzen, P. J.; Leliefeld, J.; Perner, D.; Berresheim, H.; Baboukas, E.: Role of the NO_3 Radicals in Oxidation Processes in the Eastern Mediterranean Troposphere during the MINOS Campaign. Atmos. Chem. Phys. 4 (2004) 169–182.

214 Berresheim, H.; Plass-Dülmer, C.; Elste, T.; Mihalopoulos, N.; Rohrer, F.: OH in the Coastal Boundary Layer of Crete during MINOS: Measurements and Relationship with Ozone Photolysis. Atmos. Chem. Phys. 3 (2003) 639–649.

215 Geyer, A.; Ackermann, R.; Dubois, R.; Lohrmann, B.; Müller, T.; Platt, U.: Long Term Observation of Nitrate Radicals in the Continental Layer near Berlin. Atmos. Environ. 35 (2001) 3619–3631.

216 Carslaw, N.; Plane, J. M. C.; Coe, H.; Cuevas, E.: Observation of the Nitrate Radical in the Free Troposphere. J. Geophys. Res. 102, D9 (1997) 10613–10622.
217 Heintz, F.; Platt, U.; Flentje, H.; Dubois, R.: Long Term Observation of Nitrate Radicals at the Tor Stations, Kap Arcona (Rügen). J. Geophys. Res. 101, D17 (1996) 22891–22910.
218 Allan, B. J.; Carslaw, N.; Coe, H.; Burgess, R.; Plane, J. M. C.: Observations of the Nitrate Radical in the Maritime Boundary Layer. J. Atmos. Chem. 33 (1999) 129–154.
219 Allan, B. J.; McFiggans, G.; Plane, J. M. C.; Coe, H.; McFadyen, G. G.: The Nitrate Radical in the Remote Marine Boundary Layer. J. Geophys. Res. 105, D19 (2000) 24191–24204.
220 Vingarzan, R.: A Review of Surface Ozone Background Levels and Trends. Atmos. Environ. 38 (2004) 3431–3442.
221 DMS: UV Atlas of Organic Compounds. Volume I–V. Butterworth, London and Verlag Chemie, Weinheim 1966–1971.
222 Wayne, R. P.: Principles and Applications of Photochemistry. Oxford University Press, Oxford 1988.
223 Rao, C. N. R.: Ultra-Violet and Visible Spectroscopy. Chemical Applications. Butterworth, London 1961.
224 Perkampus, H.-H.: UV-VIS-Spektroskopie und ihre Anwendungen. Anleitungen für die chemische Laboratoriumspraxis, Bd. XXI. Springer, Berlin 1986.
225 Burgess, G.; Knowly, A. (Eds.): Standards in Absorption Spectrometry. Chapman and Hall, London 1981.
226 Knowly, A.; Burgess, G. (Eds.): Practical Absorption Spectrometry. Chapman and Hall, London 1984.
227 Roehl, C. M.; Burkholder, J. B.; Moortgat, G. K.; Ravishankara, A. R.; Crutzen, P. J.: Temperature Dependence of UV Absorption Cross Sections and Atmospheric Implications of Several Alkyl Iodides. J. Geophys. Res. 102 D11 (1997) 12 819–12 829.
228 The Sadtler Handbook of Ultraviolet Spectra. Heyden, Philadelphia, since 1979.
229 Organic Electronic Spectral Data, Wiley, New York, since 1960.
230 Wendisch, M.: Absorption of Solar Radiation in the Cloudless and Cloudy Atmosphere. Wissenschaftliche Mitteilungen aus dem Institut für Meteorologie der Universität Leipzig. Bd. 31. Leipzig 2003.
231 Frank, R.; Klöpffer, W.: A Convenient Model and Program for the Assessment of Abiotic Degradation of Chemicals in Natural Waters. Ecotox. Environ. Safety 17 (1989) 323–332.
232 Herrmann, M.; Büchel, D.; Klein, M.: ABIWAS – Programm zur Berechnung des abiotischen Abbaus von Chemikalien in Gewässern UWSF-Z. Umweltchem. Ökotox. 5 (1993) 175–276.
233 Klöpffer, W.: Environmental Hazard Assessment of Chemicals and Products. Part VII. A Critical Survey of Exposure Data Requirements and Testing Methods. Chemosphere 33 (1996) 1101–1117.
234 Bidleman, T. F.: Atmospheric Processes. Wet and Dry Deposition of Organic Compounds are Controlled by their Vapor-particle Partitioning. Environ. Sci. Technol. 22 (1988) 361–367.
235 Goss, K. U.: The Air/Surface Adsorption Equilibrium of Organic Compounds under Ambient Conditions. Crit. Rev. Environ. Sci. Technol. 34 (2004) 339–389.
236 Report of the OECD/UNEP Workshop on the Use of Multimedia Models for Estimating Overall Environmental Persistence and Long Range Transport in the Context of PBTs/POPs Assessment. OECD, Series on Testing and Assessment No. 36, Annex 7 (Data Needs), Paris 2002. http://www.oecd.ehs.
237 Balmer, M. E.; Goss, K. U.; Schwarzenbach, R. P.: Photolytic Transfomation of Organic Pollutants on Soil Surfaces – An Experimental Approach. Environ. Sci. Technol. 34 (2000) 1240–1245.
238 Ohnishi, T.; Nakato, Y.; Tsubomuro, H.: The Quantum Yield of Photolysis of Water on TiO_2 Electrodes. Ber. Bunsenges. Phys. Chem. 79 (1975) 523–525.
239 Irick, Jr., G.: Determination of the Photocatalytic Activities of Titanium Dioxides and Other White Pigments. J. Appl. Polym. Sci. 16 (1972) 2387–2395.
240 Goss, K. U.: Predicting the Enrichment of Organic Compounds in Fog Caused by Adsorption on the Water Surface. Atmos. Environ. 28 (1994) 3513–3517.

Chapter 3
Table of Reaction Rate Constants of Photo-Degradation Processes

1
Content of the Table

This Table contains bimolecular reaction rate constants k_x for the atmospheric reaction of chemical substances with reactive atmospheric species: hydroxyl radical (OH), ozone (O_3) and nitrate radical (NO_3). In some instances the direct photolysis rate constant $k_{h\nu}$ and the quantum yield are also listed.

$$A + X \xrightarrow{k_x} \text{products}$$

A = chemical substance
X = reactive species
k_x = bimolecular reaction rate constant [cm^3 molecule^{-1} s^{-1}]
x = index for OH, ozone or NO_3

Gas-phase reaction rate constants are reported in most cases. Only for a few semi-volatile organic substances (SOCs) are reaction rate constants in the adsorbed state included, mimicking the adsorption to aerosols. No reaction rate constants in the aqueous medium are reported.

The Table containing most of the numerical information in this book consists of 14 columns and 1081 rows, each containing one substance. The meanings of the headings of the columns are explained below.

Atmospheric Degradation of Organic Substances. W. Klöpffer and B. O. Wagner
Copyright © 2007 WILEY-VCH Verlag GmbH & Co. KGaA, Weinheim
ISBN: 978-3-527-31606-9

2
Explanation of the Column Headings

Column	Explanation
CAS No	Chemical Abstract Service Number
Chemical name	The names are in most cases Chemical Abstract Service Index Names. This presentation should facilitate finding the chemical names in the Table quicker than with the IUPAC nomenclature.
k_OH	k_{OH} Rate constant for the reaction with the hydroxyl radical [cm^3 molecule^{-1} s^{-1}]
k_Ozone	k_{O_3} Rate constant for the reaction with ozone [cm^3 molecule^{-1} s^{-1}]
k_Nitrate	k_{NO_3} Rate constant for the reaction with the nitrate radical [cm^3 molecule^{-1} s^{-1}]
k_hν	$k_{h\nu}$ Rate constant for direct photolysis [s^{-1}]
Φ	Quantum yield for direct photolysis [dimensionless]
QI	Quality index: R = recommended value A = average value S = single value E = estimated value
Ref.	Literature reference number

There are two substances for which no CAS number is (yet) available. The nomenclature chosen facilitates the finding of substances if the parent structure of the molecule is known. Derivatives of the parent structure can be found at the same place in the Table. It should be noted, however, that several traditional names, such "toluene", "xylene(s)", "aniline", etc., are not used in this nomenclature. On the other hand, pesticides and terpenes can be found under their generic or traditional chemical names. Users knowing the CAS number of the substance can find the chemical in the CAS registry at the end of the book.

3
Content of the Footnotes

The footnotes contain additional information, most importantly on the temperature-dependence (and in a few instances on the pressure-dependence) of the rate constants.

The rate constants of SOCs may be measured in the gas phase – mostly at higher temperature – and extrapolated to room temperature by means of Eq. (2). This temperature is outside the measuring range and the resulting rate constant for 298 K is therefore labelled "E" in the Table. The footnote gives the details and also indicates whether the substance has been measured in the adsorbed state. As this was done in only a few cases, this information is not given in the main Table. The reader is advised to look at the footnote, especially for the SOCs. In many instances, the original data used for calculating the average "A" are given in the footnote.

For the quantum efficiencies, the footnotes may contain some absorption cross sections required for calculating the first-order rate constant of direct phototransformation (see Chapter 2, Section 5.3) or a reference where such information can be found.

In general the footnotes do not contain details of the measuring technique used to obtain the data. This would be difficult, especially for the recommended values "R", which are often derived from data obtained by the various methods presented in Chapter 2.

4
Completeness and Accuracy

The Table should be virtually complete up to July 2006 with respect to organic molecules and their reaction rates for indirect photochemical degradation (k_{OH}, k_{O_3} and k_{NO_3}). The authors practiced the highest care in collecting these data; however, they do not wish to take any responsibility for the completeness or accuracy.

5
Atmospheric Half-lives

The atmospheric half-lives are often of interest. The bimolecular reaction rate constant can be transformed into the half-life (see Chapter 2, Section 5.1) using the following formula (1):

$$t_{1/2} = \frac{\ln 2}{[X]_x \times k_x} \tag{1}$$

$t_{1/2}$ = the atmospheric half-life [time]
ln 2 = 0.693
$[X]_x$ = the globally averaged concentration of the reactive species (recommended conservative value)
[OH] = 5×10^5 radicals cm^{-3} (Chapter 2, Section 5.2.1)
[ozone] = 7×10^{11} molecule cm^{-3} (Chapter 2, Section 5.2.3)
[NO$_3$] = 5×10^7 radicals cm^{-3} (Chapter 2, Section 5.2.2)
k_x = the respective reaction rate constant [cm^3 molecule^{-1} s^{-1}]

For specific problems, the user is, of course, free to choose concentrations for local or regional conditions (e.g., for a smog event, tropic or arctic conditions).

Quantum efficiencies are difficult to find and have often been determined for reasons of photochemical basic research. The conditions under which Φ and or $k_{h\nu}$ have been measured may not always be relevant to the atmospheric environment. As discussed in Chapter 2, additional information is required to convert quantum efficiencies into half-lives.

6
Selection of Substances

The Table contains organic substances and a few inorganic molecules of atmospheric significance, e.g., SF$_6$ and volatile acids. As the emphasis of this book is on "real" chemicals, it does not contain unstable radicals and deuterated substances investigated for mechanistic studies only. Not all substances included are produced by the chemical industry; there are many research products, e.g., potential Freon substitution products (hydrofluorocarbons). It also contains a few transformation products of photochemical degradation processes, studied in order to elucidate degradation processes following the primary attack by radicals, ozone or photons.

No error limits have been included in the Table, in order to conform with the main aim of this data collection, which is the rapid survey of existing data and the proposition of one rate constant (or quantum efficiency) for each substance and transformation reaction. The data refer to room temperature, i.e., 293–298 K. In the great majority of cases, 298 K is the reference temperature. In some cases only an upper limit of the reaction rate constant is given in the literature.

7
Quality Index (QI)

As a rough measure of data quality, a "quality index (QI)" is given for each value. The recommended values "R" are taken mostly from the critical evaluations cited. In most instances "R" means that the reaction has been measured by more than one group as a function of temperature. The reasoning why this value has been recommended can be found in the literature cited. "A" designates an average of several values given in the cited reviews and/or original papers. Evident outliers have not been used; the same is true for very old data, if more recent values are available. "S" does not always mean that the reaction has been measured only once (in most instances this is the case), but also if a new value has been found which seemed to be more trustworthy. In the case of "S", a certain degree of subjectivity can not be avoided. Some decisions are explained in the footnotes. This unavoidable subjectivity should not be confused with arbitrariness, but is based on long experience of collecting, using and also measuring reaction rate constants. Rate constants labelled "E" are mostly extrapolated values originating from measurements at higher temperatures. There are also a few entries that have been obtained by extrapolation from another medium or by calculation. In general, values labelled "E" should be used with great care.

8
Temperature Dependence of the Rate Constant

The temperature dependence of the rate constants is important additional information, contained for many reactions in the original publications and in the critical evaluations and reviews used as sources for this compilation. Again, in order not to overload the Table, this information is given in the footnotes for many reactions. The information given there is based on the Arrhenius equation, Eq. (2)

$$k_x(T) = A \times \exp(-E_a/RT) \tag{2}$$

k_x = k_{OH}, k_{O_3}, k_{NO_3} [cm^3 molecule^{-1} s^{-1}]
A = pre-exponential factor [cm^3 molecule^{-1} s^{-1}]
E_a = activation energy [J mol^{-1}]
R = gas constant (8.314 J mol^{-1} K^{-1})
T = absolute temperature [K]

In order to use the data given in the footnotes correctly, please note that for the sake of brevity the units for A are omitted in the footnotes; they are the same as those given for the rate constants at room temperature in the main part of the Table [cm^3 molecule^{-1} s^{-1}]. Furthermore, although kJ is used as the unit for E_a, this should read as kJ mol^{-1}. Equation (2) should only be used within the temperature range used for measuring A and E_a given in the footnotes. This is especially important for reactions of OH and NO_3 with aromatic molecules, as the overall rate constant measured may depend on the association and/or dissociation of the radical forming a primary complex, which reacts in a second step to give the reaction products. In many cases, therefore, the activation energy measured may be negative – i.e., the reaction becomes faster at lower temperature.

Not all reaction rate parameters reported in the literature are included in the footnotes. One reason is the formulation of fairly complex temperature functions going beyond the simple Arrhenius equation, Eq. (2). These complex equations are not well suited for reproduction in the footnotes and can be found in the respective evaluated data collection and review cited. This is especially true for the reactions labelled "R". The reaction rate constant given in the Table is frequently – but not always – identical with k_x (298 K) calculated from the Arrhenius parameters and Eq. (2). In some instances the authors recommend a value measured at 298 or 296 K, which may deviate slightly from the calculated one. Given the large uncertainties (typically for recommended values ±30%), these deviations are of no practical importance. This is also the reason why a uniform presentation of the results, such as for example, 7.3E-12 has been chosen. Only if an author gives an upper limit of the type $k_x < 1E-19$, was this value used instead of the standard presentation, in order to avoid the impression of an unreasonably high precision.

9
Pressure Dependence of the Rate Constant

Some gas-phase reactions depend on the total pressure used during the experiment; if known, the pressure used is given in the footnote and the value cited in the Table refers to conditions near to the tropospheric ones.

10
Direct Photolysis

Finally, it should be noted that the art of measuring k_{OH}, k_{NO_3} and k_{O_3} (going back to the 1970s) has reached a high standard so that noticeable discrepancies between publications in the last ten years are only found infrequently. Unfortunately, this cannot be said with respect to the quantum efficiencies and rate constants of direct photochemical transformation. One reason may be that many substances absorbing in the near-UV and the visible part of the solar radiation belong to the semi-volatile organic substances (SOCs, see Chapters 1 and 2), for which there are also only few rate constants because of the well-known experimental difficulties. The other reason may be the doctrine that the reaction with OH-radicals is the one and only important tropospheric sink. This may be true for simple, saturated hydrocarbons, but certainly not for the complex world of SOCs, the biggest challenge for the atmospheric degradation measuring community in the near future.

Table: Reaction Rate Constants and Quantum Efficiencies for Atmospheric Photo-degradation of Chemicals

CAS No.	Chemical name	k_OH [cm^3 molecule^{-1} s^{-1}]	QI	Ref.
208-96-8	Acenaphthylene	1.1E-10	A	9, 477
83-32-9	Acenaphthylene, 1,2-dihydro-	7.2E-11	A	43, 44, 297
75-07-0	Acetaldehyde	1.6E-11	R	32, 218, 416[1]
107-20-0	Acetaldehyde, chloro-	3.1E-12	R	218
811-96-1	Acetaldehyde, chlorodifluoro-	8.3E-14	A	291, 416
138689-24-4	Acetaldehyde, chlorofluoro-	2.1E-12	A	285, 416
63034-44-6	Acetaldehyde, dichlorofluoro-	1.2E-12	A	291
79-02-7	Acetaldehyde, dichloro-	2.5E-12	R	218
430-69-3	Acetaldehyde, difluoro-	1.6E-12	R	220, 285, 416
141-46-8	Acetaldehyde, hydroxy-	1.0E-11	R	416, 65
107-29-9	Acetaldehyde, oxime	2.2E-12	S	9
75-87-6	Acetaldehyde, trichloro-	1.3E-12	R	218
75-90-1	Acetaldehyde, trifluoro-	6.0E-13	R	416
127-19-5	Acetamide, N,N-dimethyl-	1.4E-11	S	220[4]
79-16-3	Acetamide, N-methyl-	5.2E-12	S	220[5]
64-19-7	Acetic acid	8.0E-13	R	218[6]
123-86-4	Acetic acid, butyl ester	4.2E-12	R	9, 346, 510[7]
540-88-5	Acetic acid, 1,1-dimethylethyl ester	5.3E-13	A	264, 322, 512[8]
108-05-4	Acetic acid, ethenyl ester			
141-78-6	Acetic acid, ethyl ester	1.6E-12	R	9
79-20-9	Acetic acid, methyl ester	2.6E-13	A	77, 168
108-21-4	Acetic acid, 1-methylethyl ester	3.4E-12	R	9, 512
105-46-4	Acetic acid, 1-methylpropyl ester	5.5E-12	R	9, 220, 322, 512
110-19-0	Acetic acid, 2-methylpropyl ester	6.3E-12	A	322, 330[11]
628-63-7	Acetic acid, pentyl ester	7.6E-12	S	282
109-60-4	Acetic acid, propyl ester	3.4E-12	R	9, 282, 512
76-05-1	Acetic acid, trifluoro-	1.4E-13	R	220, 285
431-47-0	Acetic acid, trifluoro-, methyl ester	5.2E-14	S	9, 77
67-64-1	Acetone	2.2E-13	R	32
75-05-8	Acetonitrile	2.6E-14	R	32, 218, 227
32388-55-9	Acetyl cedrene	7.7E-11	S	482
75-36-5	Acetyl chloride	9.1E-15	S	228
6090-09-1	4-Acetyl-1-methylcyclohexene	1.3E-10	S	259
309-00-2	Aldrin	2.6E-11	A	476[14]
7664-41-7	Ammonia	1.6E-13	R	220, 285[15]
62-53-3	Aniline	1.1E-10	R	9
120-12-7	Anthracene	1.9E-10	E	297[16]
613-31-0	Anthracene, 9,10-dihydro-	2.3E-11	S	298
1912-24-9	Atrazine	1.4E-11	E	405[17]

Table of Reaction Rate Constants of Photo-Degradation Processes | 115

k_Ozone [cm³ molecule⁻¹ s⁻¹]	QI	Ref.	k_Nitrate [cm³ molecule⁻¹ s⁻¹]	QI	Ref.	k_hν/Φ [s⁻¹] / [–]	QI	Ref.
3.5E-16	A	218, 477	5.4E-12	S	8			
< 5E-19	S	218						
6.0E-21	S	4	2.7E-15	R	32, 416[2]	Φ = 0.06	S	450
						Φ = 1.32[3]	S	453
						Φ = 1.00	S	503
3.2E-18	S	220, 291[9]						
			1.2E-17	A	242[10]			
			7.2E-17	S	242			
			5.4E-17	S	218, 242[12]			
1.0E-20	E	2	< 1.1E-17	S	218, 217	k_hν = 7.85E-7 / Φ = 0.15	S	11 / 32[13]
1.5E-19	S	4	5.0E-19	S	8			
< 2.2E-18	S	482	4.1E-15	S	482			
1.5E-16	S	259	1.0E-11	S	259			
1.1E-18	S	218						
9.0E-19	S	298	1.2E-12	S	298			

CAS No.	Chemical name	k_OH [cm^3 molecule^{-1} s^{-1}]	QI	Ref.
151-56-4	Aziridine	6.1E-12	S	9
275-51-4	Azulene	2.7E-10	S	229
56-55-3	Benz[a]anthracene	3.5E-12	E	504[18]
100-52-7	Benzaldehyde	1.3E-11	R	9
5779-93-1	Benzaldehyde, 2,3-dimethyl-	2.6E-11	S	233
15764-16-6	Benzaldehyde, 2,4-dimethyl-	3.5E-11	A	338, 233
5779-94-2	Benzaldehyde, 2,5-dimethyl-	3.6E-11	A	338, 233
1123-56-4	Benzaldehyde, 2,6-dimethyl	3.1E-11	S	233
5973-71-7	Benzaldehyde, 3,4-dimethyl-	2.3E-11	A	338, 233
5779-95-3	Benzaldehyde, 3,5-dimethyl-	2.8E-11	S	233
106-47-8	Benzenamine, 4-chloro-	8.3E-11	S	9, 40
121-69-7	Benzenamine, N,N-dimethyl-	1.5E-10	S	9, 195
101-77-9	Benzenamine, 4,4'-methylenebis-	3.0E-11	S	9, 41
71-43-2	Benzene	1.4E-12	R	9[20]
108-86-1	Benzene, bromo-	7.7E-13	R	9
108-90-7	Benzene, chloro-	7.7E-13	R	9
100-44-7	Benzene, (chloromethyl)-	2.9E-12	R	9
98-56-6	Benzene, 1-chloro-4-(trifluoromethyl)-	2.4E-13	S	9, 39
823-40-5	1,3-Benzenediamine, 2-methyl-	1.0E-10	S	9, 39
95-80-7	1,3-Benzenediamine, 4-methyl-	1.9E-10	S	9, 39
117-81-7	1,2-Benzenedicarboxylic acid, bis(2-ethylhexyl) ester	1.4E-11	A	401, 402[22]
95-50-1	Benzene, 1,2-dichloro-	4.2E-13	S	9, 40
541-73-1	Benzene, 1,3-dichloro-	7.2E-13	S	9, 40
106-46-7	Benzene, 1,4-dichloro-	3.8E-13	S	218, 243[23]
584-84-9	Benzene, 2,4-diisocyanato-1-methyl-	7.1E-12	S	9, 41
91-16-7	Benzene, 1,2-dimethoxy-	3.5E-11	S	220, 314
150-78-7	Benzene, 1,4-dimethoxy-			
151-10-0	Benzene, 1,3-dimethoxy-			
95-47-6	Benzene, 1,2-dimethyl-	1.4E-11	R	9
108-38-3	Benzene, 1,3-dimethyl-	2.2E-11	A	9, 335
106-42-3	Benzene, 1,4-dimethyl-	1.4E-11	R	9, 21, 26
98-06-6	Benzene, (1,1-dimethylethyl)-	4.6E-12	S	9, 35
120-80-9	1,2-Benzenediol			
488-17-5	1,2-Benzenediol, 3-methyl-	2.0E-10	S	425
452-86-8	1,2-Benzenediol, 4-methyl-	1.6E-10	S	425
100-41-4	Benzene, ethyl-	7.1E-12	R	9
611-14-3	Benzene, 1-ethyl-2-methyl-	1.2E-11	R	9, 497
620-14-4	Benzene, 1-ethyl-3-methyl-	1.9E-11	R	9, 497
622-96-8	Benzene, 1-ethyl-4-methyl-	1.2E-11	R	9, 497
462-06-6	Benzene, fluoro-	6.9E-13	R	9

k_Ozone [cm^3 molecule^{-1} s^{-1}]	QI	Ref.	k_Nitrate [cm^3 molecule^{-1} s^{-1}]	QI	Ref.	$k\_h\nu/\Phi$ [s^{-1}] / [–]	QI	Ref.
< 7E-17	S	229	3.9E-10	S	229			
			1.5E-15	A	200, 16, 212	$\Phi \ll 1$	S	7[19]
1.7E-22	S	230[21]	< 3E-17	R	8			
< 6E-20	S	218	< 5.7E-16	S	8			
			3.4E-17	S	8			
			1.0E-14	S	218			
			8.8E-15	S	218[24]			
			1.0E-14	S	218			
8.5E-22	S	4, 230	3.8E-16	S	8, 16			
< 1E-20	R	218						
1.4E-21	E	218, 230	2.3E-16	R	8			
9.6E-18	S	491	9.8E-11	S	492			
2.8E-17	S	491						
2.6E-17	S	491						
			5.7E-16	S	8, 212			

CAS No.	Chemical name	k_OH [cm^3 molecule^{-1} s^{-1}]	QI	Ref.
118-74-1	Benzene, hexachloro-	2.7E-14	E	454[25]
392-56-3	Benzene, hexafluoro-	1.7E-13	R	218[26]
87-85-4	Benzene, hexamethyl-	1.1E-10	S	200
591-50-4	Benzene, iodo-	1.1E-12	R	9
103-72-0	Benzene, isothiocyanato-	2.2E-11	S	530
2284-20-0	Benzene, 1-isothiocyanato-4-methoxy-	4.5E-11	S	530
614-69-7	Benzene, 1-isothiocyanato-2-methyl	2.9E-11	S	530
100-51-6	Benzenemethanol	2.3E-11	S	9
100-66-3	Benzene, methoxy-	1.7E-11	R	9, 19
108-88-3	Benzene, methyl-	6.0E-12	R	218
98-83-9	Benzene, (1-methylethenyl)-	5.2E-11	S	9, 196
98-82-8	Benzene, (1-methylethyl)-	6.5E-12	R	9
99-87-6	Benzene, 1-methyl-4-(1-methylethyl)-	1.5E-11	S	218
88-72-2	Benzene, 1-methyl-2-nitro-	7.0E-13	S	9
99-08-1	Benzene, 1-methyl-3-nitro-	1.1E-12	A	9, 156
768-49-0	Benzene, (2-methyl-1-propenyl)-	3.3E-11	S	9, 37
98-95-3	Benzene, nitro-	1.5E-13	A	218, 154[28]
101-84-8	Benzene, 1,1'-oxybis-	9.6E-12	S	220, 314
52144-69-1	Benzene, pentafluoropropyl-	3.1E-12	S	9, 25
873-66-5	Benzene, 1-propenyl-, (E)-	5.9E-11	S	9, 196
103-65-1	Benzene, propyl-	6.0E-12	R	9
108-98-5	Benzenethiol	1.1E-11	S	9, 38
120-82-1	Benzene, 1,2,4-trichloro-	5.5E-13	A	3, 10, 44, 155
98-08-8	Benzene, (trifluoromethyl)-	4.6E-13	S	39
526-73-8	Benzene, 1,2,3-trimethyl-	3.3E-11	R	9
95-63-6	Benzene, 1,2,4-trimethyl-	3.2E-11	R	9
108-67-8	Benzene, 1,3,5-trimethyl-	6,0E-11	A	335[29], 533, 300
50-32-8	Benzo[a]pyrene	> 4.3E-12	E	504[30]
493-09-4	1,4-Benzodioxin, 2,3-dihydro-	2.5E-11	S	215
192-97-2	Benzo[e]pyrene	> 4.3E-12	E	504[31]
271-89-6	Benzofuran	3.7E-11	S	9, 215
496-16-2	Benzofuran, 2,3-dihydro-	3.7E-11	S	9, 215
191-24-2	Benzo[ghi]perylene	> 5.9E-12	E	504[32]
207-08-9	Benzo[k]fluoranthene	> 3.5E-12	E	504[33]
100-47-0	Benzonitrile	3.3E-13	S	3
119-61-9	Benzophenone			
493-05-0	1H-2-Benzopyran, 3,4-dihydro-	3.7E-11	S	482
532-55-8	Benzoyl isothiocyanate	3.6E-11	S	530
4551-51-3	cis-Bicyclo[4.3.0]nonane	1.7E-11	S	9, 100
3296-50-2	trans-Bicyclo[4.3.0]nonane	1.8E-11	S	9, 100

k_Ozone [cm^3 molecule^{-1} s^{-1}]	QI	Ref.	k_Nitrate [cm^3 molecule^{-1} s^{-1}]	QI	Ref.	k_hν/Φ [s^{-1}] / [–]	QI	Ref.
			1.5E-16	A	8, 16			
4.1E-22	S	218[27)]	6.8E-17	R	8			
< 5E-20	S	218	1.0E-15	S	8			
< 7E-21	S	218						
						k_hν = 8.3E-8 / Φ = 0.47	S	193
1.6E-21	S	4, 24	1.9E-15	S	8			
1.3E-21	S	4, 24	1.8E-15	S	8, 16			
2.9E-21	S	335	8.0E-16	S	8, 16			
< 1.2E-20	S	218, 215	5.0E-16	A	8, 212			
1.8E-18	S	218, 215						
< 1.0E-19	S	215	1.1E-13	S	8, 212			
						k_hν = 5.6E-5 / Φ = 0.003	S	10[34)]

CAS No.	Chemical name	k_OH [cm^3 molecule^{-1} s^{-1}]	QI	Ref.
280-33-1	Bicyclo[2.2.2]octane	1.5E-11	S	9
931-64-6	Bicyclo[2.2.2]oct-2-ene	4.1E-11	S	9, 90
694-72-4	Bicyclo[3.3.0]octane	1.1E-11	S	9, 100
92-52-4	1,1'-Biphenyl	7.2E-12	R	9, 116
2051-60-7	1,1'-Biphenyl, 2-chloro-	2.8E-12	S	9, 42[35]
2051-61-8	1,1'-Biphenyl, 3-chloro-	5.4E-12	S	42, 220[36]
2051-62-9	1,1'-Biphenyl, 4-chloro-	3.9E-12	S	9, 42[37]
2050-68-2	1,1'-Biphenyl, 4,4'-dichloro-	2.0E-12	E	220, 294[38]
34883-41-5	1,1'-Biphenyl, 3,5-dichloro-	4.2E-12	S	314[39]
33284-50-3	1,1'-Biphenyl, 2,4-dichloro-	2.6E-12	E	294
13029-08-8	1,1'-Biphenyl, 2,2'-dichloro-	2.0E-12	S	220, 314[40]
34883-43-7	1,1'-Biphenyl, 2,4'-dichloro-	1.4E-12	E	220, 290[41]
2050-67-1	1,1'-Biphenyl, 3,3'-dichloro-	4.1E-12	S	220, 314
38379-99-6	1,1'-Biphenyl, 2,2',3,5',6-pentachloro-	4.0E-13	E	294[42]
38380-03-9	1,1'-Biphenyl, 2,3,3',4',6-pentachloro-	6.0E-13	E	294[43]
18259-05-7	1,1'-Biphenyl, 2,3,4,5,6-pentachloro-	9.0E-13	E	294[44]
41464-39-5	1,1'-Biphenyl, 2,2',3,5'-tetrachloro-	8.0E-13	E	294[45]
2437-79-8	1,1'-Biphenyl, 2,2',4,4'-tetrachloro-	1.0E-12	E	294[46]
7012-37-5	1,1'-Biphenyl, 2,4,4'-trichloro-	1.1E-12	E	294[47]
16606-02-3	1,1'-Biphenyl, 2,4',5-trichloro-	1.2E-12	E	294[48]
15862-07-4	1,1'-Biphenyl, 2,4,5-trichloro-	1.3E-12	E	294[49]
38444-86-9	1,1'-Biphenyl, 2',3,4-trichloro-	1.9E-12	E	294
37680-68-5	1,1'-Biphenyl, 2',3,5-trichloro-	1.4E-12	E	220[50]
38444-87-0	1,1'-Biphenyl, 3,3',5-trichloro-	1.6E-12	E	220[51]
38444-88-1	1,1'-Biphenyl, 3,4',5-trichloro-	1.4E-12	E	220[52]
2568-89-0	Bis-isopropoxymethane	3.7E-11	A	398
2568-91-4	Bis-(2-methylpropoxy)-methane	3.7E-11	S	432
2568-92-5	Bis-(1-methylpropoxy)-methane	4.5E-11	A	398, 432
10294-34-5	Borane, trichloro-	4.6E-12	S	312
76-49-3	Bornyl acetate	1.4E-11	S	393
7726-95-6	Bromine	4.4E-11	R	220[53]
590-19-2	1,2-Butadiene	2.6E-11	S	9, 84
106-99-0	1,3-Butadiene	6.7E-11	R	9[54]
513-81-5	1,3-Butadiene, 2,3-dimethyl-	1.2E-10	S	9, 84
598-25-4	1,2-Butadiene, 3-methyl-	5.7E-11	S	9, 84
78-79-5	1,3-Butadiene, 2-methyl-	1.0E-10	R	9
460-12-8	1,3-Butadiyne	1.9E-11	R	9
123-72-8	Butanal	2.4E-11	R	393, 413, 414, 416[57, 58]

$k\_$Ozone [cm³ molecule⁻¹ s⁻¹]	QI	Ref.	$k\_$Nitrate [cm³ molecule⁻¹ s⁻¹]	QI	Ref.	$k\_h\nu/\Phi$ [s⁻¹] / [–]	QI	Ref.
7.3E-17	S	4, 108	1.4E-13	S	8, 203			
< 2E-19	S	4, 116						
						$k\_h\nu$ = 8.3E-6 / Φ = 0.32	S	193
						$k\_h\nu$ = 1.4E-6 / Φ = 0.056	S	193
						$k\_h\nu$ = 3.7E-5 / Φ = 0.078	S	193
< 2.0E-20	S	220, 314						
6.3E-18	R	218[55)]	1.0E-13	R	8, 122, 124, 125, 127, 436			
2.7E-17	S	218, 221	2.1E-12	R	8			
1.3E-17	R	438, 439[56)]	6.8E-13	R	218, 122, 127, 128, 129, 499			
			1.4E-14	A	393, 356, 413, 414, 415, 416[59)]	Φ = 0.21	S	450[60)]

CAS No.	Chemical name	k_OH [cm³ molecule⁻¹ s⁻¹]	QI	Ref.
2987-16-8	Butanal, 3,3-dimethyl-			
97-96-1	Butanal, 2-ethyl-	4.0E-11	S	393
96-17-3	Butanal, 2-methyl-			
590-86-3	Butanal, 3-methyl-	2.7E-11	R	9, 414
106-97-8	Butane	2.5E-12	R	9
109-65-9	Butane, 1-bromo-	2.5E-12	S	218, 279
109-69-3	Butane, 1-chloro-	1.5E-12	A	218, 255
13179-96-9	Butane, 1,4-dimethoxy-	2.9E-11	S	448
75-83-2	Butane, 2,2-dimethyl-	2.3E-12	R	9
79-29-8	Butane, 2,3-dimethyl-	6.0E-12	R	218
584-03-2	1,2-Butanediol	2.7E-11	S	207
107-88-0	1,3-Butanediol	3.3E-11	S	207
513-85-9	2,3-Butanediol	2.4E-11	S	207
20280-41-1	1,2-Butanediol, dinitrate	1.7E-12	S	218, 269
106-65-0	Butanedioic acid, dimethyl ester	1.6E-12	A	321, 486
6423-45-6	2,3-Butanediol, dinitrate	1.1E-12	S	218, 269
431-03-8	2,3-Butanedione	2.4E-13	R	9
628-81-9	Butane, 1-ethoxy-	2.1E-11	R	218, 72, 73
163702-05-4	Butane, 1-ethoxy-1,1,2,2,3,3,4,4,4-nonafluoro-	7.0E-14	S	407, 516
161791-33-9	Butane, 1,1,1,2,2,4-hexafluoro-	1.5E-14	S	220, 382[63]
407-59-0	Butane, 1,1,1,4,4,4-hexafluoro-	7.1E-15	R	220, 237[64]
628-28-4	Butane, 1-methoxy-	1.5E-11	R	218, 72
994-05-8	Butane, 2-methoxy-2-methyl-	5.5E-12	S	218, 72, 240
163702-07-6	Butane, 1,1,1,2,2,3,3,4,4-nonafluoro-4-methoxy-	7.2E-15	S	508
78-78-4	Butane, 2-methyl-	3.9E-12	R	9
2568-90-3	Butane, 1,1'-[methylenebis(oxy)]bis-	3.5E-11	S	432
627-05-4	Butane, 1-nitro-	1.5E-12	R	218, 64, 159
377-36-6	Butane, 1,1,2,2,3,3,4,4-octafluoro-	4.2E-15	S	218
142-96-1	Butane, 1,1'-oxybis-	2.9E-11	R	218, 72, 73, 75, 487
406-58-6	Butane, 1,1,1,3,3-pentafluoro-	5.6E-15	R	220, 301[65]
34454-97-2	1-Butanesulfonamide, 1,1,2,2,3,3,4,4,4-nonafluoro-N-(2-hydoxyethyl)-N-methyl-	5.8E-12	S	517
40630-67-9	1-Butanesulfonamide, 1,1,2,2,3,3,4,4,4-nonafluoro-N-ethyl-	3.7E-13	S	520
594-82-1	Butane, 2,2,3,3-tetramethyl-	1.1E-12	R	218[66]
544-40-1	Butane, 1,1'-thiobis-	3.7E-11	S	218, 265
109-79-5	1-Butanethiol	5.1E-11	R	9
513-53-1	2-Butanethiol	4.0E-11	R	9
464-06-2	Butane, 2,2,3-trimethyl-	4.2E-12	R	9
107-92-6	Butanoic acid	2.1E-12	A	9, 234
109-21-7	Butanoic acid, butyl ester	1.1E-11	S	9, 77

k_Ozone [cm³ molecule⁻¹ s⁻¹]	QI	Ref.	k_Nitrate [cm³ molecule⁻¹ s⁻¹]	QI	Ref.	k_$h\nu/\Phi$ [s⁻¹] / [–]	QI	Ref.
			1.9E-14	A	393, 356, 414			
			3.3E-14	S	414			
3.9E-20	S	220, 340	2.6E-14	A	393, 414, 431[61]	$\Phi = 0.72$	S	450
			2.3E-14	S	220, 356	$\Phi = 0.27$	S	450
< 9.8E-24	S	4, 101, 102	4.6E-17	R	218			
			4.1E-16	R	8			
						k_$h\nu$ = 1.7E-4	S	189[62]
			1.6E-16	A	220, 353			
			< 4.9E-17	S	220, 353			
			2.2E-16	S	220, 353			

CAS No.	Chemical name	k_OH [cm^3 molecule^{-1} s^{-1}]	QI	Ref.
105-54-4	Butanoic acid, ethyl ester	5.0E-12	S	9, 77
623-42-7	Butanoic acid, methyl ester	3.0E-12	S	9, 77
105-66-8	Butanoic acid, propyl ester	7.4E-12	S	9, 77
71-36-3	1-Butanol	8.6E-12	R	218, 71, 165[67]
78-92-2	2-Butanol	9.2E-12	S	220, 310
2517-43-3	1-Butanol, 3-methoxy-	2.4E-11	S	9
123-51-3	1-Butanol, 3-methyl-	1.3E-11	S	220, 316
598-75-4	2-Butanol, 3-methyl-	1.2E-11	S	9, 72
543-87-3	1-Butanol, 3-methyl-, nitrate	2.5E-12	S	187
127191-96-2	1-Butanol, 2-methyl-, nitrate	2.5E-12	S	187
123041-25-8	2-Butanol, 3-methyl-, nitrate	1.8E-12	A	9, 60, 187
78-93-3	2-Butanone	1.1E-12	R	9, 416[69]
75-97-8	2-Butanone, 3,3-dimethyl-	1.2E-12	S	9, 67
5077-67-8	2-Butanone, 1-hydroxy-	7.7E-12	S	409
513-86-0	2-Butanone, 3-hydroxy-	1.0E-11	S	409
590-90-9	2-Butanone, 4-hydroxy-	8.1E-12	S	409
115-22-0	2-Butanone, 3-hydroxy-3-methyl-	9.4E-13	S	409
3393-64-4	2-Butanone, 4-hydroxy-3-methyl-	1.6E-11	S	409
563-80-4	2-Butanone, 3-methyl-	3.0E-12	S	220, 306
138779-12-1	2-Butanone, 1-(nitrooxy)-	8.6E-13	R	416, 269
123-73-9	2-Butenal, (E)-	3.6E-11	R	9, 524
29343-64-4	2-Butenal, 4-hydroxy-	5.7E-11	S	518
3675-14-7	Butenedial, (E)-	5.2E-11	S	220, 303
106-98-9	1-Butene	3.2E-11	R	9
5162-44-7	1-Butene, 4-bromo-			
13294-71-8	2-Butene, 2-bromo-			
22037-73-6	1-Butene, 3-bromo-	4.0E-15	S	218
4461-42-1	1-Butene, 1-chloro-			
2211-70-3	1-Butene, 2-chloro-			
563-52-0	1-Butene, 3-chloro-			
591-97-9	2-Butene, 1-chloro-			
4461-41-0	2-Butene, 2-chloro-			
503-60-6	2-Butene, 1-chloro-3-methyl-			
13602-13-6	2-Butene, 1,2-dichloro-			
764-41-0	2-Butene, 1,4-dichloro-			
563-78-0	1-Butene, 2,3-dimethyl-			
558-37-2	1-Butene, 3,3-dimethyl-	2.8E-11	S	9, 87
563-79-1	2-Butene, 2,3-dimethyl-	1.1E-10	R	9
22615-23-3	1-Butene-3,4-diol, dinitrate	1.9E-11	S	218, 269
19931-40-9	2-Butene-1,4-diol, dinitrate	1.5E-11	S	218, 269

k_Ozone [cm^3 molecule^{-1} s^{-1}]	QI	Ref.	k_Nitrate [cm^3 molecule^{-1} s^{-1}]	QI	Ref.	k_hν/Φ [s^{-1}] / [–]	QI	Ref.
			4.8E-17	S	218, 242			
			2.1E-15	R	416			
						$\Phi = 1$	E	184, 185[68)]
						k_hν = 1.4E-5	S	189
< 1.1E-19	S	409	< 9.0E-16	S	409			
< 1.1E-19	S	409	6.5E-16	S	409			
< 1.1E-19	S	409	< 2.2E-15	S	409			
< 1.1E-19	S	409	< 2.0E-16	S	409			
< 1.1E-19	S	409	< 2.2E-15	S	409			
6.3E-21	S	220, 340						
1.1E-18	A	4, 117, 376	1.6E-14	S	430[70)]	$\Phi = 0.03$	S	453
1.6E-18	S	471						
9.6E-18	R	218	1.2E-14	R	8, 363, 370, 371[71)]			
			8.8E-16	S	363[72)]			
1.4E-17	S	218, 221	1.4E-13	S	220, 363[73)]			
			4.0E-15	S	273			
			1.2E-14	S	218, 273			
			1.9E-14	A	272, 273			
2.4E-18	S	426	3.0E-15	S	218			
2.3E-17	S	426	2.0E-14	S	218, 273			
			5.0E-15	S	218, 273			
4.4E-17	S	426						
			7.2E-14	S	273			
			1.5E-15	S	273			
1.0E-17	S	220, 379						
3.9E-18	S	218, 379						
1.1E-15	R	218	5.7E-11	R	8			

CAS No.	Chemical name	k_OH [cm^3 molecule^{-1} s^{-1}]	QI	Ref.
624-64-6	2-Butene, (E)-	6.4E-11	R	9
111823-35-9	1-Butene, 2-isopropyl-3-methyl-			
563-46-2	1-Butene, 2-methyl-	7.0E-11	A	9, 85, 86, 87
563-45-1	1-Butene, 3-methyl-	3.2E-11	R	9
513-35-9	2-Butene, 2-methyl-	8.7E-11	R	9
360-89-4	2-Butene, 1,1,1,2,3,4,4,4-octafluoro-			
594-56-9	1-Butene, 2,3,3-trimethyl-			
590-18-1	2-Butene, (Z)-	5.6E-11	R	9
623-43-8	2-Butenoic acid, methyl ester, (E)-			
6117-91-5	2-Buten-1-ol			
627-27-0	3-Buten-1-ol	5.5E-11	S	209
598-32-3	3-Buten-2-ol	5.9E-11	S	209
556-82-1	2-Buten-1-ol, 3-methyl-	1.5E-10	S	493
763-32-6	3-Buten-1-ol, 3-methyl-	9.7E-11	S	493
115-18-4	3-Buten-2-ol, 2-methyl-	5.0E-11	A	209, 308, 485
78-94-4	3-Buten-2-one	1.9E-11	R	9, 483
51731-17-0	3-Buten-2-one, 4-methoxy-, (E)-			
273223-83-9	Methanol, butoxy-, formate	8.0E-12	S	463, 464
94-80-4	2,4-D-butyl ester	1.5E-11	S	511
107-00-6	1-Butyne	8.0E-12	R	9
503-17-3	2-Butyne	2.7E-11	R	9
79-92-5	Camphene	5.3E-11	S	218, 244
76-22-2	Camphor	4.6E-12	S	205
13211-15-9	Camphenilone	5.1E-12	S	218
6642-30-4	Carbamic acid, methyl-, methyl ester	4.3E-12	S	220, 325
3013-02-3	Carbamothioic acid, dimethyl-, S-methyl ester	1.3E-11	S	218, 287
63-25-2	Carbaryl	3.3E-11	S	511
75-15-0	Carbon disulfide	8.0E-12	R	429
542-52-9	Carbonic acid, dibutyl ester	7.1E-12	S	463, 464
105-58-8	Carbonic acid, diethyl ester	1.8E-12	S	463, 464
616-38-6	Carbonic acid, dimethyl ester	3.1E-13	S	220, 336
623-96-1	Carbonic acid, dipropyl ester	3.4E-11	A	467, 468
463-58-1	Carbon oxide sulfide (COS)	2.0E-15	R	32, 285, 292
554-61-0	2-Carene	8.0E-11	S	218
13466-78-9	3-Carene	8.8E-11	S	9, 89
87-44-5	Caryophyllene	2.0E-10	S	220, 332
469-61-4	α-Cedrene	6.7E-11	S	220, 332
6164-98-3	Chlordimeform	3.0E-10	S	511
10049-04-4	Chlorine oxide (ClO$_2$)	6.7E-12	S	220[80)]

k_Ozone [cm³ molecule⁻¹ s⁻¹]	QI	Ref.	$k\_Nitrate$ [cm³ molecule⁻¹ s⁻¹]	QI	Ref.	$k\_h\nu/\Phi$ [s⁻¹] / [–]	QI	Ref.
1.9E-16	R	218	3.9E-13	R	8			
3.0E-18	S	334						
9.5E-15	S	220, 378						
4.0E-15	R	218	9.4E-12	R	8			
6.8E-21	S	4, 114						
7.7E-18	S	220, 378						
1.2E-16	R	218	3.5E-13	R	8			
4.4E-18	S	220, 387						
2.5E-16	S	220, 383						
1.6E-17	S	220, 383						
9.3E-18	A	220, 383, 485	1.4E-14	A	220, 358, 359, 485[74]			
4.6E-18	R	218, 117, 146	3.2E-16	S	399	$\Phi < 0.004$	S	453
1.2E-16	S	396[75]						
						$k\_h\nu = 1.3E-6$	S	192
3.0E-20	A	4, 141, 142[76]	4.5E-16	S	8, 12, 125			
			6.7E-14	S	8, 125			
6.7E-19	A	218, 244, 426	6.2E-13	S	526[77]			
< 7E-20	S	205	< 3E-16	S	205			
			< 1.0E-16	R	220, 285, 319			
			< 1.0E-15	R	220, 285, 416[78]			
			< 1.0E-16	R	220, 285, 301			
2.3E-16	R	218	1.7E-11	S	393[79]			
3.7E-17	R	218, 10, 107	9.1E-12	R	8, 122, 128, 130, 214			
1.2E-14	S	220, 298	1.9E-11	S	220, 332			
2.8E-17	S	220, 377	8.2E-12	S	220, 332			
3.0E-19	R	220, 301[81]	< 6.0E-17	S	220, 374			

Chapter 3: Table of Reaction Rate Constants of Photo-Degradation Processes

CAS No.	Chemical name	k_OH [cm^3 molecule^{-1} s^{-1}]	QI	Ref.
67-66-3	Chloroform	9.9E-14	R	285, 301[82]
218-01-9	Chrysene	5.0E-12	E	504[83]
470-82-6	1,8-Cineol	1.1E-11	S	218, 262
106-22-9	Citronellol	1.7E-10	S	528
3856-25-5	Copaene	9.0E-11	S	220, 332
420-04-2	Cyanamide	2.3E-16	S	220, 345
1134-23-2	Cycloate	3.5E-11	S	218, 287
287-23-0	Cyclobutane	1.5E-12	R	218, 235
33689-28-0	1,2-Cyclobutanedione			
1120-56-5	Cyclobutane, methylene-			
1191-95-3	Cyclobutanone	8.7E-13	S	9, 68
4054-38-0	1,3-Cycloheptadiene	1.4E-10	S	9, 88
291-64-5	Cycloheptane	1.2E-11	A	9, 181[85]
544-25-2	1,3,5-Cycloheptatriene	9.7E-11	S	9, 88
628-92-2	Cycloheptene	7.4E-11	S	9, 90
1453-25-4	Cycloheptene, 1-methyl-			
592-57-4	1,3-Cyclohexadiene	1.6E-10	S	9, 90
628-41-1	1,4-Cyclohexadiene	1.0E-10	A	9, 84, 90
106-51-4	2,5-Cyclohexadiene-1,4-dione	4.6E-12	S	425
110-82-7	Cyclohexane	7.5E-12	R	218[87]
695-12-5	Cyclohexane, ethenyl-			
108-87-2	Cyclohexane, methyl-	1.0E-11	S	9, 96
1192-37-6	Cyclohexane, methylene-			
3073-66-3	Cyclohexane, 1,1,3-trimethyl-	8.7E-12	S	9
108-93-0	Cyclohexanol	1.9E-11	S	183
108-94-1	Cyclohexanone	6.4E-12	S	9, 68
2403-24-9	Cyclohexanone, 2-hydroxy-, nitrate	3.7E-12	S	220, 381
110-83-8	Cyclohexene	6.8E-11	R	218
1674-10-8	Cyclohexene, 1,2-dimethyl-			
591-49-1	Cyclohexene, 1-methyl-	9.4E-11	S	9, 91
591-48-0	Cyclohexene, 3-methyl-			
591-47-9	Cyclohexene, 4-methyl-			
5502-88-5	Cyclohexene, 1-methyl-4-(1-methylethyl)-			
2562-37-0	Cyclohexene, 1-nitro-	4.4E-11	S	218, 270, 278
930-68-7	2-Cyclohexen-1-one			
111-78-4	1,5-Cyclooctadiene			
3806-59-5	1,3-Cyclooctadiene, (Z,Z)-			
292-64-8	Cyclooctane	1.3E-11	S	326[89]
629-20-9	1,3,5,7-Cyclooctatetraene			
933-11-9	Cyclooctene, 1-methyl-			

Table of Reaction Rate Constants of Photo-Degradation Processes

k_Ozone [cm^3 molecule^{-1} s^{-1}]	QI	Ref.	k_Nitrate [cm^3 molecule^{-1} s^{-1}]	QI	Ref.	k_$h\nu/\Phi$ [s^{-1}] / [–]	QI	Ref.
			< 6.5E-17	S	218			
2.4E-16	S	528						
1.6E-16	S	220, 377	1.6E-11	S	220, 332			
< 3.E-19	S	218, 287	3.2E-14	S	218, 287			
						$\Phi = 0.86$		191[84)]
1.9E-17	S	426, 226	4.2E-13	S	220, 360			
1.5E-16	S	218, 108	6.5E-12	S	8, 128			
5.4E-17	S	4, 108	1.2E-12	S	8, 203			
2.9E-16	R	218, 108, 225[86)]	4.8E-13	S	8, 203			
9.3E-16	S	500						
1.3E-15	S	218, 108	1.2E-11	R	8			
4.6E-17	S	218, 108	6.6E-13	R	8			
1.0E-23	R	2	7.2E-17	R	2			
7.5E-18	S	220, 384						
1.8E-17	A	384, 226	8.5E-16	S	325			
7.2E-17	R	218, 105, 108, 112	5.3E-13	S	8, 128			
			5.1E-11	S	360			
1.6E-16	S	380[88)]	1.3E-11	A	360, 361			
9.0E-17	S	380						
7.3E-17	S	500						
5.2E-16	S	4, 107						
1.1E-18	S	380						
< 1.9E-18	S	218, 117						
1.4E-16	S	500						
2.0E-17	S	500						
2.6E-18	S	500						
1.5E-15	S	500						

CAS No.	Chemical name	k_OH [cm^3 molecule^{-1} s^{-1}]	QI	Ref.
13152-05-1	Cyclooctene, 3-methyl-			
931-87-3	Cyclooctene, (Z)-			
4125-18-2	1,3-Cyclopentadiene, 5,5-dimethyl-	1.3E-10	S	220, 300
287-92-3	Cyclopentane	5.1E-12	R	218
1528-30-9	Cyclopentane, methylene-			
96-41-3	Cyclopentanol	1.1E-11	S	9, 72
120-92-3	Cyclopentanone	2.9E-12	S	9, 68
541-02-6	Cyclopentasiloxane, decamethyl-	1.5E-12	S	237
142-29-0	Cyclopentene	6.0E-11	A	9, 90, 136
930-29-0	Cyclopentene, 1-chloro-			
693-89-0	Cyclopentene, 1-methyl-			
1120-62-3	Cyclopentene, 3-methyl-			
75-19-4	Cyclopropane	8.4E-14	R	218, 181
6142-73-0	Cyclopropane, methylene-			
3638-35-5	Cyclopropane, (1-methylethyl)-	2.8E-12	S	9, 98
556-67-2	Cyclotetrasiloxane, octamethyl-	1.0E-12	S	218, 237
541-05-9	Cyclotrisiloxane, hexamethyl-	5.2E-13	S	218, 237
6753-98-6	1,4,8-Cycloundecatriene, 2,6,6,9-tetramethyl-, (E,E,E)-	2.9E-10	S	332
50-29-3	4,4'-DDT	5.4E-12	S	401[92]
1163-19-5	Decabromodiphenylether			
112-31-2	Decanal			
124-18-5	Decane	1.2E-11	R	9
693-54-9	2-Decanone	1.3E-11	S	9, 67
872-05-9	1-Decene			
7433-78-5	5-Decene, (Z)-			
262-12-4	Dibenzo[b,e][1,4]dioxin	1.3E-11	A	297, 343[94]
39227-53-7	Dibenzo[b,e][1,4]dioxin, 1-chloro-	4.7E-12	S	314
39227-54-8	Dibenzo[b,e][1,4]dioxin, 2-chloro-	3.9E-11	E	452[95]
29446-15-9	Dibenzo[b,e][1,4]dioxin, 2,3-dichloro-	2.2E-11	E	452[96]
33857-26-0	Dibenzo[b,e][1,4]dioxin, 2,7-dichloro-	5.9E-12	E	297, 452[97]
38964-22-6	Dibenzo[b,e][1,4]dioxin, 2,8-dichloro-	7.1E-12	E	452[98]
3268-87-9	Dibenzo[b,e][1,4]dioxin, octachloro-	5.2E-12	E	405, 452[99]
30746-58-8	Dibenzo[b,e][1,4]dioxin, 1,2,3,4-tetrachloro-	8.5E-13	E	462[100]
1746-01-6	Dibenzo[b,e][1,4]dioxin, 2,3,7,8-tetrachloro-	9.0E-12	E	290[101]
132-64-9	Dibenzofuran	3.5E-12	S	297
5409-83-6	Dibenzofuran, 2,8-dichloro-	2.2E-12	E	297[102]
19287-45-7	Diborane(6)	5.0E-13	S	320
62-73-7	Dichlorvos	2.3E-11	A	511, 525
115-32-2	Dicofol	2.0E-12	A	476[103]
462-95-3	Diethoxymethane	1.7E-11	S	432, 467

k_Ozone [cm³ molecule⁻¹ s⁻¹]	QI	Ref.	$k\_Nitrate$ [cm³ molecule⁻¹ s⁻¹]	QI	Ref.	$k\_h\nu/\Phi$ [s⁻¹] / [–]	QI	Ref.
1.4E-16	S	500						
3.9E-16	A	380, 500[90]						
9.0E-17	S	426	1.5E-12	S	360			
< 3E-20	S	237	< 3E-16	S	237			
6.3E-16	R	218, 97, 236	4.6E-13	S	8			
1.5E-17	S	380						
7.5E-16	A	380, 527[91]						
3.3E-16	S	500						
2.8E-18	S	426	1.4E-14	S	220			
< 3E-20	S	218, 237	1.4E-16	S	218, 237			
< 3E-20	S	218, 237	< 2E-16	S	218, 237			
1.2E-14	S	377	3.5E-11	S	332			
						$k\_h\nu$ = 2.2E-8	A/E	245[93]
			2.2E-14	S	488			
			2.6E-16	S	353			
8.0E-18	S	379						
1.1E-16	S	334						
< 5.0E-20	S	343	< 8E-15	S	343			
< 7.0E-20	S	314						
< 8.0E-20	S	343	< 1.6E-15	S	343			

CAS No.	Chemical name	k_OH [cm^3 molecule^{-1} s^{-1}]	QI	Ref.
111-46-6	Diethylene glycol	3.0E-11	S	9
112-34-5	Diethylene glycol butyl ether	7.4E-11	S	509
112-36-7	Diethylene glycol diethyl ether	2.7E-11	S	9
111-96-6	Diethylene glycol dimethyl ether	1.7E-11	S	9
111-90-0	Diethylene glycol ethyl ether	5.7E-11	S	509
60-29-7	Diethylether	1.3E-11	R	218[104)]
598-26-5	Dimethylketene	1.1E-10	S	9, 69
67-71-0	Dimethyl sulfone	< 3E-13	S	523
67-68-5	Dimethylsulfoxide	7.7E-11	A	9, 62, 523, 364
505-22-6	1,3-Dioxane	9.1E-12	S	9, 390
123-91-1	1,4-Dioxane	1.2E-11	A	317, 467, 390[106)]
1120-97-4	1,3-Dioxane, 4-methyl-	1.1E-11	S	9, 390
646-06-0	1,3-Dioxolane	8.8E-12	S	465
96-49-1	1,3-Dioxolane-2-one	5.0E-13	E	465
107-46-0	Disiloxane, hexamethyl-	1.4E-12	S	218
110-06-5	Disulfide, bis(1,1-dimethylethyl)-	4.1E-11	S	9
624-92-0	Disulfide, dimethyl-	2.3E-10	R	218
6163-66-2	Di-*tert*-butyl ether	3.8E-12	A	241, 145
112-40-3	Dodecane	1.4E-11	R	9
759-94-4	Eptam	3.3E-11	A	218, 287
75-04-7	Ethanamine	2.8E-11	S	9, 53
3710-84-7	Ethanamine, *N*-ethyl-*N*-hydroxy-	1.0E-10[107)]	S	9[108)]
753-90-2	Ethanamine, 2,2,2-trifluoro-	9.4E-13	S	220, 323
74-84-0	Ethane	2.6E-13	R	218
188690-78-0	Ethane, 1,2-bis(difluoromethoxy)-1,1,2,2-tetrafluoro-	4.7E-15	S	508
74-96-4	Ethane, bromo-	2.7E-13	R	218
151-67-7	Ethane, 1-bromo-1-chloro-2,2,2-trifluoro-	6.0E-14	S	218
107-04-0	Ethane, 1-bromo-2-chloro-	2.2E-13	E	220
354-06-3	Ethane, 1-bromo-2-chloro-1,1,2-trifluoro-	1.4E-14	R	301[109)]
151-67-7	Ethane, 2-bromo-2-chloro-1,1,1-trifluoro-	4.6E-14	R	218
762-49-2	Ethane, 1-bromo-2-fluoro-	9.3E-14	E	220, 319
6482-24-2	Ethane, 1-bromo-2-methoxy-	6.6E-12	S	218, 283[110)]
124-72-1	Ethane, 1-bromo-1,2,2,2-tetrafluoro-	1.7E-14	S	218, 258
421-06-7	Ethane, 2-bromo-1,1,1-trifluoro-	1.6E-14	R	218[111)]
75-00-3	Ethane, chloro-	4.1E-13	R	218
81729-06-8	Ethane, 2-chloro-1,1-difluoro-1-(1-fluoroethoxy)-	< 3E-13	S	218[112)]
75-68-3	Ethane, 1-chloro-1,1-difluoro-	3.1E-15	R	218[113)]
13838-16-9	Ethane, 2-chloro-1-(difluoromethoxy)-1,1,2-trifluoro-	1.6E-14	R	416, 261[114)]
26675-46-7	Ethane, 2-chloro-2-(difluoromethoxy)-1,1,1-trifluoro-	2.1E-14	R	416, 283
1615-75-4	Ethane, 1-chloro-1-fluoro-	1.7E-13	E	220, 319

Table of Reaction Rate Constants of Photo-Degradation Processes | 133

k_Ozone [cm^3 $molecule^{-1}$ s^{-1}]	QI	Ref.	k_Nitrate [cm^3 $molecule^{-1}$ s^{-1}]	QI	Ref.	k_$h\nu/\Phi$ [s^{-1}] / [–]	QI	Ref.
			3.5E-15	A	351, 357[105]			
< 4E-17	S	218						
< 1E-19	S	523	< 2E-15	S	523			
< 1E-19	S	523	3.3E-13	A	62, 523			
			< 8E-17	S	218, 237			
			7.0E-13	R	32, 416			
			2.8E-16	S	241			
< 1.3E-19	S	218	9.0E-15	S	218, 287			
2.8E-20	S	4						
< 1E-23	E	4	< 1.0E-17	R	416			

CAS No.	Chemical name	k_OH [cm^3 molecule^{-1} s^{-1}]	QI	Ref.
627-42-9	Ethane, 1-chloro-2-methoxy-	4.7E-12	S	218, 283[115]
2837-89-0	Ethane, 2-chloro-1,1,1,2-tetrafluoro-	9.5E-15	R	218[116]
75-88-7	Ethane, 2-chloro-1,1,1-trifluoro-	1.3E-14	R	416[117]
421-04-5	Ethane, 1-chloro-1,1,2-trifluoro-	1.4E-14	S	220
431-07-2	Ethane, 1-chloro-1,2,2-trifluoro-	1.7E-14	S	220
73602-63-8	Ethane, 2-chloro-1,1,1-trifluoro-2-(fluoromethoxy)-	1.5E-14	S	437[118]
5073-63-2	Ethane, 2-chloro-1,1,2-trifluoro-1-(fluoromethoxy)-	1.2E-14	S	437[119]
425-87-6	Ethane, 2-chloro-1,1,2-trifluoro-1-methoxy-	3.8E-14	S	393[120]
107-22-2	Ethanedial	1.1E-11	R	32, 416
557-91-5	Ethane, 1,1-dibromo-	2.8E-13	E	220, 319
106-93-4	Ethane, 1,2-dibromo-	2.2E-13	R	218, 95
75-81-0	Ethane, 1,2-dibromo-1,1-dichloro-	1.4E-13	E	220, 319
75-82-1	Ethane, 1,2-dibromo-1,1-difluoro-	2.4E-14	E	220, 319
359-19-3	Ethane, 1,1-dibromo-2,2-difluoro-	2.9E-14	E	220, 319
124-73-2	Ethane, 1,2-dibromo-1,1,2,2-tetrafluoro-	< 1E-16	R	218, 231[123]
354-04-1	Ethane, 1,2-dibromo-1,1,2-trifluoro-	1.5E-14	E	220, 319
75-34-3	Ethane, 1,1-dichloro-	2.7E-13	R	218, 95[124]
107-06-2	Ethane, 1,2-dichloro-	2.5E-13	R	218, 95
1649-08-7	Ethane, 1,2-dichloro-1,1-difluoro-	1.5E-14	R	32[125]
431-06-1	Ethane, 1,2-dichloro-1,2-difluoro-	4.3E-14	E	220, 319
1717-00-6	Ethane, 1,1-dichloro-1-fluoro-	6.0E-15	R	218, 416[126]
430-57-9	Ethane, 1,2-dichloro-1-fluoro-	1.1E-13	E	220, 319
34862-07-2	Ethane, 1,1-dichloro-2-methoxy-	2.3E-12	S	218, 283
76-14-2	Ethane, 1,2-dichloro-1,1,2,2-tetrafluoro-	5.0E-16	S	9, 95
354-23-4	Ethane, 1,2-dichloro-1,1,2-trifluoro-	1.2E-14	S	218, 232
354-15-4	Ethane, 1,2-difluoro-1,2,2-trichloro-	1.5E-14	R	220, 301[127]
306-83-2	Ethane, 2,2-dichloro-1,1,1-trifluoro-	3.6E-14	R	218, 32
75-37-6	Ethane, 1,1-difluoro-	3.5E-14	R	220, 513
624-72-6	Ethane, 1,2-difluoro-	1.1E-13	S	9, 94
69948-24-9	Ethane, 1-difluoromethoxy-1,1,2-trifluoro-	5.6E-15	S	490[128]
1885-48-9	Ethane, 2-(difluoromethoxy)-1,1,1-trifluoro-	1.2E-14	A	218, 288
534-15-6	Ethane, 1,1-dimethoxy-	8.9E-12	S	9, 210
110-71-4	Ethane, 1,2-dimethoxy-	2.8E-11	S	317
460-19-5	Ethanedinitrile	2.5E-15	S	9
629-15-2	1,2-Ethanediol, diformate	4.7E-13	S	467, 469
512-51-6	Ethane, 2-ethoxy-1,1,2,2-tetrafluoro-	3.8E-14	A	410, 433, 434[129]
353-36-6	Ethane, fluoro-	2.0E-13	R	416, 93
75-03-6	Ethane, iodo-			
542-85-8	Ethane, isothiocyanato-	5.5E-11	S	530
540-67-0	Ethane, methoxy-	6.6E-12	S	219

k_Ozone [cm³ molecule⁻¹ s⁻¹]	QI	Ref.	k_Nitrate [cm³ molecule⁻¹ s⁻¹]	QI	Ref.	k_hν/Φ [s⁻¹] / [−]	QI	Ref.
3.0E-21	S	4				$k\_h\nu = 1.1\text{E-}5 \,/\, \Phi = 0.033$	S/A	34[121]/32[122]
5.4E-25	E	4, 103						
						$k\_h\nu = 3\text{E-}6 \,/\, \Phi = 1$	E	484[130]

CAS No.	Chemical name	k_OH [cm^3 molecule^{-1} s^{-1}]	QI	Ref.
624-89-5	Ethane, (methylthio)-	8.5E-12	S	9, 63
79-24-3	Ethane, nitro-	1.5E-13	R	218, 286
333-36-8	Ethane, 1,1'-oxybis[2,2,2,-trifluoro-	1.6E-13	S	407[131]
75-95-6	Ethane, pentabromo-	1.2E-13	E	220, 319
76-01-7	Ethane, pentachloro-	2.3E-13	S	218, 254
354-33-6	Ethane, pentafluoro-	1.9E-15	R	218, 285, 301[132]
22410-44-2	Ethane, pentafluoromethoxy-	1.2E-14	S	435[133]
630-16-0	Ethane, 1,1,1,2-tetrabromo-	9.8E-14	E	220, 319
79-27-6	Ethane, 1,1,2,2-tetrabromo-	1.2E-13	E	220, 319
630-20-6	Ethane, 1,1,1,2-tetrachloro-	2.0E-14	S	218, 277
79-34-5	Ethane, 1,1,2,2-tetrachloro-	1.0E-13	S	218[134]
354-14-3	Ethane, 1,1,2,2-tetrachloro-1-fluoro-	2.0E-13	A	254, 277, 420[135]
354-11-0	Ethane, 1,1,1,2-tetrachloro-2-fluoro-	1.2E-13	E	220, 319
811-97-2	Ethane, 1,1,1,2-tetrafluoro-	4.2E-15	R	285, 301[136]
359-35-3	Ethane, 1,1,2,2-tetrafluoro-	5.8E-15	R	285, 301[137]
425-88-7	Ethane, 1,1,2,2-tetrafluoro-1-methoxy-	2.2E-14	S	410[138]
406-78-0	Ethane, 1,1,2,2-tetrafluoro-1-(2,2,2-trifluoroethoxy)-	9.0E-15	S	393[139]
2356-62-9	Ethane, 1,1,1,2-tetrafluoro-2-(trifluoromethoxy)-	5.0E-15	S	444
2356-61-8	Ethane, 1,1,2,2-tetrafluoro-2-(trifluoromethoxy)-	2.3E-15	S	515
352-93-2	Ethane, 1,1'-thiobis-	1.5E-11	R	9, 218
75-08-1	Ethanethiol	4.7E-11	R	9, 178, 179
79-00-5	Ethane, 1,1,2-trichloro-	2.0E-13	R	410[140]
71-55-6	Ethane, 1,1,1-trichloro-	9.7E-15	R	285, 301[141]
354-21-2	Ethane, 1,2,2-trichloro-1,1-difluoro-	5.3E-14	S	232
354-12-1	Ethane, 1,1,1-trichloro-2,2-difluoro-	5.1E-14	E	220, 319
76-13-1	Ethane, 1,1,2-trichloro-1,2,2-trifluoro-	3.0E-16	S	9, 95
420-46-2	Ethane, 1,1,1-trifluoro-	1.3E-15	R	218, 94
430-66-0	Ethane, 1,1,2-trifluoro-	1.5E-14	R	285, 301[143]
460-43-5	Ethane, 1,1,1-trifluoro-2-methoxy-	6.3E-13	A	218, 74, 288, 407
64-17-5	Ethanol	3.4E-12	R	32
107-07-3	Ethanol, 2-chloro-	1.3E-12	A	9, 182
359-13-7	Ethanol, 2,2-difluoro-	4.5E-13	S	505
108-01-0	Ethanol, 2-(dimethylamino)-	7.5E-11	A	9, 54, 157[144]
371-62-0	Ethanol, 2-fluoro-	1.4E-12	S	505
628-82-0	Ethanol, 2-methoxy-, formate	5.1E-12	S	463, 464
115-20-8	Ethanol, 2,2,2-trichloro-	2.4E-13	S	9, 182
75-89-8	Ethanol, 2,2,2-trifluoro-	1.0E-13	A	9, 182[145]
54464-57-2	Ethanone, 1-(1,2,3,4,5,6,7,8-octahydro-2,3,8,8-tetramethyl-2-naphthalenyl)-	9.8E-11	S	482
98-86-2	Ethanone, 1-phenyl-	2.7E-12	S	9

k_Ozone [cm³ molecule⁻¹ s⁻¹]	QI	Ref.	k_Nitrate [cm³ molecule⁻¹ s⁻¹]	QI	Ref.	k_hν/Φ [s⁻¹] / [–]	QI	Ref.
			4.8E-12	S	8, 121			
			1.2E-12	S	8			
1.5E-25	E	103[142)]						
			≤ 2.0E-15	R	32, 225, 416			
6.7E-18	S	375						
2.1E-18	S	482	1.7E-11	S	482			

CAS No.	Chemical name	k_OH [cm³ molecule⁻¹ s⁻¹]	QI	Ref.
74-85-1	Ethene	8.5E-12	R	218[146]
593-60-2	Ethene, bromo-	6.8E-12	S	9, 13
75-01-4	Ethene, chloro-	7.0E-12	R	218
75-35-4	Ethene, 1,1-dichloro-	1.1E-11	R	218[150]
79-35-6	Ethene, 1,1-dichloro-2,2-difluoro-	7.5E-12	A	218, 257[252]
156-60-5	Ethene, 1,2-dichloro-, (E)-	2.3E-12	R	218, 82
156-59-2	Ethene, 1,2-dichloro-, (Z)-	2.6E-12	R	218, 88
75-38-7	Ethene, 1,1-difluoro-			
1630-78-0	Ethene, 1,2-difluoro-, (E)-			
1630-77-9	Ethene, 1,2-difluoro-, (Z)-			
109-92-2	Ethene, ethoxy-			
75-02-5	Ethene, fluoro-	5.6E-12	S	9
107-25-5	Ethene, methoxy-	3.3E-11	S	9, 167
3638-64-0	Ethene, nitro-	1.2E-12	S	218, 278
127-18-4	Ethene, tetrachloro-	1.7E-13	R	32, 218
116-14-3	Ethene, tetrafluoro-	1.1E-11	S	394
79-01-6	Ethene, trichloro-	2.1E-12	R	218
359-29-5	Ethene, trichlorofluoro-	7.6E-12	S	218, 257
359-11-5	Ethene, trifluoro-			
1187-93-5	Ethene, trifluoro-(trifluoromethoxy)-	3.0E-12	A	410, 411, 412[159]
463-51-4	Ethenone	1.5E-11	R	218
20334-52-5	Ethylketene	1.2E-10	S	9, 69
107-21-1	Ethylene glycol	1.1E-11	A	9, 481
111-76-2	Ethylene glycol butyl ether	2.1E-11	A	9, 327, 481
629-14-1	Ethylene glycol diethyl ether	5.7E-11	S	317
110-80-5	Ethylene glycol ethyl ether	1.5E-11	A	9
111-15-9	Ethylene glycol ethyl ether acetate	1.0E-11	S	282
109-59-1	Ethylene glycol isopropyl ether	1.8E-11	S	317
109-86-4	Ethylene glycol methyl ether	1.2E-11	S	9
2807-30-9	Ethylene glycol propyl ether	1.6E-11	S	327
74-86-2	Ethyne	8.15E-13	R	218[160]
206-44-0	Fluoranthene	1.1E-11	S	297[161]
86-73-7	9H-Fluorene	1.4E-11	A	3, 10, 297, 298[162]
50-00-0	Formaldehyde	9.4E-12	R	218, 1, 32, 301
6629-91-0	Formaldehyde hydrazone			
75-17-2	Formaldehyde, oxime	6.3E-13	S	9
64-18-6	Formic acid	4.8E-13	R	32[165]
592-84-7	Formic acid, butyl ester	3.1E-12	S	299[166]
762-75-4	Formic acid, 1,1-dimethylethyl ester	7.7E-13	A	220

k_Ozone [cm^3 molecule^{-1} s^{-1}]	QI	Ref.	k_Nitrate [cm^3 molecule^{-1} s^{-1}]	QI	Ref.	$k\_h\nu/\Phi$ [s^{-1}] / [−]	QI	Ref.
1.7E-18	R	4, 32	2.1E-16	R	32, 218, 285[147]			
2.4E-19	A	4, 14, 15, 112[148]	4.9E-16	S	362[149]			
3.7E-21	S	4, 113	1.8E-15	A	357, 362[151]			
2.0E-19	A	4, 15, 138, 140	1.1E-16	A	220, 357, 362, 363, 367[153]			
6.2E-20	S	4[154]	1.2E-16	A	220, 363, 366[155]			
1.3E-19	A	4, 112, 137[156]						
2.1E-18	S	4, 112						
2.6E-19	S	4, 112						
1.5E-16	S	220, 376						
7.0E-19	S	4, 13, 112						
< 1.0E-21	R	285	9.6E-17	S	362, 363[157]			
2.8E-20	S	218, 428	3.0E-15	S	393			
< 5.0E-20	R	32, 285	2.9E-16	R	32, 285[158]			
1.4E-19	S	4, 112						
< 1E-21	S	218	1.1E-13	S	368			
< 1E-18	S	218						
< 1.1E-19	S	481	3.0E-15	S	481			
1.0E-20	R	32, 416	< 1.0E-16	R	32, 416			
−								
< 2.0E-19	S	298	3.5E-14	S	298			
≤ 2.1E-24	S	4, 197	5.8E-16	R	8	7.9E-14 / Φ = 1.0	R	32 / 5[163]
2.5E-17	S	4						

[164]

CAS No.	Chemical name	k_OH [cm^3 molecule^{-1} s^{-1}]	QI	Ref.
109-94-4	Formic acid, ethyl ester	1.0E-12	S	9, 442
107-31-3	Formic acid, methyl ester	2.0E-13	A	299, 442[167]
625-55-8	Formic acid, 1-methylethyl ester	2.1E-12	S	330
110-74-7	Formic acid, propyl ester	2.1E-12	A	77, 299[168]
2757-23-5	Formylchloride	< 3.2E-13	S	218
1493-02-3	Formylfluoride	≤ 1.0E-14	S	416
110-00-9	Furan	4.0E-11	R	9, 27, 218
98-01-1	2-Furancarboxaldehyde	3.5E-11	S	296
498-60-2	3-Furancarboxaldehyde	4.9E-11	S	296
620-02-0	2-Furancarboxaldehyde, 5-methyl-	5.1E-11	S	296
1708-29-8	Furan, 2,5-dihydro-			
3710-43-8	Furan, 2,4-dimethyl-			
625-86-5	Furan, 2,5-dimethyl-	1.3E-10	S	218, 263
14920-89-9	Furan, 2,3-dimethyl-			
108-31-6	2,5-Furandione	1.4E-12	S	303
3208-16-0	Furan, 2-ethyl-	1.1E-10	S	218, 263
930-27-8	Furan, 3-methyl-	9.3E-11	S	9, 76
534-22-5	Furan, 2-methyl-	6.2E-11	S	218, 263
20825-71-2	2(3H)-Furanone	4.4E-11	S	303
591-12-8	2(3H)-Furanone, 5-methyl-	6.9E-11	S	303
109-99-9	Furan, tetrahydro-	1.6E-11	R	9, 390
10599-58-3	Furan, tetramethyl-			
1222-05-5	Galoxolide	2.6E-11	S	482
111-71-7	Heptanal	2.9E-11	A	470, 524
142-82-5	Heptane	7.2E-12	R	9
111-70-6	1-Heptanol	1.4E-11	R	218, 72
82944-61-4	3-Heptanol, nitrate	3.7E-12	S	9
110-43-0	2-Heptanone	1.2E-11	S	424
108-83-8	4-Heptanone, 2,6-dimethyl-	2.7E-11	R	9
18829-55-5	2-Heptenal, (E)-	2.4E-11	S	524
929-22-6	4-Heptenal, (E)-			
592-76-7	1-Heptene	4.0E-11	A	9, 131
592-77-8	2-Heptene	6.8E-11	A	218[173]
14686-13-6	2-Heptene, (E)-	6.7E-11	S	208
110-93-0	5-Hepten-2-one, 6-methyl-	1.6E-10	S	479, 522
319-84-6	α-Hexachlorocyclohexane	1.4E-13	E	454[174], [175]
544-76-3	Hexadecane	2.5E-11	S	9, 97
142-83-6	2,4-Hexadienal, (E,E)-	7.6E-11	S	393
53398-76-8	2,4-Hexadienal, (2E,4Z)-	7.4E-11	S	392
592-42-7	1,5-Hexadiene	6.2E-11	A	9, 84

k_Ozone [cm³ molecule⁻¹ s⁻¹]	QI	Ref.	k_Nitrate [cm³ molecule⁻¹ s⁻¹]	QI	Ref.	k_hν/Φ [s⁻¹] / [–]	QI	Ref.
			2.0E-17	A	218, 242			
			4.1E-18	S	218, 242			
9.1E-18	S	218						
2.4E-18	S	4, 118	1.2E-12	A	8, 17, 365			
1.6E-17	S	4, 112						
			5.7E-11	S	365[169)]			
			5.8E-11	A	365			
2.0E-17	S	388	2.1E-11	A	365, 399			
			2.6E-11	S	365			
			1.8E-13	S	471			
			4.9E-15	S	8			
			1.2E-10	S	365			
			2.4E-14	S	415[170)]			
			1.4E-16	R	8			
						Φ = 1		184
			9.8E-14	S	430[171)]			
≤ 1.5E-16	S	396[172)]	2.2E-14	S	430			
1.4E-17	S	218, 115						
8.8E-17	S	4						
3.9E-16	A	479, 385, 522	7.5E-12	S	479			

CAS No.	Chemical name	k_OH [cm^3 molecule^{-1} s^{-1}]	QI	Ref.
592-46-1	2,4-Hexadiene	1.3E-10	S	9, 84
592-48-3	cis,trans-1,3-Hexadiene	1.1E-10	S	9, 84
53042-85-6	2,4-Hexadienedial, (E,Z)-	7.4E-11	S	216
121020-77-7	2,4-Hexadienedial, 2-methyl, (2E,4E)-	1.2E-10	S	531
18409-46-6	2,4-Hexadienedial, (E.E)-	7.6E-11	S	216
627-58-7	1,5-Hexadiene, 2,5-dimethyl-	1.2E-10	S	9, 84
764-13-6	2,4-Hexadiene, 2,5-dimethyl-	2.1E-10	S	9, 84
1515-79-3	1,3-Hexadiene, 5,5-dimethyl-			
7319-00-8	1,4-Hexadiene, (E)-	9.1E-11	A	9, 84
5194-51-4	2,4-Hexadiene, (E,E)-			
5194-50-3	2,4-Hexadiene, (E,Z)-			
4049-81-4	1,5-Hexadiene, 2-methyl-	9.6E-11	S	9, 84
66-25-1	Hexanal	3.0E-11	A	413, 414, 524
110-54-3	Hexane	5.6E-12	R	9
111-25-1	Hexane, 1-bromo-	5.8E-12	S	218, 279
544-10-5	Hexane, 1-chloro-	3.8E-12	A	218, 255
95576-25-3	Hexane, 1,1,1,2,2,5,5,6,6,6-decafluoro-	8.3E-15	S	382
590-73-8	Hexane, 2,2-dimethyl-	5.0E-12	A	9, 393
627-93-0	1,6-Hexanedioic acid, dimethyl ester	3.6E-12	S	486
110-13-4	2,5-Hexanedione	7.1E-12	S	9, 68
297730-93-9	Hexane, 3-ethoxy-1,1,1,2,3,4,4,5,5,6,6,6- dodecafluoro-2-(trifluoromethyl)-	2.6E-14	S	507
No CAS No.	Hexane, 1,1,1,2,3-pentafluoro-2-(trifluoromethyl)-3-ethoxy-	2.6E-14	S	507
355-37-3	Hexane, 1,1,1,2,2,3,3,4,4,5,5,6,6-tridecafluoro-	1.7E-15	S	494[179)]
1069-53-0	Hexane, 2,3,5-trimethyl-	7.9E-12	S	9
111-27-3	1-Hexanol	1.2E-11	R	218, 72[180)]
626-93-7	2-Hexanol	1.2E-11	S	9, 72
82944-60-3	3-Hexanol, nitrate	2.7E-12	S	9, 59
21981-49-7	2-Hexanol, nitrate	3.2E-12	S	9, 59
591-78-6	2-Hexanone	9.1E-12	R	9
589-38-8	3-Hexanone	6.9E-12	S	9, 66
4984-85-4	3-Hexanone, 4-hydroxy-	1.5E-11	S	409
110-12-3	2-Hexanone, 5-methyl-	1.2E-11	S	306
2235-12-3	1,3,5-Hexatriene			
821-07-8	1,3,5-Hexatriene, (E)-	1.1E-10	S	9, 88
2612-46-6	1,3,5-Hexatriene, (Z)-	1.1E-10	S	9, 88
592-47-2	3-Hexen (Z)			
505-57-7	2-Hexenal			
6728-26-3	2-Hexenal, (E)-	3.7E-11	A	333, 524
592-41-6	1-Hexene	3.5E-11	A	9, 131

k_Ozone [cm^3 molecule^{-1} s^{-1}]	QI	Ref.	k_Nitrate [cm^3 molecule^{-1} s^{-1}]	QI	Ref.	k_$h\nu$/Φ [s^{-1}] / [–]	QI	Ref.
								176)
								177)
3.1E-15	S	440						
2.4E-17	S	440						
3.7E-16	S	4, 88	1.6E-11	S	218, 284			
3.1E-16	S	4, 88						
			1.7E-14	A	356, 413, 414, 415[178]	Φ = 0.30	A	453, 470
			1.0E-16	S	8, 201			
						Φ = 1		184
						Φ = 1		184
< 1.1E-19	S	409	1.2E-15	S	409			
2.6E-17	S	4, 88						
1.4E-16	S	334						
2.0E-18	S	313[181]						
			3.2E-14	A	333, 313, 430[182]			
1.1E-17	R	218[183]	8.0E-14	R	354[184]			

CAS No.	Chemical name	k_OH [cm^3 molecule^{-1} s^{-1}]	QI	Ref.
59643-69-5	2-Hexene, 3,4-diethyl-, (Z)-			
690-93-7	3-Hexene, 2,2-dimethyl-, (E)-			
692-70-6	3-Hexene, 2,5-dimethyl-, (E)-			
820-69-9	3-Hexene-2,5-dione, (E)-	4.6E-11	A	9, 303
17559-81-8	3-Hexene-2,5-dione, (Z)-	6.3E-11	S	220
10153-61-4	3-Hexene-2,5-dione, 3,4-dihydroxy-	2.7E-10	S	311
13269-52-8	3-Hexene, (E)-			
19430-93-4	1-Hexene, 3,3,4,4,5,5,6,6,6-nonafluoro-	1.4E-12	S	408[187]
7642-09-3	3-Hexene, (Z)-			
2497-18-9	2-Hexen-1-ol, acetate, (E)-			
3681-71-8	3-Hexen-1-ol, acetate, (Z)-	7.8E-11	S	313
928-96-1	3-Hexen-1-ol, (Z)-	1.1E-10	S	313
50396-87-7	4-Hexen-3-one, (4E)-			
693-02-7	1-Hexyne	1.3E-11	S	9, 80
764-35-2	2-Hexyne			
107-54-0	1-Hexyn-3-ol, 3,5-dimethyl-	2.9E-11	S	498
302-01-2	Hydrazine	3.6E-11	S	350
57-14-7	Hydrazine, 1,1-dimethyl-			
60-34-4	Hydrazine, methyl-	6.5E-11	S	9, 56
6415-12-9	Hydrazine, tetramethyl-			
10035-10-6	Hydrobromic acid	1.1E-11	R	285, 301
7647-01-0	Hydrochloric acid	8.0E-13	R	220[188]
74-90-8	Hydrocyanic acid	3.1E-14	R	285, 301[189]
7722-84-1	Hydrogen peroxide (H$_2$O$_2$)	1.7E-12	R	285, 301, 99
7783-06-4	Hydrogen sulfide (H$_2$S)	4.7E-12	R	285, 301
75-91-2	Hydroperoxide, 1,1-dimethylethyl-	3.8E-12	S	9, 79
3031-73-0	Hydroperoxide, methyl-	1.1E-11	R	32
288-32-4	1H-Imidazole	3.6E-11	S	9
95-13-6	1H-Indene	7.8E-11	S	298
496-11-7	1H-Indene, 2,3-dihydro-	9.2E-12	S	9
120-72-9	1H-Indole	1.5E-10	S	307
119-65-3	Isoquinoline	8.5E-12	S	307
78-00-2	Lead, tetraethyl-	6.3E-11	R	218, 151, 152
75-74-1	Lead, tetramethyl-	4.6E-12	R	218, 151, 152
138-86-3	Limonene			
5989-27-5	d-Limonene	1.4E-10	S	130
78-70-6	Linalool	1.6E-10	S	342
58-89-9	Lindane	1.9E-13	E	454[194]
		3.4E-13	S	458[195]
475-20-7	(+)-Longifolene	4.7E-11	S	332

Table of Reaction Rate Constants of Photo-Degradation Processes

k_Ozone [cm³ molecule⁻¹ s⁻¹]	QI	Ref.	k_Nitrate [cm³ molecule⁻¹ s⁻¹]	QI	Ref.	k_hν/Φ [s⁻¹] / [−]	QI	Ref.
4.0E-18	S	334						
4.0E-17	S	334						
3.8E-17	S	334						
8.3E-18	S	4				k_hν ca. 5E-4	S	70[185)]
1.8E-18	S	471				k_hν ca. 5E-4	S	70[186)]
1.6E-16	S	334						
1.4E-16	S	334						
2.2E-17	S	334						
			2.5E-13	S	313			
6.4E-17	S	313	2.7E-13	S	313			
6.4E-17	S	396						
			1.6E-15	S	8, 12			
			2.8E-14	S	218, 274			
> 1.0E-15	S	4						
> 1.0E-15	S	4						
5.2E-18	S	218, 375						
< 3.0E-20	S	386	< 1.0E-16	R	285, 301			
< 4.7E-24	S	391	< 5.0E-17	R	285, 301			
			< 2.0E-15	S	372	Φ = 1.0	R	342[190)]
< 2.0E-20	R	220	< 9.0E-16	R	285, 301			
1.7E-16	S	298	4.1E-12	S	298			
< 3.0E-19	S	298	6.6E-15	S	298			
4.9E-17	S	307	1.3E-10	S	307			
< 1.1E-19	S	307						
1.1E-17	A	4, 151						
8.7E-19	A	4, 151[191)]						
2.3E-16	A	214, 438[192)]						
2.0E-16	R	218	1.2E-11	A	122, 130[193)]			
4.3E-16	S	342	1.1E-11	S	342			
< 7.0E-18	S	332	6.8E-13	S	377			

CAS No.	Chemical name	k_OH [cm^3 molecule^{-1} s^{-1}]	QI	Ref.
593-74-8	Mercury, dimethyl-	1.9E-11	A	9, 153
74-89-5	Methanamine	2.2E-11	S	9, 52
75-50-3	Methanamine, N,N-dimethyl-	6.1E-11	S	9, 53
124-40-3	Methanamine, N-methyl-	6.5E-11	S	9, 53
4164-28-7	Methanamine, N-methyl-N-nitro-	3.8E-12	S	9, 55
62-75-9	Methanamine, N-methyl-N-nitroso-	3.1E-12	A	9, 55[196]
74-82-8	Methane	6.4E-15	R	363[197]
78522-47-1	Methane, bis(difluoromethoxy)difluoro-	2.4E-15	S	220
74-83-9	Methane, bromo-	3.0E-14	R	218
74-97-5	Methane, bromochloro-	1.2E-13	S	339
353-59-3	Methane, bromochlorodifluoro-	< 1E-16	R	218, 231, 416
75-27-4	Methane, bromodichloro-	1.2E-13	S	339
1511-62-2	Methane, bromodifluoro-	9.6E-15	R	218
75-63-8	Methane, bromotrifluoro-	< 1E-16	R	218, 231, 232, 416
74-87-3	Methane, chloro-	4.3E-14	R	218
75-45-6	Methane, chlorodifluoro-	4.7E-15	R	218, 232[198]
593-70-4	Methane, chlorofluoro-	4.4E-14	R	416[199]
593-71-5	Methane, chloroiodo-			
75-72-9	Methane, chlorotrifluoro-	< 7.0E-16	R	9
334-88-3	Methane, diazo-			
74-95-3	Methane, dibromo-	1.1E-13	R	276, 285, 301[201]
124-48-1	Methane, dibromochloro-	5.7E-14	E	220, 319
75-61-6	Methane, dibromodifluoro-	< 5E-16	R	218[202]
75-09-2	Methane, dichloro-	1.2E-13	R	220[203]
75-71-8	Methane, dichlorodifluoro-	≤ 7.0E-18	R	32
75-43-4	Methane, dichlorofluoro-	2.9E-14	R	416[204]
75-10-5	Methane, difluoro-	1.1E-14	R	9, 218
359-15-9	Methane, difluoromethoxy-	3.5E-14	S	407[205]
3822-68-2	Methane, (difluoromethoxy)trifluoro-	2.5E-14	A	218, 288
75-11-6	Methane, diiodo-			
109-87-5	Methane, dimethoxy-	4.9E-12	S	317, 469[207]
593-53-3	Methane, fluoro-	1.6E-14	R	218
74-88-4	Methane, iodo-	7.1E-14	R	416, 252
556-61-6	Methane, isothiocyanato-			
75-52-5	Methane, nitro-	1.3E-13	R	218
115-10-6	Methane, oxybis-	3.0E-12	R	218[210], 75
1691-17-4	Methane, oxybis[difluoro-	2.6E-15	A	220, 407[211]
593-79-3	Methane, selenobis-	6.8E-11	S	218
17696-73-0	Methanesulfinic acid	9.0E-11	S	449
56-23-5	Methane, tetrachloro-	< 4.0E-16	R	32

Table of Reaction Rate Constants of Photo-Degradation Processes

k_Ozone [cm^3 molecule^{-1} s^{-1}]	QI	Ref.	k_Nitrate [cm^3 molecule^{-1} s^{-1}]	QI	Ref.	k_$h\nu$/Φ [s^{-1}] / [–]	QI	Ref.
			8.7E-12	S	373			
7.4E-21	S	375						
7.8E-18	S	218, 375						
1.7E-18	S	218, 375						
3.0E-21	S	4						
< 1E-20	S	4				Φ = 1		55
≤ 1.2E-24	E	4	≤ 1.0E-18	R	218, 416			
						k_$h\nu$ = 1E-4 / Φ = 1	E	484[200)]
3.3E-17	S	4						
			< 5.8E-18	E	226			
						k_$h\nu$ = 5E-3 / Φ = 1	E	484[206)]
						k_$h\nu$ = 3E-6	E	484[208)]
						Φ = 0.98	S	457[209)]
			3.0E-16	S	220			
6.8E-17	S	218	1.4E-11	S	8			

CAS No.	Chemical name	k_OH [cm^3 molecule^{-1} s^{-1}]	QI	Ref.
75-73-0	Methane, tetrafluoro-	< 2E-18	S	218, 246
75-18-3	Methane, thiobis-	4.8E-12	R	218[212]
74-93-1	Methanethiol	3.3E-11	R	32
75-25-2	Methane, tribromo-	1.5E-13	R	301[215]
75-69-4	Methane, trichlorofluoro-	≤ 5.0E-18	R	32[216]
76-06-2	Methane, trichloronitro-			
75-46-7	Methane, trifluoro-	2.7E-16	R	285, 301[217]
2314-97-8	Methane, trifluoroiodo-	< 4.0E-14	R	220[218]
421-14-7	Methane, trifluoromethoxy-	2.1E-14	S	218, 288
149-73-5	Methane, trimethoxy-	6.0E-12	S	392
6004-38-2	4,7-Methano-1H-indene, octahydro-	1.1E-11	S	9, 100
67-56-1	Methanol	9.4E-13	R	218
32665-20-6	Methanol, ethoxy-, formate	3.7E-12	S	463, 464
4382-75-6	Methanol, methoxy-, formate	1.3E-12	S	463, 466
1493-11-4	Methanol, trifluoro-	≤ 2.0E-17	S	301
85358-65-2	Methanol, trifluoro-, formate	1.6E-14	S	494[220]
3219-63-4	Methanol, (trimethylsilyl)-	2.2E-11	S	324
26981-93-1	Methyl diazene			
6004-44-0	Methylketene			
91-20-3	Naphthalene	2.2E-11	R	9[221], 10, 20, 44, 194, 213
493-01-6	Naphthalene, decahydro-, cis-	2.0E-11	S	9, 100
493-02-7	Naphthalene, decahydro-, trans-	2.0E-11	S	9, 100
1825-31-6	Naphthalene, 1,4-dichloro-	5.8E-12	S	3, 9, 10
573-98-8	Naphthalene, 1,2-dimethyl-	6.0E-11	S	475
575-41-7	Naphthalene, 1,3-dimethyl-	7.5E-11	S	475
571-58-4	Naphthalene, 1,4-dimethyl-	5.8E-11	S	475
571-61-9	Naphthalene, 1,5-dimethyl-	6.0E-11	S	475
575-43-9	Naphthalene, 1,6-dimethyl-	6.3E-11	S	475
575-37-1	Naphthalene, 1,7-dimethyl-	6.8E-11	S	475
569-41-5	Naphthalene, 1,8-dimethyl-	6.3E-11	S	475
581-40-8	Naphthalene, 2,3-dimethyl-	6.9E-11	A	9, 31, 475[222]
581-42-0	Naphthalene, 2,6-dimethyl-	6.6E-11	S	475
582-16-1	Naphthalene, 2,7-dimethyl-	6.9E-11	S	475
130-15-4	1,4-Naphthalenedione	3.1E-12	S	9
1127-76-0	Naphthalene, 1-ethyl-	3.6E-11	S	475
939-27-5	Naphthalene, 2-ethyl-	4.0E-11	S	475
90-12-0	Naphthalene, 1-methyl-	4.7E-11	A	9, 45, 475[223]
91-57-6	Naphthalene, 2-methyl-	5.0E-11	A	9, 31, 475[224]
881-03-8	Naphthalene, 2-methyl-1-nitro-	< 8.3E-12	S	9

k_Ozone [cm³ molecule⁻¹ s⁻¹]	QI	Ref.	k_Nitrate [cm³ molecule⁻¹ s⁻¹]	QI	Ref.	$k\_h\nu/\Phi$ [s⁻¹] / [–]	QI	Ref.
< 1E-18	R	285, 301, 429	7.0E-13	R	429[213]			
			9.2E-13	R	352, 429[214]			
						$k\_h\nu$ = 5.7E-5	S	457, 461
			2.4E-16	R	285[219]			
2.0E-16	S	4						
< 7E-19	S	218						
< 3E-19	S	4, 116						
< 4E-19	S	218						
< 2.0E-19	S	218	9.5E-16	S	8			
< 1.3E-19	S	218						
< 4E-19	S	218						
< 3E-19	S	218	1.1E-14	S	8			

CAS No.	Chemical name	k_OH [cm^3 molecule^{-1} s^{-1}]	QI	Ref.
86-57-7	Naphthalene, 1-nitro-	5.4E-12	S	9
581-89-5	Naphthalene, 2-nitro-	5.6E-12	S	9
119-64-2	Naphthalene, 1,2,3,4-tetrahydro-	3.4E-11	S	9
7697-37-2	Nitric acid	1.3E-13	S	220
928-45-0	Nitric acid, butyl ester	1.7E-12	R	218, 59, 159
2108-66-9	Nitric acid, cyclohexyl ester	3.3E-12	S	9, 60
426-05-6	Nitric acid, 1,1-dimethylethyl ester			
625-58-1	Nitric acid, ethyl ester	3.7E-13	A	58, 267, 369[227]
598-58-3	Nitric acid, methyl ester	3.0E-14	A	158, 289
1712-64-7	Nitric acid, 1-methylethyl ester	2.8E-13	R	416[229]
543-29-3	Nitric acid, 2-methylpropyl ester	8.6E-13	R	416
924-52-7	Nitric acid, 1-methylpropyl ester	9.0E-13	R	218, 159
1002-16-0	Nitric acid, pentyl ester	3.3E-12	S	267
627-13-4	Nitric acid, propyl ester	7.3E-13	R	218, 58, 159
544-16-1	Nitrous acid, butyl ester	2.7E-12	R	218, 270
540-80-7	Nitrous acid, 1,1-dimethylethyl ester	1.4E-12	S	9, 57
109-95-5	Nitrous acid, ethyl ester	7.0E-13	R	218, 270
188479-35-8	Nitrous acid, 2-hydroxycyclopentyl ester	3.9E-12	S	381
624-91-9	Nitrous acid, methyl ester	2.2E-13	R	218, 57, 148–150
542-56-3	Nitrous acid, 2-methylpropyl ester	5.3E-12	S	9, 57
924-52-7	Nitrous acid, 1-methylpropyl ester	6.0E-12	S	9, 57
463-04-7	Nitrous acid, pentyl ester	4.2E-12	R	218, 270
543-67-9	Nitrous acid, propyl ester	1.2E-12	R	218, 57, 149
124-19-6	Nonanal	2.9E-11	S	470
111-84-2	Nonane	1.0E-11	R	9
821-55-6	2-Nonanone	1.2E-11	S	9, 67
24903-95-5	Nopinone	1.4E-11	S	218, 259
121-46-0	2,5-Norbornadiene	1.2E-10	S	9, 90
279-23-2	Norbornane	5.5E-12	S	9, 100
498-66-8	2-Norbornene	4.9E-11	S	9, 90
2436-90-0	1,6-Octadiene, 3,7-dimethyl-			
123-35-3	1,6-Octadiene, 7-methyl-3-methylene-	2.1E-10	S	9, 89
124-13-0	Octanal	2.9E-11	S	470[233]
111-65-9	Octane	8.7E-12	R	9
3221-61-2	Octane, 2-methyl-	1.0E-11	S	9
2216-34-4	Octane, 4-methyl-	9.7E-12	S	9
111-87-5	1-Octanol	1.4E-11	S	218, 59
82944-62-5	3-Octanol, nitrate	3.9E-12	S	9
111-13-7	2-Octanone	1.1E-11	S	9, 67

k_Ozone [cm³ molecule⁻¹ s⁻¹]	QI	Ref.	k_Nitrate [cm³ molecule⁻¹ s⁻¹]	QI	Ref.	k_$h\nu$/Φ [s⁻¹] / [–]	QI	Ref.
< 1.3E-19	S	218	7.2E-15	S	8			
< 6E-19	S	218	7.3E-15	S	8			
			9.9E-15	A	8, 43, 212[225)]			
						Φ = 1	E	184
						k_$h\nu$ = 2.0E-6 / Φ = 1	S	184[226)]
						k_$h\nu$ = 1.4E-6 / Φ = 1	S	186 / 185[228)]
						k_$h\nu$ = 4.6E-7 / Φ = 1	A	184
						k_$h\nu$ = 4.7E-7 / Φ = 1	A	184, 185, 187[230)]
						k_$h\nu$ = 6.9E-7 / Φ = 1	S	184
						k_$h\nu$ = 1.3E-6 / Φ = 1	S	186 / 184[231)]
1.2E-19	S	4, 147						
1.3E-20	S	4, 147						
						Φ = 0.23	S	453
			1.9E-16	S	353			
3.5E-15	S	218, 203	1.0E-12	S	8, 225			
1.5E-15	S	218, 225	2.5E-13	S	8, 203			
6.8E-16	S	4, 107						
4.7E-16	S	218, 224	1.1E-11	S	8, 130[232)]			
			1.8E-16	S	8, 201			
						Φ = 1		184

CAS No.	Chemical name	k_OH [cm^3 molecule^{-1} s^{-1}]	QI	Ref.
13877-91-3	1,3,6-Octatriene, 3,7-dimethyl-	2.5E-10	S	9, 89
3338-55-4	1,3,6-Octatriene, 3,7-dimethyl-, (Z)-			
111-66-0	1-Octene			
14850-23-8	4-Octene, (E)-	6.9E-11	A	218, 208
25291-17-2	1-Octene, 3,3,4,4,5,5,6,6,7,7,8,8,8-tridecafluoro-	1.4E-12	S	408[234)]
7642-15-1	4-Octene, (Z)-			
18479-58-8	7-Octen-2-ol, 2,6-dimethyl-	3.8E-11	S	211
288-42-6	Oxazole	9.5E-12	A	9
592-90-5	Oxepane	1.5E-11	S	9, 390
291-70-3	Oxepin	1.0E-10	S	329
16479-77-9	Oxepin, 2-methyl-			
503-30-0	Oxetane	1.0E-11	S	9, 390
425-82-1	Oxetane, hexafluoro-	< 2E-16	S	218, 288
144109-03-5	Oxetane, 2,2,3,4,4-pentafluoro-	2.5E-15	A	218, 288
75-21-8	Oxirane	7.6E-14	R	9
106-89-8	Oxirane, (chloromethyl)-	4.4E-13	S	9
930-22-3	Oxirane, ethenyl-			
86777-83-5	Oxirane, 2-ethenyl-2-methyl-			
106-88-7	Oxirane, ethyl-	2.0E-12	A	9, 78, 182[235)]
75-56-9	Oxirane, methyl-	5.2E-13	R	9
56-38-2	Parathion			
629-62-9	Pentadecane	2.2E-11	S	9, 97
591-95-7	1,2-Pentadiene	3.5E-11	S	9, 84
591-93-5	1,4-Pentadiene	5.3E-11	S	9, 84
2004-70-8	1,3-Pentadiene, (E)-			
1118-58-7	1,3-Pentadiene, 2-methyl-			
926-56-7	1,3-Pentadiene, 4-methyl-	1.3E-10	S	9, 84
763-30-4	1,4-Pentadiene, 2-methyl-	7.9E-11	S	9, 84
4549-74-0	1,3-Pentadiene, 3-methyl-	1.4E-10	S	9, 84
1574-41-0	1,3-Pentadiene, (Z)-	1.0E-10	S	9, 84
110-62-3	Pentanal	2.8E-11	R	9, 524[237)]
32749-94-3	Pentanal, 2,3-dimethyl-	4.2E-11	S	447
123-15-9	Pentanal, 2-methyl-	2.0E-11	R	413, 414, 416
15877-57-3	Pentanal, 3-methyl-			
1119-16-0	Pentanal, 4-methyl-	2.6E-11	S	414
626-96-0	Pentanal, 4-oxo-	2.0E-11	S	522
109-66-0	Pentane	3.9E-12	R	9, 218
110-53-2	Pentane, 1-bromo-	4.0E-12	S	218, 279[239)]
543-59-9	Pentane, 1-chloro-	3.0E-12	A	218, 255
138495-42-8	Pentane, 1,1,1,2,3,4,4,5,5,5-decafluoro-	3.9E-15	S	218, 281

k_Ozone [cm^3 molecule^{-1} s^{-1}]	QI	Ref.	k_Nitrate [cm^3 molecule^{-1} s^{-1}]	QI	Ref.	$k\_h\nu/\Phi$ [s^{-1}] / [–]	QI	Ref.
5.4E-16	A	218, 214	2.2E-11	S	8, 130			
2.0E-15	S	4, 107	2.2E-11	S	8, 130			
1.2E-17	S	379						
1.3E-16	S	220						
9.0E-17	S	389						
			9.2E-12	S	329			
			1.3E-11	S	417			
1.6E-18	S	427	6.1E-15	S	365			
2.5E-18	S	218						
						$k\_h\nu$ = 6E-3	S	457
1.5E-17	S	218						
4.3E-17	R	275, 440[236)]	1.6E-12	S	218, 284			
8.0E-17	S	218, 221						
1.3E-17	S	218, 221						
2.8E-17	S	440	1.4E-12	S	218, 233			
			1.6E-14	A	414, 421[238)]	Φ = 0.29	S	450
			2.7E-14	S	414			
			2.4E-14	S	414			
			1.7E-14	S	414			
< 5E-20	S	522	3.1E-15	S	522			
1.0E-23	E	2	4.3E-17	R	2			

CAS No.	Chemical name	k_OH [cm^3 molecule^{-1} s^{-1}]	QI	Ref.
111-30-8	Pentanedial	2.4E-11	A	9, 136
1067-20-5	Pentane, 3,3-diethyl-	4.9E-12	A	218[240)]
590-35-2	Pentane, 2,2-dimethyl-	3.4E-12	S	9
108-08-7	Pentane, 2,4-dimethyl-	5.2E-12	S	9, 96
1119-40-0	1,5-Pentanedioic acid, dimethyl ester	2.1E-12	S	486
107-41-5	2,4-Pentanediol, 2-methyl-	1.5E-12	S	423
123-54-6	2,4-Pentanedione	1.1E-12	S	9, 68
628-80-8	Pentane, 1-methoxy-	6.2E-12	S	305
107-83-5	Pentane, 2-methyl-	5.6E-12	R	9
96-14-0	Pentane, 3-methyl-	5.7E-12	R	9
760-21-4	Pentane, 3-methylene-			
628-05-7	Pentane, 1-nitro-	3.2E-12	R	218, 64
102526-10-3	Pentane, 1,1,1,3,3,5,5,5-octafluoro-	2.6E-15	S	382
693-65-2	Pentane, 1,1'-oxybis-	3.3E-11	R	218, 74
540-84-1	Pentane, 2,2,4-trimethyl-	3.7E-12	R	218
565-75-3	Pentane, 2,3,4-trimethyl-	6.9E-12	A	9, 180, 501
624-24-8	Pentanoic acid, methyl ester	5.3E-12	S	347[242)]
71-41-0	1-Pentanol	1.1E-11	R	218, 71
6032-29-7	2-Pentanol	1.2E-11	S	9, 72
584-02-1	3-Pentanol	1.2E-11	S	9, 72
123024-70-4	2-Pentanol, 3-methyl-, nitrate	3.0E-12	S	9, 60
133764-33-7	2-Pentanol, 2-methyl-, nitrate	1.7E-12	S	9, 60
21981-48-6	2-Pentanol, nitrate	1.8E-12	S	9
82944-59-0	3-Pentanol, nitrate	1.1E-12	S	9, 60, 187[244)]
107-87-9	2-Pentanone	4.9E-12	R	9, 424[245)]
96-22-0	3-Pentanone	2.0E-12	R	9
565-80-0	3-Pentanone, 2,4-dimethyl-	5.4E-12	S	9, 66
123-42-2	2-Pentanone, 4-hydroxy-4-methyl-	3.8E-12	A	220[247)]
108-10-1	2-Pentanone, 4-methyl-	1.4E-11	R	9, 218
756-13-8	3-Pentanone, 1,1,1,2,2,4,5,5,5-nonafluoro-4-(trifluoromethyl)-	< 5.0E-16	S	451
1576-87-0	2-Pentenal, (E)-	2.3E-11	S	524
5729-47-5	2-Pentenal, 4-oxo-			
109-67-1	1-Pentene	3.1E-11	R	9
109-68-2	2-Pentene	9.0E-11	S	9, 85
10574-37-5	2-Pentene, 2,3-dimethyl-	1.0E-10	A	9, 86
690-08-4	2-Pentene, 4,4-dimethyl-, (E)-	5.4E-11	S	9, 86
646-04-8	2-Pentene, (E)-	6.7E-11	S	9
763-29-1	1-Pentene, 2-methyl-	6.3E-11	S	9, 86
760-20-3	1-Pentene, 3-methyl-			
691-37-2	1-Pentene, 4-methyl-			

k_Ozone [cm^3 molecule^{-1} s^{-1}]	QI	Ref.	k_Nitrate [cm^3 molecule^{-1} s^{-1}]	QI	Ref.	$k\_h\nu/\Phi$ [s^{-1}] / [–]	QI	Ref.
			1.4E-16	S	353			
			1.7E-16	S	353			
8.1E-18	S	387[241]	4.5E-13	S	360			
			7.5E-17	S	353			
						$\Phi = 1$	E	184
						$\Phi = 1$	E	186[243]
						$\Phi = 1$		184
						$\Phi = 0.07$	S	450[246]
< 4.0E-22		451						
			3.1E-14	S	430[248]			
4.8E-18	S	471						
1.0E-17	R	218[249]	6.8E-14	S	354[250]			
1.6E-16	S	4	4.8E-12	S	273			
3.1E-16	S	4, 104						
1.3E-17	S	390[251]						
3.8E-18	S	218, 379						
1.1E-17	S	4, 104						

CAS No.	Chemical name	k_OH [cm³ molecule⁻¹ s⁻¹]	QI	Ref.
625-27-4	2-Pentene, 2-methyl-	8.9E-11	A	9, 86
616-12-6	2-Pentene, 3-methyl-, (E)-			
674-76-0	2-Pentene, 4-methyl-, (E)-	6.0E-11	S	9, 86
922-62-3	2-Pentene, 3-methyl-, (Z)-			
107-40-4	2-Pentene, 2,4,4-trimethyl-			
627-20-3	2-Pentene, (Z)-	6.6E-11	A	9, 86, 87
616-25-1	1-Penten-3-ol			
1576-85-8	4-Penten-1-ol, acetate	4.3E-11	A	218, 282
1576-95-0	2-Penten-1-ol, (Z)-			
625-33-2	3-Penten-2-one			
627-19-0	1-Pentyne	1.1E-11	S	9, 80
No CAS No.	Perfluoropolymethylisopropyl ethers	< 6.8E-16	S	139
2278-22-0	Peroxide, acetyl nitro	1.1E-13	A	9, 163, 198[252]
198-55-0	Perylene	5.0E-12	E	504[253]
555-10-2	β-Phellandrene	1.7E-10	S	218, 249
4221-98-1	(−)-α-Phellandrene	3.1E-10	S	9
85-01-8	Phenanthrene	3.0E-11	A,E	297[255], 534[256]
832-69-9	Phenanthrene, 1-methyl-	2.9E-11	E	534[257]
2531-84-2	Phenanthrene, 2-methyl-	6.5E-11	E	534[258]
832-71-3	Phenanthrene, 3-methyl-	6.6E-11	E	534[259]
883-20-5	Phenanthrene, 9-methyl-	7.6E-11	E	534[260]
108-95-2	Phenol	2.6E-11	R	3, 9, 218
576-24-9	Phenol, 2,3-dichloro-	1.7E-12	S	9
120-83-2	Phenol, 2,4-dichloro-	1.1E-12	S	9
526-75-0	Phenol, 2,3-dimethyl-	8.0E-11	S	9
105-67-9	Phenol, 2,4-dimethyl-	7.1E-11	S	9
95-87-4	Phenol, 2,5-dimethyl-	8.0E-11	S	9
576-26-1	Phenol, 2,6-dimethyl-	6.6E-11	S	9
108-68-9	Phenol, 3,5-dimethyl-	1.1E-10	S	9
95-65-8	Phenol, 3,4-dimethyl-	8.1E-11	S	9
95-48-7	Phenol, 2-methyl-	4.2E-11	R	9, 202, 333
108-39-4	Phenol, 3-methyl-	6.3E-11	A	202, 333
106-44-5	Phenol, 4-methyl-	4.9E-11	A	202, 333
88-75-5	Phenol, 2-nitro-	9.0E-13	S	9, 3
697-82-5	Phenol, 2,3,5-trimethyl-	1.2E-10	S	338
2416-94-6	Phenol, 2,3,6-trimethyl-	1.2E-10	S	338
298-02-2	Phorate			
75-44-5	Phosgene	< 5.0E-15	R	416
7803-51-2	Phosphine	1.6E-11	S	220, 160[261]
868-85-9	Phosphonic acid, dimethyl ester	5.0E-12	A	532, 292

k_Ozone [cm³ molecule⁻¹ s⁻¹]	QI	Ref.	$k\_Nitrate$ [cm³ molecule⁻¹ s⁻¹]	QI	Ref.	$k\_h\nu/\Phi$ [s⁻¹] / [–]	QI	Ref.
5.6E-16	S	4, 105						
4.6E-16	S	4, 97						
1.4E-16	S	378						
2.1E-16	S	4, 104						
1.8E-14	S	383						
1.7E-16	S	383						
3.5E-17	R	218, 225						
			7.5E-16	S	8, 12			
4.7E-17	R	218, 249	8.9E-12	S	218, 249			
1.8E-15	S	218, 107, 224	4.9E-11	A	355[254)]			
			3.8E-12	R	218, 519			
2.5E-19	S	4, 36	1.4E-11	R	218			
1.9E-19	S	4, 36	9.7E-12	R	8, 218			
4.7E-19	S	4, 36	1.1E-11	R	218, 16			
			< 1.2E-14	S	218			
						$k\_h\nu$ = 2E-3	S	457, 460
< 8E-20	S	292	< 1.4E-16	S	292			

CAS No.	Chemical name	k_OH [cm^3 molecule^{-1} s^{-1}]	QI	Ref.
78-38-6	Phosphonic acid, ethyl-, diethyl ester	6.1E-11	A	533, 300[262]
6163-75-3	Phosphonic acid, ethyl-, dimethyl ester	1.6E-11	S	300[263]
683-08-9	Phosphonic acid, methyl-, diethyl ester	5.7E-11	A	533, 300[264]
756-79-6	Phosphonic acid, methyl-, dimethyl ester	1.1E-11	A	300[265], 292
676-97-1	Phosphonic dichloride, methyl-	6.4E-14	S	532
597-07-9	Phosphoramidic acid, dimethyl-, dimethyl ester	3.2E-11	S	9, 18
17321-47-0	Phosphoramidothioic acid, O,O-dimethyl ester	2.4E-10	S	9, 18
28167-51-3	Phosphoramidothioic acid, dimethyl-, O,O-dimethyl ester	4.7E-11	S	9, 18
31464-99-0	Phosphoramidothioic acid, methyl-, O,O-dimethyl ester	2.3E-10	S	9, 18
78-40-0	Phosphoric acid, triethyl ester	5.3E-11	A	533, 300[266]
512-56-1	Phosphoric acid, trimethyl ester	7.4E-12	S	9, 248
10025-87-3	Phosphoric trichloride	< 4E-14	S	532
2524-03-0	Phosphorochloridothioic acid, O,O-dimethyl ester	5.9E-11	S	9, 50
2953-29-9	Phosphorodithioic acid, O,O,S-trimethylester	5.6E-11	S	9, 51
22608-53-3	Phosphorodithioic acid O,S,S-trimethylester			
152-20-5	Phosphorothioic acid, O,O,S-trimethyl ester			
152-18-1	Phosphorothioic acid, O,O,O-trimethyl ester	7.0E-11	S	9, 51
121-45-9	Phosphorous acid, trimethyl ester	7.1E-10	S	532
7719-12-2	Phosphorous trichloride	3.8E-11	S	220
10281-53-5	(–)-trans-Pinane	1.3E-11	S	218, 251
7785-26-4	(–)-α-Pinene	5.8E-11	R	9, 506
18172-67-3	(–)-β-Pinene	7.9E-11	R	9
2704-78-1	Pinonaldehyde	9.1E-11	S	340
463-49-0	1,2-Propadiene	9.8E-12	R	9, 133–135
123-38-6	Propanal	2.0E-11	R	393[270]
630-19-3	Propanal, 2,2-dimethyl-	2.6E-11	R	218[272]
20818-81-9	Propanal, 2-hydroxy-2-methyl-	1.5E-11	S	529
78-84-2	Propanal, 2-methyl-	4.6E-11	R	218, 223[274]
78-98-8	Propanal, 2-oxo-	1.7E-11	R	32
460-40-2	Propanal, 3,3,3-trifluoro-	2.6E-12	S	248
758-96-3	Propanamide, N,N-dimethyl-	2.1E-11	S	302
1187-58-2	Propanamide, N-methyl-	7.6E-12	S	302
75-64-9	2-Propanamine, 2-methyl-	1.2E-11	S	323[276]
74-98-6	Propane	1.1E-12	R	218[277]
106-94-5	Propane, 1-bromo-	1.2E-12	S	218, 279
75-26-3	Propane, 2-bromo-	8.8E-13	S	218, 279
540-54-5	Propane, 1-chloro-	1.0E-12	R	218, 255
75-29-6	Propane, 2-chloro-	9.0E-13	R	218
507-20-0	Propane, 2-chloro-2-methyl-	4.1E-13	S	218, 279

k_Ozone [cm³ molecule⁻¹ s⁻¹]	QI	Ref.	k_Nitrate [cm³ molecule⁻¹ s⁻¹]	QI	Ref.	k_hν/Φ [s⁻¹] / [–]	QI	Ref.
< 6E-20	S	533	3.4E-16	S	533			
< 6E-20	S	292	3.4E-16	S	292			
< 6E-20	S	533	3.7E-16	S	533			
< 6E-20	S	292	2.0E-16	S	292			
< 2E-19	S	218, 18	< 3.6E-14	S	8, 18			
< 4E-19	S	218, 18	4.0E-13	S	8, 18			
< 2E-19	S	218, 18	3.2E-14	S	8, 18			
< 2E-19	S	218, 18	3.0E-13	S	8, 18			
< 6E-20	S	533	2.4E-16	S	533			
< 6E-20	S	218, 248						
< 2E-19	S	218, 50	< 2.8E-14	S	8, 50			
< 2E-19	S	218, 51	< 2.8E-14	S	8, 51			
< 1E-19	S	218, 51	< 2.4E-15	S	8, 51			
< 2E-19	S	218, 51	< 1.1E-15	S	8, 51			
< 3E-19	S	218, 51	1.6E-14	S	8, 51			
8.7E-17	R	218, 214, 109–111	6.2E-12	R	6, 128, 129, 130, 526[267)]			
1.5E-17	R	218	2.5E-12	R	2, 6, 122, 128, 130, 526[268)]			
8.9E-20	S	340	5.4E-14	S	340	Φ = 0.14	S	453[269)]
1.8E-19	S	218, 132						
4.7E-16	S	218, 224	5.7E-15	S	356, 416	k_hν = 3.2E-5 / Φ = 0.5	S	189 / 32[271)]
			2.5E-14	A	393[273)]	Φ = 0.56	S	453
			1.2E-14	S	356	k_hν = 2.7E-5 / Φ = 0.70	S	189 / 450
1.1E-21	S	4				k_hν = 2.7E-15 / Φ = 0.107	S	34 / 32[275)]
7.0E-24	E	4	2.2E-17	S	226			

CAS No.	Chemical name	k_OH [cm³ molecule⁻¹ s⁻¹]	QI	Ref.
96-12-8	Propane, 1,2-dibromo-3-chloro-	4.3E-13	S	9, 49
78-87-5	Propane, 1,2-dichloro-	4.4E-13	S	9, 83
142-28-9	Propane, 1,3-dichloro-	7.8E-13	S	218, 279
507-55-1	Propane, 1,3-dichloro-1,1,2,2,3-pentafluoro-	8.8E-15	R	218, 256
422-56-0	Propane, 3,3-dichloro-1,1,1,2,2-pentafluoro-	2.5E-14	R	218
64712-27-2	Propane, 1,1-dichloro-1,3,3,3-tetrafluoro-	8.0E-16	E	220, 319
7125-99-7	Propane, 1,1-dichloro-1,2,2-trifluoro-	2.3E-15	A	416[278]
126-84-1	Propane, 2,2-diethoxy-	1.2E-11	S	218, 210
35042-99-0	Propane, 3-(difluoromethoxy)-1,1,2,2-tetrafluoro-	1.6E-14	S	410[279]
77-76-9	Propane, 2,2-dimethoxy-	3.9E-12	S	9, 210
7778-85-0	Propane, 1,2-dimethoxy-	2.4E-11	A	210, 317
17081-21-9	Propane, 1,3-dimethoxy-	5.1E-11	S	448
463-82-1	Propane, 2,2-dimethyl-	8.5E-13	R	9[280]
57-55-6	1,2-Propanediol	1.6E-11	A	9, 481
6423-43-4	1,2-Propanediol, dinitrate	< 3.2E-13	S	218, 269
163702-06-5	Propane, 2-(ethoxydifluoromethyl)-1,1,1,2,3,3,3-heptafluoro-	7.7E-14	S	516
637-92-3	Propane, 2-ethoxy-2-methyl-	8.8E-12	R	218, 72, 73
420-26-8	Propane, 2-fluoro-	5.6E-13	S	535[281]
431-89-0	Propane, 1,1,1,2,3,3,3-heptafluoro-	1.6E-15	R	285[282]
375-03-1	Propane, 1,1,1,2,2,3,3-heptafluoro-3-methoxy-	1.2E-14	A	435, 389
22052-84-2	Propane, 1,1,1,2,3,3,3-heptafluoro-2-methoxy-	1.5E-14	S	435[283]
690-39-1	Propane, 1,1,1,3,3,3-hexafluoro-	3.2E-16	R	301[284]
431-63-0	Propane, 1,1,1,2,3,3-hexafluoro-	5.3E-15	R	301[285]
677-56-5	Propane, 1,1,1,2,2,3-hexafluoro-	4.2E-15	R	301
56860-85-6	Propane, 1,1,1,2,3,3-hexafluoro-(3-difluoromethoxy)-	1.8E-15	S	514
13171-18-1	Propane, 1,1,1,3,3,3-hexafluoro-2-methoxy-	1.3E-13	S	502
382-34-3	Propane, 1,1,1,2,3,3-hexafluoro-3-methoxy	2.1E-14	S	407
28523-86-6	Propane, 1,1,1,3,3,3-hexafluoro-2-propyl-2-(fluoromethoxy)-	7.3E-14	S	218
65064-78-0	Propane, 1,1,1,2,3,3-hexafluoro-3-(2,2,3,3-tetrafluoropropoxy)-	1.3E-14	S	434[286]
428454-68-6	Propane, 1,1,1,2,3,3-hexafluoro-3-(trifluoromethoxy)-	1.4E-15	S	514
162401-05-0	Propane, 1,1,1,3,3,3-hexafluoro-2-(trifluoromethoxy)-	3.3E-16	S	515
107-08-4	Propane, 1-iodo-	1.5E-12	S	478
75-30-9	Propane, 2-iodo-	1.2E-12	S	478
1118-00-9	Propane, 1-methoxy-2,2-dimethyl-			
1634-04-4	Propane, 2-methoxy-2-methyl-	2.9E-12	R	218, 512
75-28-5	Propane, 2-methyl-	2.3E-12	R	218[288]
42125-48-4	1,2-Propanediol, 2-methyl-, 1-acetate	9.5E-12	S	330
78448-33-6	Propane, 2-methyl-1-(1-methylethoxy)-	1.9E-11	S	220, 330
107-12-0	Propanenitrile	1.9E-13	S	9, 162
108-03-2	Propane, 1-nitro-	4.4E-13	R	218, 64

k_Ozone [cm³ molecule⁻¹ s⁻¹]	QI	Ref.	k_Nitrate [cm³ molecule⁻¹ s⁻¹]	QI	Ref.	k_hν/Φ [s⁻¹] / [–]	QI	Ref.
			5.4E-13	S	360			
			2.3E-15	S	536[287]			
< 2.0E-23	S	4, 101, 102	1.1E-16	R	218			
1.0E-19	S	4						

CAS No.	Chemical name	k_OH [cm³ molecule⁻¹ s⁻¹]	QI	Ref.
79-46-9	Propane, 2-nitro-	2.5E-13	S	218, 271[289]
111-43-3	Propane, 1,1'-oxybis-	1.8E-11	R	218, 487
108-20-3	Propane, 2,2'-oxybis-	1.0E-11	R	218
628-55-7	Propane, 1,1'-oxybis[2-methyl-	2.6E-11	S	9, 73
1814-88-6	Propane, 1,1,1,2,2-pentafluoro-	1.5E-15	S	220, 315[291]
679-86-7	Propane, 1,1,2,2,3-pentafluoro-	7.2E-15	R	285[292]
24270-66-4	Propane, 1,1,2,3,3-pentafluoro-	1.6E-14	S	382
431-31-2	Propane, 1,1,1,2,3-pentafluoro-	1.5E-14	S	382
460-73-1	Propane, 1,1,1,3,3-pentafluoro-	7.0E-15	S	301, 407[293]
56860-81-2	Propane; 1,1,1,2,2-pentafluoro-3-(difluoromethoxy)-	1.0E-14	S	410[294]
378-16-5	Propane, 1,1,1,2,2-pentafluoro-3-methoxy-	8.7E-13	S	502
60598-17-6	Propane, 1,1,2,2-tetrafluoro-3-methoxy-	8.4E-13	S	502
111-47-7	Propane, 1,1'-thiobis-	2.1E-11	R	218
107-03-9	1-Propanethiol	4.8E-11	R	9
75-33-2	2-Propanethiol	4.2E-11	R	9
513-44-0	1-Propanethiol, 2-methyl-	4.5E-11	R	9
75-66-1	2-Propanethiol, 2-methyl-	3.3E-11	R	9
421-07-8	Propane, 1,1,1-trifluoro-	4.2E-14	R	301
993-95-3	Propane, 1-(2,2,2-trifluoroethoxy)-1,1,2,3,3,3-hexafluoro-	8.9E-15	S	490
14315-97-0	Propane, 1,1,3-trimethoxy-	1.8E-11	A	9, 210
79-09-4	Propanoic acid	1.2E-12	R	218, 169
763-69-9	Propanoic acid, 3-ethoxy-, ethyl ester	2.3E-11	S	341
105-37-3	Propanoic acid, ethyl ester	1.9E-12	A	9, 77, 168
79-31-2	Propanoic acid, 2-methyl-	2.0E-12	S	9, 170
554-12-1	Propanoic acid, methyl ester	9.6E-13	A	77, 299[295]
54396-97-3	Propanoic acid, 2-methyl-, 2-ethoxyethyl ester	1.4E-11	S	218, 282
547-63-7	Propanoic acid, 2-methyl-, methyl ester	1.7E-12	S	418
617-50-5	Propanoic acid, 2-methyl-, 1-methylethyl ester	6.5E-12	S	330
127-17-3	Propanoic acid, 2-oxo-	1.2E-13	S	445[296]
106-36-5	Propanoic acid, propyl ester	4.0E-12	S	9, 442
71-23-8	1-Propanol	5.5E-12	R	218, 489[298]
67-63-0	2-Propanol	5.1E-12	R	218
124-68-5	1-Propanol, 2-amino-2-methyl-	2.8E-11	S	9, 54
5131-66-8	2-Propanol, 1-butoxy-	3.8E-11	S	509
75-84-3	1-Propanol, 2,2-dimethyl-	5.3E-12	S	218
926-42-1	1-Propanol, 2,2-dimethyl-, nitrate	8.5E-13	S	9, 60
111-35-3	1-Propanol, 3-ethoxy-	2.2E-11	S	9
920-66-1	2-Propanol, 1,1,1,3,3,3-hexafluoro-	2.5E-14	S	392[299]
107-98-2	2-Propanol, 1-methoxy-	2.0E-11	A	317, 481
75-65-0	2-Propanol, 2-methyl-	1.1E-12	R	9

k_Ozone [cm^3 molecule^{-1} s^{-1}]	QI	Ref.	k_Nitrate [cm^3 molecule^{-1} s^{-1}]	QI	Ref.	$k\_h\nu/\Phi$ [s^{-1}] / [–]	QI	Ref.
			5.5E-15	A	220[290)]			
			3.5E-17	A	242			
			3.3E-17	S	218, 242			
						$\Phi = 0.43$	S	453[297)]
			< 2.1E-15	S	357			
			1.4E-15	R	416			
						$\Phi = 1$		184
< 1.1E-19	S	481	1.7E-15	S	481			

CAS No.	Chemical name	k_OH [cm^3 molecule^{-1} s^{-1}]	QI	Ref.
196881-04-6	2-Propanol, 2-methyl-1-(1-methylethoxy)-, nitrate	1.6E-11	S	330
422-05-9	1-Propanol, 2,2,3,3,3-pentafluoro-	1.0E-13	A	392[300)]
2240-88-2	1-Propanol, 3,3,3-trifluoro-	6.9E-13	S	248
116-09-6	2-Propanone, 1-hydroxy-	3.0E-12	R	416, 443[301)]
5878-19-3	2-Propanone, 1-methoxy-	6.8E-12	S	9, 260
6745-71-7	2-Propanone, 1-(nitrooxy)-	< 4.4E-13	S	218, 269
421-50-1	2-Propanone, 1,1,1-trifluoro-	1.5E-14	S	9, 182
107-02-8	2-Propenal	2.0E-11	R	9
78-85-3	2-Propenal, 2-methyl-	2.8E-11	S	483
14371-10-9	2-Propenal, 3-phenyl-, (E)-	4.8E-11	S	479
115-07-1	1-Propene	3.0E-11	R	33
590-14-7	1-Propene, 1-bromo-			
106-95-6	1-Propene, 3-bromo-	1.5E-11	A	328, 527[305)]
107-05-1	1-Propene, 3-chloro-	1.7E-11	R	9, 206, 527[307)]
1871-57-4	1-Propene, 3-chloro-2-(chloromethyl)-	3.7E-11	A	9, 82, 83
513-37-1	1-Propene, 1-chloro-2-methyl-	9.0E-14	S	218
563-47-3	1-Propene, 3-chloro-2-methyl-			
6065-93-6	1-Propene, 1,1-dichloro-2-methyl-			
563-58-6	1-Propene, 1,1-dichloro-			
542-75-6	1-Propene, 1,3-dichloro-	1.3E-11	A	9[309)]
78-88-6	1-Propene, 2,3-dichloro-			
10061-02-6	1-Propene, 1,3-dichloro-, (E)-			
10061-01-5	1-Propene, 1,3-dichloro-, (Z)-	8.4E-12	S	9, 82, 83
818-92-8	1-Propene, 3-fluoro-	1.6E-11	S	527[311)]
116-15-4	1-Propene, 1,1,2,3,3,3-hexafluoro-	2.2E-12	A	239, 315[313)]
556-56-9	1-Propene, 3-iodo-			
57-06-7	1-Propene, isothiocyanato	9.9E-11	S	530
115-11-7	1-Propene, 2-methyl-	5.2E-11	R	9
107-13-1	2-Propenenitrile			
625-46-7	1-Propene, 3-nitro-	1.2E-11	S	218, 278
754-12-1	1-Propene, 2,3,3,3-tetrafluoro-	1.1E-12	S	315[315)]
677-21-4	1-Propene, 3,3,3-trifluoro-	2.2E-11	S	220
79-10-7	2-Propenoic acid			
1001-26-9	2-Propenoic acid, 3-ethoxy-, ethyl ester	3.3E-11	S	218, 282
140-88-5	2-Propenoic acid, ethyl ester	1.6E-11	S	530[316)]
34846-90-7	2-Propenoic acid, 3-methoxy-, methyl ester			
79-41-4	2-Propenoic acid, 2-methyl-			
96-33-3	2-Propenoic acid, methyl ester	E-11	S	530[317)]
2370-63-0	2-Propenoic acid, 2-methyl-, 2-ethoxyethyl ester	2.8E-11	A	218, 282
80-62-6	2-Propenoic acid, 2-methyl-, methyl ester	3.2E-11	A	316, 530[318)]

k_Ozone [cm^3 $molecule^{-1}$ s^{-1}]	QI	Ref.	k_Nitrate [cm^3 $molecule^{-1}$ s^{-1}]	QI	Ref.	k_hν/Φ [s^{-1}] / [–]	QI	Ref.
2.9E-19	R	218, 117	8.9E-15	S	399[302]	Φ < 0.004	S	453
1.1E-18	R	218, 117, 146	6.4E-15	A	220, 357	Φ < 0.004	S	453, 483
2.2E-18	S	479	1.9E-14	S	479			
1.0E-17	R	220, 32	9.5E-15	R	218[303]			
			2.2E-14	S	363[304]			
			3.7E-15	S	328[306]			
1.5E-18	R	218	5.3E-16	S	8, 204			
3.9E-19	S	4, 83						
			9.0E-14	S	273			
3.7E-18	S	426	2.5E-14	S	218, 273			
			7.2E-14	S	292			
			1.4E-14	S	328[308]			
			1.0E-14	S	220, 318[310]			
			1.3E-14	S	328			
6.7E-19	S	4, 83						
1.5E-19	S	4						
			3.8E-15	S	442[312]			
< 3.0E-21	S	428						
1.2E-17	R	4	3.3E-13	R	218[314]			
1.0E-19	S	4, 36						
6.5E-18	S	81						
5.7E-18	S	218						
1.1E-17	S	396						
4.1E-18	S	81						
2.9E-19	S	218	1.9E-16	S	399			

CAS No.	Chemical name	k_OH [cm³ molecule⁻¹ s⁻¹]	QI	Ref.
107-18-6	2-Propen-1-ol	5.5E-11	S	209
108-22-5	1-Propen-2-ol, acetate	6.3E-11	S	349
591-87-7	1-Propen-3-ol, acetate	3.0E-11	A	349, 473
74-99-7	1-Propyne	5.9E-12	R	9
142-68-7	2H-Pyran, tetrahydro-	1.4E-11	S	9, 390
129-00-0	Pyrene	5.0E-11	S	220, 331
100-69-6	Pyridine, 2-ethenyl-	5.7E-11	S	218, 250
88283-41-4	Pyrifenox	1.8E-11	S	456[321]
109-97-7	1H-Pyrrole	1.1E-10	R	9, 30
872-50-4	2-Pyrrolidinone, 1-methyl-	2.1E-11	S	397
91-22-5	Quinoline	1.2E-11	S	307
3387-41-5	Sabinene	1.2E-10	S	218, 244
7803-62-5	Silane	1.2E-11	A	9, 47, 446
1111-74-6	Silane, dimethyl-	4.4E-11	S	446
1066-42-8	Silanediol, dimethyl-	8.1E-13	S	480
992-94-9	Silane, methyl-	3.3E-11	S	446
75-76-3	Silane, tetramethyl-	9.2E-13	A	218[324]
993-07-7	Silane, trimethyl-	3.1E-11	S	446
1066-40-6	Silanol, trimethyl-	7.2E-13	S	480
100-42-5	Styrene	5.8E-11	R	9
7446-09-5	Sulfur dioxide	1.1E-12	S	220
64-67-5	Sulfuric acid, diethyl ester	1.8E-12	A	218, 266
77-78-1	Sulfuric acid, dimethyl ester	< 5E-13	S	218, 266
5915-41-3	Terbuthylazine	1.1E-11	A	401[326]
99-86-5	α-Terpinene	3.6E-10	S	1
99-85-4	γ-Terpinene	1.8E-10	S	9, 89
98-55-5	α-Terpineol	1.9E-10	S	521
586-62-9	Terpinolene	2.2E-10	S	218, 245
278-06-8	Tetracyclo[3.2.0.0²,⁷.0⁴,⁶]heptane	1.8E-12	S	218, 251
629-59-4	Tetradecane	1.9E-11	S	9, 97
1120-36-1	1-Tetradecene			
141-62-8	Tetrasiloxane, decamethyl-	2.7E-12	S	220, 318
288-47-1	Thiazole	1.4E-12	S	9
420-12-2	Thiirane			
110-02-1	Thiophene	9.5E-12	R	1, 9, 27–29
96-43-5	Thiophene, 2-chloro-			
17249-80-8	Thiophene, 3-chloro-			
638-02-8	Thiophene, 2,5-dimethyl-			
872-55-9	Thiophene, 2-ethyl-			
554-14-3	Thiophene, 2-methyl-			

k_Ozone [cm³ molecule⁻¹ s⁻¹]	QI	Ref.	k_Nitrate [cm³ molecule⁻¹ s⁻¹]	QI	Ref.	k_hv/Φ [s⁻¹] / [–]	QI	Ref.
1.4E-17	S	387	1.4E-14	S	359[319]			
2.4E-18	S	474	5.0E-14	E	472			
1.8E-20	A	4[320]	2.3E-16	R	8			
1.5E-17	S	218, 250						
2.0E-19	S	456[322]						
1.6E-17	S	4, 30	4.6E-11	S	8, 17			
< 1.0E-19	S	397	1.3E-13	S	397			
< 1.0E-19	S	307						
8.6E-17	R	218	9.7E-12	S	395[323]			
< 7E-21	S	218, 237	< 8E-17	S	218, 237			
1.7E-17	R	218, 250[325]	1.5E-13	S	8, 43			
< 1.0E-22	S	200, 490	< 7.0E-21	S	220, 301			
< 5E-19	S	404						
1.5E-14	A	220, 214[327]	1.0E-10	S	220[328]			
1.4E-16	S	218, 224	2.9E-11	S	8, 130			
3.0E-16	S	521						
1.5E-15	A	218, 214, 377	5.2E-11	S	393			
2.2E-17	S	441						
1.7E-20	S	4, 119						
7.2E-20	S	4, 120	3.9E-14	R	8			
			7.3E-14	S	190[329]			
			2.1E-14	S	190			
			2.3E-12	S	190[330]			
			8.4E-13	S	190[331]			
			9.3E-13	S	496[332]			

CAS No.	Chemical name	k_OH [cm^3 molecule^{-1} s^{-1}]	QI	Ref.
616-44-4	Thiophene, 3-methyl-			
1551-27-5	Thiophene, 2-propyl-	9.0E-11	S	220, 295
110-01-0	Thiophene, tetrahydro-	2.0E-11	R	9
290-87-9	1,3,5-Triazine	1.4E-13	S	9
288-88-0	1H-1,2,4-Triazole	< 2E-13	S	9
281-23-2	Tricyclo[3.3.1.1^{3,7}]decane	2.2E-11	A	9, 100
508-32-7	Tricyclo[2.2.1.0^{2,6}]heptane, 1,7,7-trimethyl-	2.9E-12	S	218, 251
629-50-5	Tridecane	1.6E-11	R	9
1582-09-8	Trifluralin			
110-88-3	1,3,5-Trioxane	6.1E-12	A	166, 469[336)]
107-51-7	Trisiloxane, octamethyl-	1.8E-12	S	304
1120-21-4	Undecane	1.3E-11	R	9
51-79-6	Urethane			

k_Ozone [cm³ molecule⁻¹ s⁻¹]	QI	Ref.	k_Nitrate [cm³ molecule⁻¹ s⁻¹]	QI	Ref.	$k\_h\nu/\Phi$ [s⁻¹] / [–]	QI	Ref.
			1.0E-12	S	496[333]			
			8.9E-13	S	190[334]			
1.0E-17	S	4, 120						
< 4.0E-21	S	218						
						$k\_h\nu = 3.3E-4$	S	188[335]
< 4.0E-20	S	325						

Footnotes to the Table

1) $A = 4.4\text{E-}12$; $E_a = -3.1$ kJ (240–350 K) [416].
2) $A = 1.4\text{E-}12$; $E_a = 15.5$ kJ (260-380 K) [416].
3) $\Phi = 1.32 \pm 0.3$ [453].
4) Measured at 300 K [302].
5) Measured at 300 K [302].
6) $k_{OH} = f(T)$ is given in [1].
7) 5.7E-12; $E_a = -2.5$ kJ (263–372 K) [346].
8) $E_a = 3.28$ kJ (253–372 K) [322].
9) Measured at 291 K [376].
10) $f(T)$ 273–373 K [242].
11) $A = 1.11\text{E-}12$; $E_a = -4.4$ kJ (263–372 K) [322].
12) According to Atkinson [218], the low values measured by Langer et al. [242] should be considered as upper limits.
13) $\Phi(\lambda) = 0.30$ (290); 0.15 (300); 0.028 (320); 0.033 (330) [32]. UV absorption spectrum in the gas phase [32]. 1st-order rate constant calculated for $\Phi = 0.08$, zenith angle 40° [11].
14) Measured in the adsorbed state in a special aerosol chamber [476].
15) $A = 3.58\text{E-}12$; $E_a = 7.7$ kJ (230–450 K) [285].
16) Extrapolated from measurements at higher temperatures in the gas phase to 298 K [297]; measurement in the gas phase at 325 K [46]; measurement adsorbed on graphite particles [504].
17) Measured in trichlorotrifluoroethane, extrapolated to the gas phase [405, 406].
18) Adsorbed on graphite particles; k_{OH}^{II} calculated from $k_{OH}^{I} = 0.12$ s^{-1} (298 K) with the given (probably too high) [OH] [504].
19) The quantum efficiency of the photolysis, leading to either C_6H_5 + CHO or to C_6H_5CO + H, is reported to be very small [7].
20) The reaction rate constant depends on the pressure; the recommended value refers to the high pressure region (about 0.1 MPa = 1 bar). $k_{OH} = f(p)$ is given in [9] for the pressure range 8–133 hPa; $k_{OH} = f(T)$ is given in [218] for the temperature range 243–379 K.
21) O_3 reaction rate constant measured $f(T)$ 298–423 K: $A = 1.05\text{E-}11$; $E_a = 61.5$ kJ [230].
22) Measured in the adsorbed state at SiO_2; $k_{OH,ads}$ at 280 K = 2.0E-11 [401].
23) For older measurements see [3, 9, 40, 44].
24) NO_3 reaction rate constant is valid for H-abstraction only [218] (addendum).
25) OH reaction rate constant measured $f(T)$ 347–386 K: $A = 4.9\text{E-}10$; $E_a = 24.3$ kJ; k_{OH} (277 K) = 1.3E-14, corresponding $\tau_{OH} = 940$ d (with <[OH]> = 9.7E5 cm^{-3} [455]).
26) New $k_{OH}(T)$ recommended in [218], but value for 298 K unchanged relative to [9]; $k_{OH}f(T)$ 234–438 K: $A = 1.3\text{E-}12$; $E_a = 5.07$ kJ [48].
27) One higher value of k_{O_3} (1.2E-20) seems to be erroneous [4]; Atkinson [218] recommends < 1E-20 for methylated benzenes.
28) OH reaction rate constant studied as a function of T, including the dissociation of primary adducts [154].
29) This average of ten measured values [335] is very near to the value previously recommended by Atkinson [9].
30) Adsorbed on graphite particles; k_{OH}^{II} calculated from $k_{OH}^{I} = 0.145$ s^{-1} (298 K) with the given (probably too high) [OH] [504].
31) Adsorbed on graphite particles; k_{OH}^{II} calculated from $k_{OH}^{I} = 0.145$ s^{-1} (298 K) with the given (probably too high) [OH] [504].
32) Adsorbed on graphite particles; k_{OH}^{II} calculated from $k_{OH}^{I} = 0.20$ s^{-1} (298 K) with the given (probably too high) [OH] [504].
33) Adsorbed on graphite particles; k_{OH}^{II} calculated from $k_{OH}^{I} = 0.12$ s^{-1} (298 K) with the given (probably too high) [OH] [504].
34) Calculated from the average of two measured $t_{1/2}$ values of 200 and 400 min in a smog chamber experiment with known irradiation intensity [10].
35) Measurement of $k_{OH}(T)$ 343–363 K [294]; extrapolation to 298 K: 2.7E-12.
36) Measurement of $k_{OH}(T)$ 343–363 K [294]; $E_a = 1.66$ kJ; extrapolation to 298 K: 2.5E-12.
37) Measurement of $k_{OH}(T)$ 327–363 K [294]; extrapolation to 298 K: 3.4E-12.

Footnotes to the Table

38) Measurement of $k_{OH}(T)$ 329–363 K [294]; $A = 1.3E-11$; $E_a = 4.66$ kJ; extrapolation to 298 K: $(2.0 \pm 0.5)E-12$.
39) Measurement of $k_{OH}(T)$ 323–363 K [294]; $A = 1.5E-11$; $E_a = 4.32$ kJ; extrapolation to 298 K: 2.6E-12.
40) Measurement of $k_{OH}(T)$ 343–364 K [294]; $E_a = 0.98$ kJ; extrapolation to 298 K: 2.2E-12.
41) Estimated for OH-addition at 298 K [290].
42) Measurement of $k_{OH}(T)$ 343–363 K [294]; $A = 2.5E-10$; $E_a = 16.2$ kJ.
43) Measurement of $k_{OH}(T)$ 343–363 K [294]; $A = 2.8E-11$; $E_a = 9.39$ kJ.
44) Measurement of $k_{OH}(T)$ 343–363 K [294]; $A = 1.1E-10$; $E_a = 12.0$ kJ.
45) Measurement of $k_{OH}(T)$ 323–363 K [294]; $A = 2.9E-10$; $E_a = 14.5$ kJ.
46) Measurement of $k_{OH}(T)$ 323–363 K [294]; $A = 1.9E-11$; $E_a = 7.23$ kJ.
47) Measurement of $k_{OH}(T)$ 322–363 K [294]; $A = 2.7E-10$; $E_a = 13.7$ kJ.
48) Measurement of $k_{OH}(T)$ 323–363 K [294]; $A = 1.4E-11$; $E_a = 6.07$ kJ.
49) Measurement of $k_{OH}(T)$ 323–363 K [294]; $A = 1.9E-10$; $E_a = 5.57$ kJ.
50) Calculated for OH-addition [290].
51) Calculated for OH-addition [290].
52) Calculated for OH-addition [290].
53) The recommended value is the average of two evaluated reviews: 4.2E-11 [301] and 4.6E-11 [285]; $k_{OH}(T)$ 260–360 K: $E_a = -3.3$ kJ [285].
54) OH reaction rate constant $f(T)$ in [9].
55) O_3 reaction rate constant $f(T)$ in [4].
56) O_3 reaction rate constant $f(T)$ 242–363 K: $A = 1.09E-14$; $E_a = 16.6$ kJ [438].
57) OH reaction rate constant $f(T)$ 250–430 K: $A = 6.0E-12$; $E_a = -3.4$ kJ [416].
58) Albaladejo et al. (2002) measured $k_{OH} = 2.9E-11$ [524].
59) NO_3 reaction rate constant $f(T)$ 298–433 K: $A = 7.6E-11$; $E_a = 20.5$ kJ [415].
60) A lower value of $\Phi = 0.03$ has been reported by Moortgat [453].
61) NO_3 reaction rate constant $f(T)$ 298–433 K: $A = 5.5E-11$; $E_a = 19.1$ kJ [431].
62) A lower $k_{hv} = 5.0E-5$ s^{-1} has been measured at low irradiance [$k_1(NO_2) = 0.084$ min^{-1}] [34].
63) OH reaction rate constant $f(T)$ 252–346 K: $E_a = 9.2$ kJ [382].
64) OH reaction rate constant $f(T)$ 200–300 K: $E_a = 15.0$ kJ [237].
65) OH reaction rate constant $f(T)$ 200–300 K: $E_a = 14.5$ kJ [416].
66) OH reaction rate constant $f(T)$ 290–1180 K [218].
67) OH reaction rate constant $f(T)$ 260–380 K: $A = 5.3E-12$; $E_a = -1.17$ kJ [71].
68) Measured value of ethyl nitrate [184, 185].
69) OH reaction rate constant $f(T)$ 240–300 K: $A = 1.3E-12$; $E_a = 0.21$ kJ [416].
70) NO_3 reaction rate constant $f(T)$ 298–433 K: $A = 5.5E-11$; $E_a = 20.1$ kJ [204].
71) NO_3 reaction rate constant $f(T)$ 232–401 K: $E_a = 8.9$ kJ [370]; 270–330 K: $E_a = 12.1$ kJ [363]; 299–473 K: $E_a = 7.8$ kJ [371].
72) NO_3 reaction rate constant $f(T)$ 270–330 K: $A = 3.3E-12$; $E_a = 20.4$ kJ [363].
73) NO_3 reaction rate constant $f(T)$ 270–330 K: $E_a = 3.29$ kJ [363].
74) NO_3 reaction rate constant $f(T)$ 267–400 K: $E_a = 3.3$ kJ [358].
75) It is not clear wheather the rate constant k_{O_3}, measured by Grosjeans and Grosjeans 1999 [396], really refers to the *trans* isomer.
76) A much higher value reported in [143] has been disregarded in calculating the average k_{O_3}.
77) NO_3 reaction rate constant $f(T)$ 298–433 K: $A = 3.1E-12$; $E_a = 4.0$ kJ [526].
78) This upper limit of k_{NO_3} is recommended [285, 416] despite one experimental value of 1.8E-15 above this limit has been reported [220].
79) NO_3 reaction rate constant $f(T)$ 298–433 K: $A = 1.4E-12$; $E_a = -6.16$ kJ [395].
80) OH reaction rate constant $f(T)$ 293–473 K: $E_a = 6.68$ kJ.
81) O_3 reaction rate constant $f(T)$ 293–473 K: $A = 2.1E-12$; $E_a = 39.1$ kJ [301].
82) OH reaction rate constant $f(T)$ 200–300 K: $E_a = 7.48$ kJ [301]; 240–300 K: $E_a = 5.49$ kJ [285]; see also [32, 218, 247].
83) Adsorbed onto graphite particles; calculated from first measured 1st-order rate constant 0.19 s^{-1} and estimated [OH] = 3.4E10 [504], this concentration is probably too high and, thus, k_{OH} is a lower limit of the true 2nd-order rate constant in the adsorbed state.
84) Average of six values measured between 418 and 493 nm; Φ increases towards shorter wavelengths from 0.69 to 0.99 in the range given.

85) OH reaction rate constant $f(T)$ 300–390 K: E_a = 3.1 kJ [326].
86) O_3 reaction rate constant $f(T)$ measured for *cis* (Z) cycloheptene in the range 240–331 K: A = 1.3E-15; E_a = 4.1 kJ [380].
87) OH reaction rate constant $f(T)$ 298–368 K: A = 3.59E-12; E_a = 4.15 kJ [400].
88) O_3 reaction rate constant $f(T)$ 240–331 K: E_a = 8.64 kJ [380].
89) OH reaction rate constant $f(T)$ 300–390 K: E_a = 3.42 kJ [326]; k_{OH} (298 K) is in excellent agreement with Behnke et al. (1988), cited in [9].
90) O_3 reaction rate constant $f(T)$ 240–331 K: E_a = 1.80 kJ [380].
91) O_3 reaction rate constant $f(T)$ 240–331 K: A = 1.8E-15; E_a = 2.47 kJ [380].
92) Bimolecular reaction rate constant calculated from 1st-order decay and [OH] by Zetzsch et al.: DDT adsorbed onto aerosil particles; same group (2005) at 280 K: $k_{OH,ads.}$ = 5.2E-12.
93) The 1st-order rate constant of direct photochemical degradation refers to an average of 3 adsorbents (kaolonite, montmorillonite and sediment) at sunlight [550].
94) OH reaction rate constant $f(T)$ 326–907 K: A = 1.7E-12; E_a = 8.14 kJ [452].
95) OH reaction rate constant $f(T)$ 346–905 K: A = 2.79E-12; E_a = –6.52 kJ [452].
96) OH reaction rate constant $f(T)$ 400–927 K: A = 1.83E-12; E_a = –6.17 kJ [452].
97) OH reaction rate constant $f(T)$ 355–395 K: A = 6.6E-11; E_a = 6.7 kJ [297] and 390–769 K; A = 1.1E-12; E_a = –4.73 kJ [452]; the k_{OH} given is the average of the two extrapolated values (298 K); it should be noted that E_a has a different sign in the two measurements.
98) OH reaction rate constant $f(T)$ 379–931 K: A = 1.02E-12; E_a = –4.81 kJ [452].
99) OH reaction rate constant $f(T)$ 514–928 K: A = 3.18E-11; E_a = 5.54 kJ [452]; the extrapolated value (298 K) coincides with the relative value measured in Freon 113 [405, 406].
100) OH reaction rate constant $f(T)$ 373–432 K [462]; another measurement of $k_{OH}(T)$ over the range 409 936 K yielded A = 1.66E-12, E_a = –5.93 kJ [452]; due to the negative activation energy, the room temperature value obtained by extrapolation is much higher than the value given in the table.
101) This estimated value [290] is included here, as it is unlikely that the extremely toxic 2,3,7,8-TCDD ("Seveso-Dioxin") will be measured directly.
102) OH reaction rate constant $f(T)$ 355–365 K: A = 2.1E-11; E_a = 5.6 kJ [297].
103) Measured adsorbed onto SiO_2 artificial aerosol particles; value given is the average of several measurements at 280 K: k_{OH} = (2 ± 1)E-12.
104) The new R is slightly higher than the previous one: 1.31 (± 35%)E-11 [218,9]; $k_{OH}(T)$ 240–442 K [218].
105) NO_3 reaction rate constant $f(T)$ 258–373 K: E_a = 17.1 kJ [351].
106) The OH reaction rate is the average of three values: Dagaut et al. 1990 [542], 1.09E-11; Porter et al. 1997 [317], 1.26E-11; and Maurer et al. 1999 [467], 1.24E-11.
107) OH reaction rate constant $f(T)$ 296–401 K: E_a = 3.74 kJ [323].
108) Measured at 308 K [9].
109) OH reaction rate constant $f(T)$ 200-300 K: E_a = 10.4 kJ [301].
110) Measured at 300 K [283].
111) Recommended $k_{OH}(T)$ for the temperature range 280–460 K [218].
112) Measured at 300 K [283].
113) Recommended $k_{OH}(T)$ for the range 223–427 K [218]; confirmed by DeMore et al. [301] for the range 200–300 K: E_a = 15.0 kJ [301].
114) OH reaction rate constant $f(T)$ 300–400 K: A = 6.1E-13; E_a = 9.0 kJ [416].
115) OH reaction rate constant measured at 300 K [283].
116) OH reaction rate constant $f(T)$ 297–867 K: A = 5.02E-14; E_a = 3.81 kJ [422]; 240–300 K: A = 5.4E-13; E_a = 10.0 kJ [416].
117) OH reaction rate constant $f(T)$ 260–380 K: A = 5.22E-13; E_a = 9.2 kJ [416]; see also $k_{OH}f(T)$ 295–866 K [392].
118) OH reaction rate constant $f(T)$ 250–430 K: A = 1.12E-12; E_a = 10.6 kJ [437].
119) OH reaction rate constant $f(T)$ 250–430 K: A = 7.46E-13; E_a = 10.2 kJ [437].
120) OH reaction rate constant $f(T)$ 250–430 K: A = 2.59E-12; E_a = 10.5 kJ [393].
121) 1st-order rate constant of photolysis measured at low irradiation intensity: k^l (NO_2) = 0.084 min^{-1} [34].
122) Φ is the average of two effective quantum efficiencies [32, 453]; preferred absorption cross sections for calculating the photolytic lifetime given in [32].
123) Upper limit of k_{OH} at 298 and 460 K: ≤ 4E-16 [232]; at room temperature ≤ 1.3E-16 [416].

124) Recommended $k_{OH} = f(T)$ 294–800 K [218]; there is one much lower value of k_{OH} (1.22E-15) [392].
125) OH reaction rate constant $f(T)$ 200–300 K: E_a = 13.3 kJ, k_{OH}(298 K) = 1.7E-14 [301].
126) OH reaction rate constant $f(T)$ 240–300 K: A = 7.0E-13; E_a = 11.9 kJ [416].
127) OH reaction rate constant $f(T)$ 200–300 K: A = 1.0E-12; E_a = 10.4 kJ [301]; 270–340 K: A = 4.1E-12; E_a = 7.86 kJ [285].
128) OH reaction rate constant $f(T)$ A = 1.68E-12; E_a = 14.1 kJ [490].
129) The value of k_{OH} given is the average of 3 diverging values: (1) [410] 9.0E-15, calculated from $f(T)$ 250–430 K; A = 1.49E-12; E_a = 12.65 kJ; 10–50 hPa. (2) [433] 9.4E-14, measured at 298 K and 1000 hPa. (3) [434] 9.8E-15, calculated from $f(T)$ 268–308 K; A = 1.36E-12; E_a = 12.23 kJ; 130 hPa. The different pressures may be one, but perhaps the only reason for the divergence observed.
130) Calculated for a zenith angle of 40° and Φ = 1 (assumed).
131) OH reaction rate constant $f(T)$ 277–370 K: A = 2.32E-12; E_a = 6.58 kJ [407].
132) OH reaction rate constant $f(T)$ 200–300 K: E_a = 14.1 kJ [301]; 240–300 K: E_a = 13.5 kJ [285].
133) OH reaction rate constant $f(T)$ 250–430 K: A = 1.9E-12; E_a = 12.5 kJ [435].
134) OH reaction rate constant $f(T)$ 292–365 [218]; 295–882 [277]; no recommendation due to discrepancies in different labs [218] (addendum).
135) OH reaction rate constant $f(T)$ 292–366 K: A = 3.70E-12; E_a = 6.78 kJ [254]; 293–418 K: A = 5.01E-12; E_a = 7.63 kJ [420]; k_{OH}(298 K) = 1.26E-13 [420].
136) The recommended (R) value is the average of two evaluated data sets: (1) [301] k_{OH} (T) 200–300 K; E_a = 14.5 kJ; k_{OH}(298) = 4.23E-15. (2) [285]: k_{OH}(T) 240–300 K; E_a = 12.8 kJ; k_{OH} (298) = 4.16E-15; see also [9, 32, 218].
137) OH reaction rate constant $f(T)$ 200–300 K: E_a = 14.0 kJ [301]; 270–340 K: E_a = 13.7 kJ [285].
138) OH reaction rate constant $f(T)$ 250–430 K: A = 2.6E-12; E_a = 11.8 kJ [410].
139) OH reaction rate constant $f(T)$ 250–430 K: A = 1.49E-12; E_a = 12.6 kJ; p = 10–50 hPa [410]; see also [490].
140) OH reaction rate constant $f(T)$ 250–430 K: A = 2.6E-12; E_a = 11.8 kJ [410].
141) The recommended (R) values is the average of two evaluated data sets: (1) [301] k_{OH}(T) 200–300 K; E_a = 12.9 kJ; k_{OH}(298) = 9.92E-15. (2) [285]: k_{OH}(T) 240–300 K; E_a = 12.0 kJ; k_{OH}(298) = 9.6E-15.
142) O_3 reaction rate constant $f(T)$ 307–359 K: A = 1.05E-12; E_a = 73.2 kJ [103].
143) The recommended (R) values is the average of two two evaluated data sets: (1) [301] k_{OH}(T) 200–300 K; E_a = 13.7 kJ. (2) [285] k_{OH}(T) 270–300 K; E_a = 13.4 kJ.
144) The k_{OH} value given is the average of 7.5E-11 [54] and 10.3E-11 [157].
145) OH reaction rate constant $f(T)$ 250–430 K: A = 2.01E-12; E_a = 7.42 kJ [392], k_{OH}(298 K) = 1.0E-13; a similar value for the room-temperature OH reaction rate constant has been reported [505]: (1.23 ± 0.06)E-16.
146) High pressure limit [218].
147) NO_3 reaction rate constant $f(T)$ 270–340 K: A = 3.30E-12; E_a = 23.95 kJ [285]: k_{OH} (298 K) = 2.1E-16.
148) The ozone reaction rate constant given is the average of 2.3E-19 [14] and 2.45E-19 [15]; a much lower value cited in [4] has been disregarded.
149) This value of k_{NO_3} replaces an older (R); $f(T)$ 266–367 K: E_a = 14.7 kJ [362].
150) R (k_{OH}) [218] essentially confirms the average of older values (1.48E-11; 1.46E-11 [164]; 8.11E-12 [82]).
151) Average of two nearly identical values: k_{NO_3} (298 K) = 1.80E-15 [357] and 1.88E-15 [362].
152) Average of two values measured at 297 K: k_{OH} = 7.63E-12 and 7.35E-12 [257].
153) Average of four values: k_{NO} = 1.3E-16 [357], 0.90E-16 [362], 0.797E-16 [363], and 1.43 [367]; $f(T)$ 293–473 K: E_a = 19.0 kJ [367]; 270–330 K: E_a = 19.4 kJ [363].
154) Average of two values: k_{O_3} = 8.0E-20 [137] and 19E-20 [112].
155) A is the average of k_{NO_3} = 1.40E-16 [366] and 9.26E-17 [363]; $f(T)$ 270–330 K: E_a = 18.9 kJ.
156) Average of two values: k_{NO_3} = 1.40E-16 [366] and 0.926E-16 [363], $f(T)$ 270–330 K: E_a = 18.9 kJ [363].
157) NO_3 reaction rate constant $f(T)$ 270–330 K: E_a = 15.4 kJ [363].
158) NO_3 reaction rate constant $f(T)$ 278–368 K: E_a = 16.9 kJ [362].
159) OH reaction rate constant $f(T)$ 253–348 K: A = 6.41E-11; E_a = 7.2 kJ [412]; 250–430 K: A = 1.1E-12; E_a = 2.68 kJ [410].
160) OH reaction rate constant $k_{OH} = f(T)$ and $f(p)$ given in [32].

161) A measurement adsorbed onto grapite particles (k^I_{OH} = 0.115 s^{-1}) has been reported [504].
162) The reaction rate constant given is the average of 3 values measured at 297–300 K: 1.2E-11 [3, 10, 44], 1.6E-11 [298] and 1.3E-11 [297].
163) The total quantum efficiency of photolysis is Φ = 1 below 330 nm. There are two channels of photolysis. Channel I leads to H$_2$ + CO over the whole range of absorption in the near-UV, i.e. below 360 nm [5]. Channel II ends up in the products H + HCO. This reaction occurs below a threshold near 340 nm. The quantum yield of channel I and II depend on the wavelength of excitation. At 355 nm Φ = 0.2 (channel I) [5]. See also Seinfeld [199].
164) Photolysis see [61].
165) OH reaction rate constant $f(T)$ 290–450 K: 4.5E-13, no function of temperature in the range given [416].
166) OH reaction rate constant $f(T)$ 253–371 K: E_a = –0.125 kJ [299].
167) OH reaction rate constant $f(T)$ 233–372 K: A = 8.54E-13; E_a = 3.85 kJ [299].
168) OH reaction rate constant $f(T)$ 253–372 K: E_a = –0.35 kJ [299].
169) Also measured at 300 K [263].
170) NO$_3$ reaction rate constant $f(T)$ 298–433 K: A = 7.8E-11; E_a = 20.0 kJ [415].
171) NO$_3$ reaction rate constant $f(T)$ 298–433 K: A = 8.0E-13; E_a = 5.2 kJ [430].
172) It is not clear whether this upper limit of k_{O_3} refers to cis (Z), trans (E) or a mixture of the isomers.
173) Average of 6.7E-11 and 6.9E-11 [trans(E) 2-heptene].
174) OH reaction rate constant $f(T)$ 346–386 K: A = 1.4E-11; E_a = 11.2 kJ [454]; k_{OH} (277 K) = 1.0E-13; τ_{OH} (277 K) = 120 d (using <[OH]> = 9.7E05 cm^{-3} [455]).
175) Measured adsorbed onto SiO$_2$ particles (Aerosil) at 280 K: 2.1E-12 [476].
176) Very rapid photo-isomerization to E,E-2,4-hexadienal [216].
177) Photolysis competes with OH reaction [216].
178) NO$_3$ reaction rate constant $f(T)$ 298–433 K: A = 7.0E-11; E_a = 19.8 kJ [415].
179) OH reaction rate constant $f(T)$ 250–430 K: A = 4.71E-13; E_a = 13.5 kJ [494].
180) Aschmann and Atkinson (1998) reported 1.6E-11 [481].
181) Ozone reaction rate constant k_{O_3} (287 K) = 1.3E-18 [385].
182) Average of k_{NO_3} = 1.2E-14 [333, 313] and 5.3E-14 [430]; $f(T)$ 298–433 K: A = 1.2E-12; E_a = 1.7 kJ [430].
183) Measured by an absolute method; replaces an older, slightly higher value as R [218].
184) NO$_3$ reaction rate constant $f(T)$ 298–433 K: E_a = 6.78 kJ [354], see also [273, 363].
185) The "photolysis" rate [70] is a total photo-transformation rate and includes cis–trans isomerization (major pathway) and true photolysis (photodissociation, minor pathway). It refers to average enviromental conditions, which are not specified. The photochemical lifetime is given as 30–35 min [70].
186) See comment to trans isomer [820-69-9].
187) OH reaction rate constant $f(T)$ 233–372 K: A = 8.5E-13; E_a = –1.2 kJ [408].
188) The recommended k_{OH} is the average of two evaluated data: k_{OH} = 8.04E-13 [301] and 7.93E-13 [285]); $k_{OH} f(T)$ 200–300 K: E_a = 2.91 kJ [301]; 2.74 kJ [285].
189) OH reaction rate constant $f(T)$ 290–440 K: A = 1.2E-13; E_a = 3.33 kJ [416]; the reaction depends on pressure [160, 161].
190) Primary quantum efficiency of photolysis; for a detailed discussion see introductory chapter by Walter Klöpffer and references.
191) The ozone reaction rate constant k_{O_3} is the average of 4.8E-19 [4, 151] and 12.6E-19 [151].
192) O$_3$ reaction rate constant $f(T)$ 242–363 K: A = 2.95E-15; E_a = 6.54 kJ [438].
193) Another value k_{NO_3} = 9.4E-12 [395] coincides well with the average given.
194) Measured by Brubacker and Hites 1998 [454] $k_{OH} f(T)$ 347–385 K: A = 6.0E-11; E_a = 14.2 kJ; hence k_{OH} (298 K) = 1.95E-13 (extrapolated to room temperature); the values measured at 347 K (74 °C, the lowest temperature measured) range from k_{OH} = 4.0 to 4.7E-13; the observed thermal activation and the chemical structure of Lindane both point to H abstraction as the dominant mechanism of the OH reaction. The extrapolation seems thus to be justified. The rate constant extrapolated to 277 K, the average temperature of the troposphere, is 1.2E-13, yielding a tropospheric lifetime of 3 months (using <[OH]> = 9.7E05 cm^{-3}) for γ-HCH in the free gas phase.
195) Measured by Rühl (2004) in the adsorbed state [458]; higher values of $k_{OH,ads}$ have been measured by Zetzsch in the smog chamber [403]. Krüger et al. (2005) [476] reported $k_{OH,ads}$(280 K) = (3.0 ± 0.3)E-12.

Footnotes to the Table

196) Room temperature OH rate constant is the average of two measured values k_{OH} = 2.53E-12 [55] and 3.6E-12 [9].

197) OH reaction rate constant $f(T)$ 223–1512 K: A = 1.85E-12; E_a = 14.05 kJ [218, 285].

198) OH reaction rate constant $f(T)$ 200–300 K: k_{OH} (298 K) = 4.66E-15; E_a = 13.3 kJ [301]; 240–300 K: k_{OH} (298 K) = 4.60E-15; E_a = 12.7 kJ [285].

199) OH reaction rate constant $f(T)$ 240–300 K: A = 2.0E-12; E_a = 9.44 kJ [416].

200) UV-absorption cross section measured $f(T)$ 223–298 K in the spectral range 205 to 375 nm [484]; k_{hv} calculated for sea level.

201) OH reaction rate constant $f(T)$ 200–300 K: k_{OH} (298 K) = 1.1E-13; E_a = 7.5 kJ [301]; 240–300 K: k_{OH} (298 K) = 1.17E-13; E_a = 6.4 kJ [285].

202) 293–424 K: only upper limit [218].

203) k_{OH} = 1.2E-13 recommended here (R) is the average of two evaluated values: 1.12E-13 (DeMore et al. 1997 [301]) and 1.23E-13 (Atkinson et al. 1997 [285]); $f(T)$ 200–300 K [301] E_a = 8.73 kJ; 240–300 K [285] E_a = 7.65 kJ; these data are confirmed by Villenave et al. [348]: k_{OH} = 1.1E-13; $f(T)$ 277–370 K: E_a = 7.85 kJ.

204) OH reaction rate constant $f(T)$ 240–300 K: A = 8.8E-13; E_a = 8.42 kJ [416]; further literature [171–177].

205) OH reaction rate constant $f(T)$: A = 6.0E-12; E_a = 12.7 kJ [407].

206) Calculated for sea level, 40° zenith by Roehl et al. 1997 [484] assuming Φ = 1.

207) OH reaction rate constant $f(T)$ 230–372 K: E_a = 6.5 kJ [317]; same room temperature k_{OH} given by [469].

208) Calculated for sea level, 40° zenith by Roehl et al. 1997 [484] assuming Φ = 1.

209) Measured by Alvarez and Moore 1994 [459] Φ = 0.98 ± 0.24 at 308 nm; k_{hv} = 6.7E-6 s^{-1} calculated for a zenith angle 40° using Φ = 1 [459, 457].

210) A more recently evaluated OH-reaction constant [220] is also very near to the original one: k_{OH} = 2.9E-12; kOH = $f(T)$ given in [220]. Dito DeMore and Bayes 1999 [400]: $k_{OH}f(T)$ 263–361 K: A = 1.51E-11; E_a = 4.15 kJ, hence k_{OH} (298 K) = 2.8E-12.

211) k_{OH} (298 K) is the average of two evaluated values: DeMore et al. 1997 [301], 2.31E-15, $f(T)$ 200–300 K, E_a = 16.6 kJ mol^{-1}; and Atkinson, Baulch et al. 1997, 2001 [285, 416]: 2.90E-15, no recommended $f(T)$.

212) k_{OH} is the sum of two rate constants, $k_1 + k_2$. k_1 is independent of O_2 partial pressure, k_2 is not. The given preferred value is the sum of $k_1 + k_2$ for a total pressure of 0.1 MPa air at 298 K [9, 32, 218].

213) NO_3 reaction rate constant $f(T)$ 250–380 K: A = 1.9E-13; E_a = –4.2 kJ [429].

214) k_{NO_3} (R) is the average two evaluated values by DeMore et al. 1997 [301] (8.9E-13) and Atkinson, Baulch et al. 1997 [285] (9.2E-13); $f(T)$ 200–300 K, E_a = –1.75 kJ; "R" confirmed 2004 [429].

215) OH reaction rate constant $f(T)$ 200–300 K: E_a = 5.9 kJ [301].

216) The much lower k_{OH} (298 K) limit, compared with earlier upper limits [1], results from extrapolation of measurements performed at higher temperatures, assuming a reasonable pre-exponential factor (A) in the Arrhenius equation [32]. Atkinson [9] gives an upper limit of k_{OH} of 1E-17.

216) The recommended value ("R") is the average of two evaluated values: DeMore et al. [301] (2.78E-16) and Atkinson, Baulch et al. [285] (2.71E-16) at 298 K. k_{OH} = $f(T)$ for the range of 200–300 K (DeMore) E_a = 20.3 kJ; 240–300 K (Atkinson and Baulch) E_a = 19.1 kJ.

218) Atkinson, Baulch et al. 1997 [285] give the upper limit for k_{OH} which is valid now (evaluated = R), i.e. *no* measurable reaction with OH-degradation via photolysis is much more likely! Older measurements underestimated the role of photolysis [92].

219) NO_3 reaction rate constant $f(T)$ 290–480 K: E_a = 21.3 kJ [285].

220) OH reaction rate constant $f(T)$ 242–328 K: A = 2.33E-12; E_a = 12.3 kJ [495].

221) New k_{OH} value by Brubaker and Hites 1998 [297]: k_{OH} (298 K) = 2.3E-11 in good agreement with the recommended value (R) and Kloepffer et al. 1986 [44] (2.0E-11); $k_{OH}f(T)$: A = 2.4E-11; E_a = 0.04 ± 0.59 kJ [297] (*de facto* no thermal activation in the temperature range 306–336 K).

222) k_{OH} = 6.9E-11 is the average of two values: k_{OH} = 7.7E-11 [9] and k_{OH} = 6.1E-11 measured by Phousongphoung and Arey [475].

223) k_{OH} is the average of two values: 5.3E-11 [45] and 4.1E-11 [475].

224) k_{OH} is the average of two values: k_{OH} = 5.2E-11 [31] and 4.9E-11 [475].

225) k_{NO_3} is the average of 1.11E-14 [212] and 8.64E-15 [43].

226) k_{hv} is an average for January, April, August and October [184].

227) k_{OH} depends on pressure (p), negative $f(T)$ at room temperature [218]; $f(T)$ 298–373 K: A = 3.30E-12; E_a = 5.81 kJ [544].
228) k_{hv}, calculated for overhead sun, is an average for January, April, August and October [184, 186].
229) k_{OH} as an average of older data 1.8E-13 [59], 4.1E-13 [159] and 5.9E-13 [187] amounts to 3.9E-13. $k_{OH} = f(T)$ 230–300 K: A = 6.2E-13; E_a = 1.93 kJ yielding k_{OH} (298 K) = 2.8E-13; reviewed by Atkinson, Baulch et al. 2001 [416].
230) k_{hv} is an average from 7.2E-7 [184], 3.0E-7 [187] and 4.5E-7 [185].
231) The gas-phase UV-absorption spectrum is given in [184].
232) k_{NO_3} by Martinez et al. 1999 [395] seems to be very high: 1.3E-09 (!); $f(T)$ 298–433 K: A = 2.2E-10; E_a = –4.36 kJ; the discrepancy with [130] is too high to form an average.
233) k_{OH} measured 2001 at 1000 ± 20 hPa: k_{OH}(298 ± 2K) = (2.88 ± 0.25)E-11 [470].
234) OH reaction rate constant $k_{OH} = f(T)$ 253–372 K: A = 1.3E-12; E_a = –0.25 kJ [408].
235) k_{OH} is the average of two values: k_{OH} = 2.1E-12 [78] and 1.9E-12 [182].
236) $k_{O_3} = f(T)$ in the range 238–298 K measured by Batha et al. [275], k_{O_3} (298 K) = 4.4E-17; k_{O_3} (298 K) = 4.3E-17 measured by Lewin et al. 2001 [440]; the excellent agreement is taken as reason for the recommendation (R).
237) The recommended k_{OH} is supported by recent measurements: D'Anna et al. 2001 [414] 2.6E-11; Thevenet et al. 2000 [421] $k_{OH} = f(T)$ 243–372 K: A = 9.9E-12; E_a = –2.25 kJ; k_{OH} (298 K) = 2.8E-11.
238) k_{NO_3} is the average of three recent measurements: D'Anna et al. 2001 [414], 1.7E-14; Papagni et al. 2000 [413], 1.4E-14; Cabanas et al. 2000 [415], $k_{NO_3} = f(T)$ 298–433 K: A = 2.8E-11; E_a = 18.2 kJ, k_{NO_3} (298 K) = 1.8E-14; average (A) at 298 K: 1.6E-14.
239) Measured at 305 K [279].
240) k_{OH} is the average of 4.79 and 5.1E-12 [218], original data from Nielsen et al. 1991 [253].
241) Another value determined by the same group at 288 K: k_{O_3} = 1.4E-17 [379].
242) k_{OH} measured $f(T)$ 263–372 K: E_a = –3.33 kJ [347].
243) The estimation of Φ is based on measurements of ethyl nitrate [184, 185].
244) k_{OH} is the average of 1.72E-12 [60] and 1.95E-12 [187].
245) The more recent value by Atkinson et al. (2000) fits well: k_{OH} (298 K) = 4.6E-12 [424].
246) Φ = 0.07 reported by Wirtz 1999 [450]; effective quantum efficiency measured in the outdoor chamber EUPHOR in Valencia, Spain.
247) k_{OH} is the average of the values reported by Atkinson and Aschmann 1995 [337]: 4.0E-12 and Magneron et al. 2003 [423] 3.6E-12.
248) k_{NO_3} measured by Cabanas et al. 2001 [430] $f(T)$ 298–433 K; A = 5.4E-12; E_a = 12.8 kJ; hence k_{NO_3} (298 K) = 3.1E-14.
249) $k_{O_3} = f(T)$ 240–324 K [221]; Grosjean and Grosjean 1998 [376] measured k_{O_3} = 1.09E-17; the average of Treacy et al. [221] and Grosjean's value is 9.7E-18 and confirms the R given by Atkinson [218]. Treacy et al. also measured $k_{O_3} = f(T)$ in the range 240–324 K, E_a = 13.1 kJ.
250) k_{NO_3} = 6.8E-14 by Martinez et al. 1997 [354] is preferred over lower values published by the same group (Wayne) earlier; $f(T)$ 298–433 K: A = 2.05E-11 E_a = 14.1 kJ. Averaging seems not to be wise in that case, as $k_{NO_3} f(T)$ varies strongly.
251) Measured at 285–293 K [390].
252) k_{OH} is an average of 1.37E-13 [163] and 7.5E-14 [198].
253) Calculated from k^1_{OH} = 0.145 s^{-1} adsorbed onto graphite particles, 298 K; Esteve et al. 2004 [504]; the [OH] used for calculating the 2nd-order rate constant is probably too high and $k_{OH,ads}$ therefore too low.
254) k_{NO_3} by Berndt et al. 1996 replaces the older value (9.1E-10) by Atkinson 1985; Martinez et al. 1999 [395] measured $k_{NO_3} = f(T)$ 298–433 K; A = 1.9E-09; E_a = 9.6 kJ; thus k_{NO_3} (298 K) = 3.9E-11.
255) k_{OH} measured by Brubaker and Hites [297] 2.7E-11; older value by Biermann et al. 1985: 3.4E-11 [46].
256) OH reaction rate constant $k_{OH} = f(T)$ 323–403 K:
A = 4.46E-12; E_a = –3.97 kJ, hence k_{OH}(298 K) = 3.2E-11 (extrapolated) [534].
257) OH reaction rate constant $k_{OH} = f(T)$ 323–403 K: A = 2.46E-13; E_a = –11.8 kJ, hence k_{OH}(298 K) = 2.9E-11 (extrapolated) [534].
258) OH reaction rate constant $k_{OH} = f(T)$ 323–403 K: A = 2.546E-12; E_a = –8.0 kJ, hence k_{OH}(298 K) = 6.5E-11 (extrapolated) [534].
259) OH reaction rate constant $k_{OH} = f(T)$ 323–403 K: A = 2.97E-12; E_a = –7.7 kJ, hence k_{OH}(298 K) = 6.6E-11 (extrapolated) [534].

260) OH reaction rate constant $k_{OH} = f(T)$ 323–403 K: $A = 7.75\text{E-}12$; $E_a = -5.65$ kJ, hence $k_{OH}(298\text{ K}) = 7.6\text{E-}11$ (extrapolated) [534].
261) $k_{OH} = f(T)$ 249–438 K; $E_a = 1.29$ kJ [160].
262) OH reaction rate constant $k_{OH} = f(T)$ 278–348 K: $A = 6.46\text{E-}13$; $E_a = -11.1$ kJ [546].
263) OH reaction rate constant $k_{OH} = f(T)$ 278–348 K: $A = 9.76\text{E-}14$; $E_a = -12.6$ kJ [546].
264) OH reaction rate constant $k_{OH} = f(T)$ 278–348 K: $A = 4.20\text{E-}13$; $E_a = -12.1$ kJ [546].
265) OH reaction rate constant $f(T)$ 278–348 K: $A = 8.0\text{E-}14$; $E_a = -12.2$ kJ [546].
266) OH reaction rate constant $k_{OH} = f(T)$ 28–348 K: $A = 4.29\text{E-}13$; $E_a = -11.9$ kJ [546].
267) NO$_3$ reaction rate constant $f(T)$ 298–433 K: $A = 3.5\text{E-}13$; $E_a = -7.0$ kJ; k_{NO_3} (298 K) = 5.9E-12 [526].
268) NO$_3$ reaction rate constant $f(T)$ 298–433 K: $A = 1.6\text{E-}10$; $E_a = 10.4$ kJ; k_{NO_3} (298 K) = 2.1E-12 [526].
269) Phi $(\Phi)_{\text{effective}}$ reported by Moortgat 2001 [453], measured in the EUPHOR smog chamber in Spain (Valencia).
270) $k_{OH}(298) = 1.9\text{E-}11$ measured by D'Anna et al. 2001 [414]; $k_{OH} = f(T)$ 240–380 K; $A = 5.1\text{E-}12$; $E_a = -3.35$ kJ; k_{OH} (298 K) = 2.0E-11 according to an evaluation [416].
271) Φ (Norrish I splitting) depends on the wavelength [nm]. Preferred values are: $\Phi = 0.89$ (298), 0.85 (302), 0.50 (313), 0.26 (325), 0.15 (334) [32]. Absorption cross sections are given for the spectral range 340 to 230 nm [32].
272) Previous recommendation R of k_{OH} unchanged, although higher values have been published by Dobi et al. [223]; a similar value results from $k_{OH} = f(T)$ 243–372: K; $A = 4.7\text{E-}12$; $E_a = -4.69$ kJ; k_{OH} (298 K) = 3.1E-11 [421].
273) $k_{NO_3} = 2.3\text{E-}14$ D'Anna [356]; $k_{NO_3} = f(T)$ 298–433 K: $A = 2.7\text{E-}11$; $E_a = 17.2$ kJ; k_{NO_3} (298 K) = 2.7E-14; average: 2.5E-14.
274) OH reaction rate constant $k_{OH} = f(T)$ 298–519 K; numerical values given in [223] and [218].
275) Φ is an effective quantum efficiency, i.e., it depends on the spectral irradiance of the radiation used [32]. It may be used for calculating tropospheric photolysis rates in the range of 325–470 nm [1,2]. $k_{h\nu}$ at weak irradiation intensity of k^1 (NO$_2$) = 0.084 min^{-1} [34].
276) $k_{OH} = f(T)$ 298–420 K by Koch et al. 1996 [323]: $E_a = -1.9$ kJ.
277) Unchanged R 1994, Atkinson [218], $k_{OH} = f(T)$ 293–1220 K given. Unchanged R after DeMore et al. 1997 [301] (1.09E-12) and Atkinson and Baulch 1997 [285] (1.11E-12); $k_{OH} = f(T)$ 200–300 K: $E_a = 5.5$ kJ [301] and 240–300 K: $E_a = 4.9$ kJ [285].
278) k_{OH} (298 K) = 2.3E-15 is the average of two values: $f(T)$ 295–367K measured by Nelson et al. 1992 [256] (2.1E-15) $f(T)$ 290–400 K: $A = 7.0\text{E-}13$; $E_a = 14.1$ kJ given by Atkinson, Baulch et al. 2001 [416] (2.4E-15).
279) k_{OH} measured by Tokuhashi et al. 2000 [410] $f(T)$ 250–430 K: $A = 2.49\text{E-}12$; $E_a = 12.5$ kJ, hence k_{OH} (298 K) = 1.6E-14.
280) A more recent value, measured 1999 [392], is close to the value recommended by Atkinson [9]: k_{OH} (298 K) = 8.9E-13.
281) $k_{OH} = f(T)$ 288–394 K: $A = 3.06\text{E-}12$; $E_a = 4.18$ kJ [535].
282) $k_{OH} = f(T)$ in the range 294–369 K measured by Nelson et al. 1993 [280]. The R at 298 K is due to evaluation by Atkinson et al. 1997 [285]; unchanged 2001 [416]: $k_{OH} = f(T)$ 270–463 K: $A = 4.5\text{E-}13$; $E_a = 13.9$ kJ; k_{OH} (298 K) = 1.6E-15.
283) k_{OH} measured by Tokuhashi et al. 1999 [435] $f(T)$ 250–430 K; $A = 1.94\text{E-}12$; $E_a = 12.0$ kJ; hence k_{OH} (298 K) = 1.5E-14 [435].
284) OH reaction rate constant $f(T)$ 200–300 K: $A = 1.30\text{E-}12$; $E_a = 20.6$ kJ [301].
285) DeMore et al. 1997 [301] $k_{OH} = f(T)$ 200–300 K: $E_a = 13.2$ kJ.
286) k_{OH} measured by Chen et al. 2003 [434] $f(T)$ 268–308 K: $A = 2.39\text{E-}12$; $E_a = 13.0$ kJ; hence k_{OH} (298 K) = 1.26E-14.
287) k_{NO_3} measured by Langer et al. 1994 [536]; $f(T)$ 257–367 K: $A = 1.21\text{E-}12$; $E_a = 15.55$ kJ.
288) $k_{OH} = f(T)$ 293–1146 K given by Atkinson 1994 [218] is slightly different from 1989 [9]; k_{OH}(298 K) recommended value is the same as previously reported [9].
289) OH reaction rate constant $k_{OH} = f(T)$ 240–400 K: $A = 2.10\text{E-}12$; $E_a = 2.54$ kJ [271].
290) k_{NO_3} is the average of two values: 4.9E-15 (Chew et al. 1998 [357] and Langer 1994 [351]), $f(T)$ 258–373 K, $E_a = 15.5$ kJ.
291) $k_{OH} = f(T)$ 298–370 K measured by Orkin et al. 1997 [315]; $E_a = 14$ kJ.
292) Evaluation by Atkinson et al. 1997 $k_{OH} = f(T)$ 270–340 K: $E_a = 14.0$ kJ [285].
293) OH reaction rate constant $f(T)$: $A = 6.1\text{E-}13$; $E_a = 11.1$ kJ [301,407].

294) k_{OH} measured by Tokuhashi et al. 2000 [410] as a function of temperature 250–430 K; $A = 2.14E-12$; $E_a = 13.1$ kJ, hence k_{OH} (298 K) = 1.0E-14.

295) k_{OH} is the average of two values: LeCalve et al. 1997 [299] 7.3E-13 and Wallington et al. 1988 [77] 1.03E-12.

296) OH reaction rate constant $k_{OH} = f(T)$ 273–371 K: $A = 4.9E-14$; $E_a = -2.29$ kJ [541].

297) Effective quantum efficiency measured in Spain (EUPHOR chamber), reported by Moortgat 2001 [453]. Berges and Warneck [222] measured the quantum efficiency of disappearance of pyruvic acid with monochromatic UV-radiation: Φ (350 nm) = 0.85 ± 0.16. The same authors measured the quantum efficiency of acetaldehyde formation using the same wavelength to be $\Phi' = 0.48$. Acetaldehyde is formed as a main product of pyruvic acid photolysis together with CO_2 and acetic acid.

298) $k_{OH} = 5.5E-12$ (R) confirmed by Wu et al. 2003 [489]: 5.47 ± 0.25 at 295 K, 1000 hPa.

299) $k_{OH} = f(T)$ 250–430 K: $A = 6.99E-13$; $E_a = 8.25$ kJ [392].

300) k_{OH} measured $f(T)$ 250–430K: $A = 1.4E-12$; $E_a = 6.49$ kJ; k_{OH} (298 K) = 1.0E-13; Chen et al. 2000 [419]: $f(T)$ 298–356 K: $A = 2.27E-12$; $E_a = 7.49$ kJ; k_{OH} (298 K) = 1.1E-13.

301) $k_{OH} = 3.0E-12$ also reported in [218] and Dagaut et al. 1989 [260]; recently confirmed in a mechanistic study by Chowdhury et al. 2002 [443] (2.8E-12).

302) $k_{NO_3} = 8.9E-15$, measured by Canosa-Mas et al. 1999 [399] at 293 K, replaces an older, much lower value. There is, however, also a more recent low value reported by Cabanas et al. 2001 [430]: $k_{NO_3} = f(T)$ 298–433 K: $A = 1.7E-11$; $E_a = 26.9$ kJ, hence k_{NO_3} (298 K) = 3.2E-16. The activation energy E_a seems to be fairly high.

303) k_{NO_3} R new [218], but very near to old ABIOTIKx (R) value [537]; $k_{NO_3} = f(T)$ is given for the range of 296–423 K (R) [218]. Atkinson, Baulch et al. 1997 [285, 220] recommend 9.4E-15; $k_{NO_3} = f(T)$: 296–423 K; $A = 4.6E-13$; $E_a = 9.65$ kJ.

304) k_{NO_3} measured by Marston et al. 1993 [363]; $f(T)$ 270–330 K: $E_a = 9.7$ kJ.

305) OH reaction rate constant $k_{OH} = f(T)$ 228–388 K: $A = 6.3E-12$; $E_a = -2.37$ kJ [527].

306) k_{NO_3} measured by Martinez et al. 1996 [328]; $f(T)$ 298–428 K: $E_a = 14.4$ kJ.

307) OH reaction rate constant $k_{OH} = f(T)$ 228–388 K: $A = 3.5E-12$; $E_a = -3.94$ kJ [527].

308) k_{NO_3} measured by Martínez et al. 1996 [328] as $f(T)$ 296–419 K: $E_a = 18.9$ kJ.

309) k_{OH} is the average of 1.24E-11 [83] and 1.43E-11 [82].

310) NO_3 reaction rate constant $f(T)$ 297–432 K: $E_a = 21.4$ kJ [318].

311) OH reaction rate constant $k_{OH} = f(T)$ 228–388 K: $A = 3.6E-12$; $E_a = -3.78$ kJ [527].

312) NO_3 reaction rate constant $f(T)$ 296–430 K: $A = 7.17E-12$; $E_a = 18.7$ kJ [538].

313) k_{OH} measured $f(T)$ 293–489K: $A = 9.96E-13$; $E_a = -2.03$ kJ; k_{OH} (298 K) = 2.26E-12 [239]; $f(T)$ 252–370 K: $A = 5.67E-13$; $E_a = -3.38$ kJ; k_{OH} (298 K) = 2.22E-12 [315].

314) k_{NO_3} (R) is the average of 5 nearly identical room temperature values values [6, 122–125]. One lower value from 1975 [126] has been disregarded.

315) OH reaction rate constant $k_{OH} = f(T)$ 252–370 K: $A = 1.4E-12$; $E_a = 0.53$ kJ [315].

316) OH reaction rate constant $k_{OH} = f(T)$ 253–374 K: $A = 2.3E-12$; $E_a = -4.8$ kJ [530].

317) OH reaction rate constant $k_{OH} = f(T)$ 253–374 K: $A = 2.0E-12$; $E_a = -4.6$ kJ [530].

318) OH reaction rate constant $k_{OH} = f(T)$ 253–374 K: $A = 2.5E-12$; $E_a = -6.8$ kJ [530].

319) k_{NO_3} measured by Hallquist et al. 1996 [359]; $f(T)$ 273–363 K, $E_a = 2.83$ kJ.

320) k_{O_3} is the average of 1.43E-20 [141] and 2.2E-20 [134]. A much higher value [143] has been disregarded.

321) Measured by Palm et al. [456] in a 2400-L smog chamber; no loss by direct photolysis, O_3-reaction unimportant (see k_{O_3}).

322) Measured by Palm et al. [456] in a 2400-L smog chamber, see also k_{OH}; the O_3 reaction is judged to be unimportant under environmental conditions by the authors (OH dominating).

323) k_{NO_3} has been measured by Martinez et al. 1999 [395]; $f(T)$ 298–393 K: $A = 2.3E-10$; $E_a = 7.83$ kJ [393].

324) $k_{OH} = 9.2E-13$ is the average of two values, both measured by the Atkinson group: $k_{OH} = (1.00 ± 0.27)E-12$ [237] and $k_{OH} = (8.5 ± 0.9)E-13$ [480].

325) This R replaces the somewhat higher average of 2.6E-17 [537], based on older measurements [4], [115] and [144].

326) k_{OH} in the adsorbed state measured by Palm et al. 1997 [404] at 300 K in a special smog chamber described by Behnke et al. 1988 [403] $k_{OH,ads}$ at 280 K: 5.3E-12 Zetzsch 2004 [401]; Krüger et al. 2005 (Fig. 54 in [476]) reported $k_{OH,ads}$ at 280K = (4.6 ± 2.0)E-12.

Footnotes to the Table

327) k_{O_3} is the average of Shu and Atkinson 1994 [377] (2.11E-14) and Atkinson et al. [224] (8.65E-15).
328) k_{NO_3} measured by Berndt et al. 1996 [355] is given here and preferred over older but similar values given in [537].
329) NO$_3$ reaction rate constant $f(T)$ 263–335 K: $A = (8 \pm 2)$E-17; $E_a = -17$ kJ [552].
330) NO$_3$ reaction rate constant $f(T)$ 263–335 K: $A = (1 \pm 1)$E-14; $E_a = -13$ kJ [552].
331) NO$_3$ reaction rate constant $f(T)$ 263–335 K: $A = (4.2 \pm 0.28)$E-16; $E_a = -19.0$ kJ [552].
332) NO$_3$ reaction rate constant $f(T)$ 263–335 K: $A = (4 \pm 2)$E-16; $E_a = 19$ kJ [496].
333) NO$_3$ reaction rate constant $f(T)$ 263–335 K: $A = (3 \pm 2)$E-15; $E_a = -14$ kJ [496].
334) NO$_3$ reaction rate constant $f(T)$ 263–335 K: $A = (7.0 \pm 2)$E-18; $E_a = -29.3$ kJ [552].
335) $k_{h\nu}$ is the average of the results of 5 smog chamber experiments using outdoor irradiation: $k_{h\nu}$ (min^{-1}): 0.014, 0.009, 0.023, 0.017 and 0.035 (corrected values) [188]; measured in a 17 000-L outdoor chamber (summer mid-day) [297]. Atkinson et al. [297] quoted a second value of 6E-4 s^{-1} measured by Woodrow et al. 1978 in addition to [188] (average: 3E-4 s^{-1}).
336) k_{OH} is the average of 5.8E-12 (Zarbanick et al. 1988 [166]) and 6.4E-12 (Platz et al. 1998 [469]).

References to the Table

1 Atkinson, R.: Kinetics and Mechanisms of the Gas-Phase Reactions of the Hydroxyl Radical With Organic Compounds Under Atmospheric Conditions. Chem. Rev. 86 (1986) 69–201.

2 Finnlayson-Pitts, B. J.; Pitts, Jr., J. N.: Atmospheric Chemistry. Fundamentals and Techniques. John Wiley and Sons, New York 1986.

3 Becker, K. H.; Biehl, H. M.; Bruckmann, P.; Fink, E. H.; Führ, F.; Klöpffer, W.; Zellner, R.; Zetzsch, C. (Eds.): Methods of the Ecotoxicological Evaluation of Chemicals. Photochemical Degradation in the Gas Phase. Vol. 6: OH Reaction Rate Constants and Tropospheric Lifetimes of Selected Environmental Chemicals. Report 1980–1983. Kernforschungsanlage Jülich GmbH, Projektträgerschaft Umweltchemikalien. July–September – 279.

4 Atkinson, R.; Carter, W. P. L.: Kinetics and Mechanisms of the Gas-Phase Reactions of Ozone with Organic Compounds Under Atmospheric Conditions. Chem. Rev. 84 (1984) 437–470.

5 Warneck, P.: Chemistry of the Natural Atmosphere. International Geophysics Series, Vol. 41, Academic Press, San Diego 1988.

6 Atkinson, R.; Aschmann, S. M.; Pitts, Jr., J. N.: Rate Constants for the Gas-Phase Reactions of the NO_3 Radical with a Series of Organic Compounds at 296 ± 2 K. J. Phys. Chem. 92 (1988) 3454–3457 (with supplementary material).

7 Calvert, J. G.; Pitts, Jr., J. N.: PhotoChemistry. Wiley, New York (1966) sec. print 1967.

8 Atkinson, R.: Kinetics and Mechanisms of the Gas-Phase Reactions of the NO_3 Radical with Organic Compounds. J. Phys. Chem. Ref. Data 20 (1991), 459–507.

9 Atkinson, R.: Kinetics and Mechanism of the Gas-Phase Reactions of the Hydroxyl Radical with Organic Compounds. J. Phys. Chem. Ref. Data. Monograph 1, Am. Inst. Physics, New York 1989.

10 Klöpffer, W.; Haag, F.; Kohl, E.-G.; Frank, R.: Testing of the Abiotic Degradation of Chemicals in the Atmosphere: The Smog Chamber Approach. Ecotox. Environ. Safety 15 (1988) 298–319.

11 Gardner, E. P.; Wijayaratne, R. D.; Calvert, J. G.: Primary Quantum Yields of Photodecomposition of Acetone in Air Under Tropospheric Conditions. J. Phys. Chem. 88 (1984) 5069–5083.

12 Canosa-Mas, C.; Smith, S. J.; Toby, S.; Wayne, R. P.: Temperature Dependences of the Reactions of the Nitrate Radical with some Alkynes and with Ethylene. J. Chem. Soc., Faraday Trans. 2. 84 (1988) 263–272.

13 Perry, R. A.; Atkinson, R.; Pitts, Jr., J. N.: Rate Constants for the Reaction of OH Radicals with CH_2=CHF, CH_2=CHCl, and CH_2=CHBr over the Temperature Range 299–426 K. J. Chem. Phys. 67 (1977) 458–462.

14 Gay, Jr., B. W.; Hanst, P. L.; Bufalini, J. J.; Noonan, R. C.: Atmospheric Oxidation of Chlorinated Ethylenes. Environ. Sci. Technol. 10 (1976) 58–67.

15 Zhang, J.; Hatakeyama, S.; Akimoto, H.: Rate Constants of the Reaction of Ozone with trans-1,2-Dichlorethene and Vinyl Chloride. Int. J. Chem. Kinet. 15 (1983) 655–68.

16 Atkinson, R.; Carter, W. P. L.; Plum, C. N.; Winer, A. M.; Pitts, Jr., J. N.: Kinetics of the Gas Phase Reactions of NO_3 Radicals with a Series of Aromatics at 296 ± 2 K. Int. J. Kinet. 16 (1984) 887–898.

17 Atkinson, R.; Aschmann, S. M.; Winer, A. M.; Carter, W. P. L.: Rate Constants for the Gas-Phase Reactions of NO_3 Radicals with Furan, Thiophen, and Pyrrol at 295 ± 1 K and Atmospheric Pressure. Environ. Sci. Technol. 19 (1985) 87–90.

18 Goodman, M. A.; Aschmann, S. M.; Atkinson, R.; Winer, A. M.: Atmospheric Reactions of a Series of Dimethyl Phosphoroamidates and Dimethyl Phosphorothiamidates. Environ. Sci. Technol. 22 (1988) 578–583.

19 Perry, R. A.; Atkinson, R.; Pitts, Jr., J. N.: Kinetics and Mechanism of the Gas Phase Reaction of OH Radicals with Methoxybenzene and o-Cresol over the Temperature Range 299–435 K. J. Phys. Chem. 81 (1977) 1607–1611.

20 Atkinson, R.; Aschmann, S. M.: Rate Constants for the Gas-Phase Reactions of the OH Radical with a Series of Aromatic Hydrocarbons at 296 ± 2 K. Int. J. Chem. Kinet. 21 (1989) 355–365.

21 Doyle, G. J.; Lloyd, A. C.; Darnall, K. R.; Winer, A. M.; Pitts, Jr., J. N.: Gas Phase Kinetic Study of Relative Rates of Reaction of Selected Aromatic Compounds with Hydroxyl Radicals in an Environmental Chamber. Environ. Sci. Technol. 9 (1975) 237–241.

22 Hansen, D. A.; Atkinson, R.; Pitts, Jr., J. N.: Rate Constants for the Reaction of OH Radicals with a Series of Aromatic Hydrocarbons. J. Phys. Chem. 79 (1975) 1763–1766.

23 Perry, R. A.; Atkinson, R.; Pitts, Jr., J. N.: Kinetics and Mechanism of the Gas Phase Reaction of OH Radicals with Aromatic Hydrocarbons Over the Temperature Range 296–473 K. J. Phys. Chem. 81 (1977) 296–304.

24 Pate, C. T.; Atkinson, R.; Pitts, Jr., J. N.: The Gas Phase Reaction of Ozone with a Series of Aromatic Hydrocarbons. J. Environ. Sci. Health, Part A 11 (1976) 1–10.

25 Ravishankara, A. R.; Wagner, S.; Fischer, S.; Smith. G.; Schiff, R.; Watson, R. T.; Tesi, G.; Davis, D. D.: A Kinetic Study of the Reactions of OH with Several Aromatic and Olefinic Compounds. Int. J. Chem. Kinet. 10 (1978) 783–804.

26 Nicovich, J. M.; Thompson, R. L.; Ravishankara, A. R.: Kinetics of the Reactions of the Hydroxyl Radical with Xylenes. J. Phys. Chem. 85 (1981) 2913–2916.

27 Wine, P. H.; Thompson, R. J.: Kinetics of OH Reactions with Furan, Thiophene and Tetrahydrothiophene. Int. J. Chem. Kinet. 16 (1984) 867–878.

28 Wallington, T. J.: Kinetics of the Gas Phase Reaction of OH Radicals with Pyrrole and Thiophene. Int. J. Chem. Kinet.18 (1986) 487–496.

29 McLeod, H.; Jourdain, J. L.; Poulet, G.; Le Bras, G.: Kinetic Study of Reactions of some Organic Sulfur Compounds with OH Radicals. Atmos. Environ. 18 (1984) 2621–26.

30 Atkinson, R.; Aschmann, S. M.; Winer, A. M.; Carter, W. P. L.: Rate Constants for the Gas Phase Reactions of OH Radicals and O_3 with Pyrrole at 295 ± 1 K and Atmospheric Pressure. Atmos. Environ. 18 (1984) 2105.

31 Atkinson, G.; Aschmann, S. M.: Kinetics of the Reaction of Naphthalene, 2-Methylnaphthalene, and 2,3-Dimethyl-naphthalene with OH Radicals and with O_3 at 295 ± 1 K. Int. J. Chem. Kinet. 18 (1986) 569–573.

32 Atkinson, R.; Baulch, D. L.; Cox, R. A.; Hampson, Jr., R. F.; Kerr, J. A.; Troe, J.: Evaluated Kinetic and Photochemical Data for Atmospheric Chemistry. Supplement III IUPAC Subcommittee on Gas Kinetic Data Evaluation for Atmospheric Chemistry. J. Phys. Chem. Ref. Data 18 (1989) 881–1097.

33 Baulch, D. L.; Cox, R. A.; Hampson, Jr., R. F.; Kerr, J. A.; Troe, J.; Watson, R. T.: Evaluated Kinetic and Photochemical Data for Atmospheric Chemistry: Supplement II. CODATA Task Group on Gas Phase Chemical Kinetics. J. Phys. Ref. Data 13 (1984) 1259–1380.

34 Plum, C. N.; Sanhueza, E.; Atkinson, R.; Carter, W. P. L.; Pitts, Jr., N. J.: Radical Rate Constants and Photolysis Rates of alpha – Dicarbonyls. Environ. Sci. Technol. 17 (1983) 479–484.

35 Ohta, T.; Ohyama, T.: A Set of Rate Constants for the Reactions of Hydroxyl Radicals with Aromatic Hydrocarbons. Bull. Chem. Soc. Jpn. 58 (1985) 3029–3030.

36 Atkinson, R.; Aschmann, S. M.; Carter, W. P. L.; Pitts, Jr., J. N.: Rate Constant for the Gas-phase Reactions of O_3 with Selected Organics at 296 K. Int. J. Chem. Kinet. 14 (1982) 13–18.

37 Chiorboli, C.; Bignozzi, C. A.; Maldotti, A.; Giardini, P. F.; Rossi, A.; Carassiti, V.: Rate Constants for the Gas-Phase Reactions of OH Radicals with ββ-Dimethylstyrene and Acetone. Int. J. Chem. Kinet. 15 (1983) 579–586.

38 Barnes, I.; Bastian, V.; Becker, K. H.; Fink, E. H.; Nelsen, W.: Oxidation of Sulphur Compounds in the Atmosphere: I. Rate Constants of OH Radical Reactions with Sulphur Dioxide, Hydrogen Sulphide, Aliphatic Thiols and Thiophenol. J. Atmos. Chem. 4 (1986) 445–466.

39 Atkinson, R.; Aschmann, S. M.; Winer, A. M.; Pitts, Jr., J. N.: Atmospheric Gas Phase Loss Processes for Chlorobenzene, Benzotrifluoride, and 4-Chlorobenzotrifluoride, and Generalization of Predictive Techniqes for Atmorpheric Lifetimes or Aromatic Compounds. Arch. Environ. Contamin.Tox. 14 (1985) 417–425.

40 Wahner, A.; Zetzsch, C.: Rate Constants for the Addition of OH to Aromatics (Benzene, p-Chloroaniline, and o-, m-, and p-Dichlorobenzene) and the Unimolecular Decay of the Adduct. Kinetics into a Quasi-Equilibrium. J. Phys. Chem. 87 (1983) 4945–4951.

41 Becker, K. H.; Bastian, V.; Klein, Th.: The Reaction of OH Radicals with Toluene Diisocyanate, Toluenediamine and Methylenedianiline under Simulated Atmospheric Conditions. J. Photochem. Photobiol., A: Chemistry 45 (1988) 195–205.

42 Atkinson, R.; Aschmann, S. M.: Rate Constants for the Gas-Phase Reaction of Hydroxyl Radicals with Biphenyl and the Monochlorobiphenyls at 295 ± 1 K. Environ. Sci. Technol. 19, (1985) 462–464.

43 Atkinson, R.; Aschmann, S. M.: Kinetics of the Reactions of Acenaphthene and Acenaphthylene and Structurally-Related Aromatic Compounds with OH and NO_3 Radicals, N_2O_5 and O_3 at 296 ± 2 K. Int. J. Chem. Kinet. 20 (1988) 513–539.

44 Klöpffer, W.; Frank, R.; Kohl, E.-G; Haag, F.: Quantitative Erfassung der photochemischen Transformationsprozesse in der Troposphäre. Chem.-Ztg 110 (1986) 57–61.

45 Atkinson, R.; Aschmann, S. M.: Kinetics of the Gas-Phase Reactions of Alkylnaphthalenes with O_3, N_2O_5 and OH Radicals at 298 ± 2 K. Atmos. Environ. 21 (1987) 2323–2326.

46 Biermann, H. W.; Mac Leod, H.; Atkinson, R.; Winer, A. M.; Pitts, Jr., J. N.: Kinetics of the Gas-Phase Reactions of the Hydroxyl Radical with Naphthalene, Phenanthrene, and Anthracene.Environ. Sci. Technol. 19 (1985) 244–248.

47 Atkinson, R.; Pitts, Jr., J. N.: Absolute Rate Constants for the Reactions of Oxygen (^3P) Atoms and Hydroxyl Radicals with Silane over the Temperature Range of 297–438 K. Int J. Chem. Kinet. 10 (1978) 1151–1160.

48 Wallington, T. J.; Neumann, D. M.; Kurylo, M. J.: Kinetics of the Gas Phase Reaction of Hydroxyl Radicals with Ethane, Benzene, and a Series of Halogenated Benzenes over the Temperature Range 234–438 K. Int. J. Chem. Kinet. 19 (1987) 725–739.

49 Tuazon, E. C.; Atkinson, R.; Aschmann, S. M.; Arey, J.; Winer, A. M.; Pitts, Jr., J. N.: Atmospheric Loss Prozesses of 1,2-Dibromo-3-Chloro-propane and Trimethyl Phosphate. Environ. Sci. Technol. 20 (1986) 1043–1046.

50 Atkinson, R.; Aschmann, S. A.; Goodman, M. A.; Winer, A. M.: Kinetics of the Gas-Phase Reactions of the OH Radical with $(C_2H_5)_3PO$ and $(CH_3O)_2P(S)Cl$ at 296 ± 2 K. Int. J. Chem. Kinet. 20 (1988) 273–281.

51 Goodman, M. A.; Aschman, S. M.; Atkinson, R.; Winer, A. M.: Kinetics of the Atmospherically Important Gas-Phase Reactions of a Series of Trimethyl Phosphorothioates. Arch. Environ. Contam. Toxicol. 17 (1988) 281–288.

52 Atkinson, R.; Perry, R. A.; Pitts, Jr., J. N.: Rate Constants for the Reaction of the OH Radical with CH_3SH and CH_3NH_2 over the Temperature Range 299–426 K. J. Chem. Phys. 66 (1977) 1578–1581.

53 Atkinson, R.; Perry, R. A.; Pitts, Jr., J. N.: Rate Constants for the Reactions of the OH Radical with $(CH_3)_2NH$, $(CH_3)_3N$, and $C_2H_5NH_2$ over the Temperature Range 298–426 K. J. Chem. Phys. 68 (1978) 1850–1853.

54 Harris, G. W.; Pitts, Jr., J. N.: Rates of Reaction of Hydroxyl Radicals with 2-(Dimethylamino)ethanol and 2-Amino-2-methyl-1-propanol in the Gas Phase at 300 ± 2 K. Environ. Sci. Technol. 17 (1983) 50–51.

55 Tuazon, E. C.; Carter, W. P. L.; Atkinson, R.; Winer, A. M.; Pitts, Jr., J. N.: Atmospheric Reactions of N-Nitrosodimethyl-amine and Dimethylnitramine. Environ. Sci. Technol. 18 (1984) 49–54.

56 Harris, G. W.; Athinson, R.; Pitts, Jr., J. N.: Kinetics of the Reactions of Hydroxyl Radical with Hydrazine and Methylhydrazine. J. Phys. Chem. 83 (1979) 2557–2559.

57 Audley, G. J.; Baulch, D. L.; Campbell, I. M.; Waters, D. J.; Watling, G.: Gas-phase Reactions of Hydroxyl Radicals with Alkyl Nitrite Vapours in $H_2O_2 + NO_2 + CO$ Mixtures. J. Chem. Soc., Faraday Trans.1, 78 (1982) 611–617.

58 Kerr, J. A.; Stocker, D. W.: Kinetics of the Reactions of Hydroxyl Radicals with Alkyl Nitrates and with Some Oxygen-Containing Organic Compounds Studied under Atmospheric Conditions J. Atmos. Chem. 4 (1986) 253–262.

59 Atkinson, R.; Aschmann, S. M.; Carter, W. P. L.; Winer, A. M.: Kinetics of the Gas-Phase Reactions of OH Radicals with Alkyl Nitrates at 299 ± 2 K. Int. J. Chem. Kinet. 14 (1982) 919–926.
60 Atkinson, R.; Aschmann, S. M.; Carter, W. P. L.; Winer, A. M.; Pitts, Jr. J. N.: Formation of Alkyl Nitrates from the Reaction of Branched and Cyclic Alkyl Peroxy Radicals with NO. Int. J. Chem. Kinet. 16 (1984) 1085–1101.
61 Horne, D. G.; Norrish, R. G. W.: The Flash Photolysis of Oximes. Proc. R. Soc. London A. 315 (1970) 287–300.
62 Barnes, I.; Bastian, V.; Becker, K. H.; Martin, D.: Fourier Transform IR Studies of the Reactions of Dimethyl Sulfoxide with OH, NO_3, and Cl Radicals. Am. Chem. Soc. 393 (1989) 476–488.
63 Hynes, A. J.; Wine, P. H.; Semmes, D. H.: Kinetics and Mechanism of OH Reactions with Organic Sulfides. J. Phys. Chem. 90 (1986) 4148–4156.
64 Nielsen, O. J.: Rate Constants for the Gas-Phase Reactions of OH Radicals and Cl Atoms with $CH_3CH_2NO_2$, $CH_3CH_2CH_2NO_2$, $CH_3CH_2CH_2CH_2NO_2$, and $CH_3CH_2CH_2CH_2CH_2NO_2$. Chem. Phys. Lett. 156 (1989) 312–318.
65 Niki, H.; Maker, P. D.; Savage, C. M.; Hurley, M. D.: Fourier Transform Infrared Study of the Kinetics and Mechanisms for the Cl-Atom- and HO-Radical-Initiated Oxidation of Glycolaldehyd. J. Phys. Chem. 91 (1987) 2174–2178.
66 Atkinson, R.; Aschmann, S. M.; Carter, W. P. L.; Pitts, Jr., J. N.: Rate Constants for the Gas-Phase Reaction of Hydroxyl Radicals with a Series of Ketones at 299 ± 2 K. Int. J. Chem. Kinet. 14 (1982) 839–847.
67 Wallington, T. J.; Kurylo, M. J.: Flash Photolysis Resonance Fluorescence Investigation of the Gas-Phase Reactions of OH Radicals with a Series of Aliphatic Ketones over the Temperature Range 240–440 K. J. Phys. Chem. 91 (1987) 5050–5054.
68 Dagaut, P.; Wallington, T. J.; Liu, R.; Kurylo, M. J.: A Kinetics Investigation of the Gas-Phase Reactions of OH Radicals with Cyclic Ketones and Diones: Mechanic Insights. J. Phys. Chem. 92 (1988) 4375–4377.
69 Hatakeyama, S.; Honda, S.; Washida, N.; Akimoto, H.: Rate Constants and Mechanism for Reactions of Ketenes with OH Radicals in Air at 299 ± 2 K. Bull Chem. Soc. Jpn. 58 (1985) 2157–2162.
70 Tuazon, E. C.; Atkinson, R.; Carter, W. P. L.: Atmospheric Chemistry of cis- and trans-3-Hexene-2,5-dione. Environ. Sci. Technol. 19 (1985) 265–269.
71 Wallington, T. J.; Kurylo, M. J.: The Gas Phase Reactions of Hydroxyl Radicals with a Series of Aliphatic Alcohols Over the Temperature Range 240–440 K. Int. J. Chem. Kinet. 19 (1987) 1015–1023.
72 Wallington., T. J.; Dagaut, P.; Liu, R.: Kurylo, M. J.: Rate Constant for the Gas Phase Reactions of OH with C_5 trough C_7 Aliphatic Alcohols and Ethers: Predicted and Experimental Values. Int. J. Chem. Kinet. 20 (1988) 541–547.
73 Bennet, P. J.; Kerr, A. J.: Kinetics of the Reactions of Hydroxyl Radicals with Aliphatic Ethers Studied under Simulated Atmospheric Conditions. J. Atmos. Chem. 8 (1989) 87–94.
74 Wallington, T. J.; Liu, R.; Dagaut, P.; Kurylo, M. J.: The Gas-Phase Reactions of Hydroxyl Radicals with a Series of Aliphatic Ethers over the Temperature Range 240–440 K. Int. J. Chem. Kinet. 20 (1988) 41–49.
75 Mellouki, A.; Teton, S.; Le Bras, G.: Kinetics of OH Radical Reactions with a Series of Ethers. Int. J. Chem. Kinet. 27 (1995) 791–805.
76 Atkinson, R.; Aschmann, S. M.; Tuazon, E. C.; Arey, J.; Zielinska, B.: Formation of 3-Methylfuran from the Gas-Phase Reaction of OH Radicals with Isoprene and the Rate Constant for Its Reaction with the OH Radical. Int. J. Chem. Kinet. 21 (1989) 593–604.
77 Wallington, T. J.; Dagaut, P.; Liu, R.; Kurylo, M. J.: The Gas Phase Reactions of Hydroxyl Radicals with a Series of Esters over the Temperature Range 240–440 K. Int. J. Chem. Kinet. 20 (1988) 177–186.
78 Atkinson, R.; Darnall, K. R.; Lloyd, A. C.; Winer, A. M.; Pitts, Jr., J. N.: Kinetics and Mechanisms of the Reaction of the Hydroxyl Radical with Organic Compounds in the Gas Phase. Advances in PhotoChemistry Vol. 11, Wiley, New York (1979) 375–488.

79 Anastasi, C.; Smith, I. W. M.; Parkes, D. A.: Flash Photolysis Study of the Spectra of CH_3O_2 and $C(CH_3)O_2$ Radicals and the Kinetics of their Mutual Reactions with NO. J. Chem. Soc. Faraday Trans. 1, 74 (1978) 1693–1701.

80 Boodaghians, R. B.; Hall, I. W.; Toby, F. S.; Wayne, R. P.: Absolute Determinations of the Kinetics and Temperature Dependences of the Reactions of OH with a Series of Alkynes. J. Chem. Soc., Faraday Trans. 2, 83 (1987) 2073–2080.

81 Neeb, B.; Kolloff, A.; Koch, S.; Moortgat, G. K.: Rate Constants for Reactions of Methylvinyl Ketone, Methacrolein, Methacrylic Acid, and Acrylic Acid with Ozone. Int. J. Chem. Kinet. 30 (1998) 769–776.

82 Tuazon, E. C.; Atkinson, R.; Aschmann, S. M.; Goodman, M. A.; Winer, A. M.: Atmospheric Reactions of Chloroethenes with the OH Radical. Int. J. Chem. Kinet. 20 (1988) 241–265.

83 Tuazon, E. C.; Atkinson, R.; Winer, A. M.; Pitts, Jr., J. N.: A Study of the Atmospheric Reactions of 1,3-Dichloropropene and Other Selected Organochlorine Compounds. Arch. Environ. Toxicol. 13 (1984) 691–700.

84 Ohta, T.: Rate Constants for the Reactions of Diolefins with OH Radicals in the Gas Phase. J. Phys. Chem. 87 (1983) 1209–1213.

85 Morris, Jr., E. D.; Niki, H.: Reactivity of Hydroxyl Radicals with Olefins. J. Phys. Chem. 75 (1971) 3640–3641.

86 Ohta, T.: Rate Constants for the Reactions of OH Radicals with Alkyl Substituted Olefins. Int. J. Chem. Kinet. 16 (1984) 879–886.

87 Wu, C. H.; Japar, M.; Niki, H.: Relative Reactivities of HO-Hydrogencarbon Reactions from Smog Reactor Studies. J. Environ. Sci. Health – Environ. Sci. Eng. A11 (2), (1976) 191–200.

88 Atkinson, R.; Aschmann, S. A.; Carter, W. P. L.: Kinetics of the Reactions of O_3 and OH Radicals with a Series of Dialkenes and Trialkenes at 294 ± 2 K. Int. J. Chem. Kinet. 16 (1984) 967–975.

89 Atkinson, R.; Aschmann, S. M.; Pitts, Jr., J. N.: Rate Constants for the Gas-Phase Reactions of the OH Radical with a Series of Monoterpenes at 294 ± 1 K. Int. J. Chem. Kinet. 18 (1990) 287–299.

90 Atkinson, R.; Aschmann, S. M.; Carter, W. P. L.: Effects of Ring Strain on Gas-Phase Rate Constants. 2. Hydroxyl Radical Reactions with Cycloalkenes. Int. J. Chem. Kinet. 15 (1983) 1161–1177.

91 Darnall, K. R.; Winer, A. M.; Lloyd, A. C.; Pitts, Jr., J. N.: Relative Rate Constants for the Reaction of Hydroxyl Radicals with Selected C_6 and C_7 Alkanes and Alkenes at 305 ± K. Chem. Phys. Lett. 44 (1976) 415–418.

92 Garraway, J.; Donovan, R. J.: Gas-Phase Reaction of Hydroxyl Radical with Alkyl Iodides. J. Chem. Soc., Chem. Commun. 23 (1979) 1108.

93 Nip, W. S.; Singleton, D. L.; Overend, R.; Paraskevopoulos, G.: Rates of Hydroxyl Radical Reactions. 5. Reactions with Fluoromethane, Difluoromethane, Trifluoromethane, Fluoroethane, and 1,1-Difluoroethane at 297 K. J. Phys. Chem. 83 (1979) 2440–2443.

94 Martin, J. P.; Paraskevopoulos, G.: A Kinetic Study of the Reactions of Hydroxyl Radicals with Fluoroethanes. Estimates of Carbon–Hydrogen Bond Strengths in Fluoroalkanes. Can. J. Chem. 61 (1983) 861–865.

95 Howard, C. J.; Evenson, K. M.: Rate Constants for the Reactions of OH with Ethane and some Halogene Substituted Ethanes at 296 K. J. Chem. Phys. 64 (1976) 4303–4306.

96 Atkinson, R.; Carter, W. P.; Aschmann, S. M.; Winer, A. M.; Pitts, Jr., J. N.: Kinetics of the Reaction of OH Radicals with a Series of Branched Alkanes at 297 ± 2 K. Int. J. Chem. Kinet. 16 (1984) 469–481.

97 Nolting, F.; Behnke, W.; Zetzsch, C.: A Smog Chamber for Studies of the Reaction of Terpenes and Alkanes with Ozone and OH. J. Atmos. Chem. 6 (1988) 47–59.

98 Atkinson, R.; Aschmann, S. M.: Rate Constant for the Reaction of OH Radicals with Isopropyl-cylopropane at 298 ± 2 K: Effect of Ring Strain on Substituted Cycloalkanes. Int. J. Chem. Kinet. 20 (1988) 339–342.

99 Gorse, R. A.; Volmann, D. H.: PhotoChemistry of the Gaseous Hydrogen Peroxide–Carbon Monoxide System. II: Rate Constants for Hydroxyl Radical Reactions with Hydrocarbons and for Hydrogen Atom Reactions with Hydrogen Peroxide. J. Photochem. 3 (1974) 115–122.

100 Atkinson, R; Aschmann, S. M.; Carter, W. P. L.: Rate Constants for the Gas-Phase Reactions of Hydroxyl Radicals with a Series of Bi- and Tricycloalkanes at 299 ± 2 K: Effects of Ring Strain. Int. J. Chem. Kinet. 15 (1983) 37–50.

101 Schubert, C. C.; Pease, S. J.; Pease, N.: The Oxidation of Lower Paraffin Hydrocarbons. I. Room Temperature Reaction of Methane, Propane, n-Butane and Isobutane with Ozonized Oxygen. J. Am. Chem. Soc. 78 (1956) 2044–2048.

102 Schubert, C. C.; Pease, S. J.; Pease, N.: The Oxidation of Lower Paraffin Hydrocarbons. II. Observations on the Role of Ozone in the Slow Combustion of Isobutane. J. Am. Chem. Soc. 78 (1956) 5553–5556.

103 Dillemuth, F. J.; Drescher Lalancette, B.; Skidmore, D. R.: Reaction of Ozone with 1,1-Difluoroethane and 1,1,1-Trifluoroethane. J. Phys. Chem. 80 (1976) 571–575.

104 Cox, R. A.; Penkett, S. A.: Aerosol Formation from Sulphur Dioxide in the Presence of Ozone and Olefinic Hydrocarbons. J. Chem. Soc. Faraday Trans. 1, 68 (1972) 1735–1753.

105 Japar, S. M.; Wu, C. H.; Niki, H.: Rate Constants of the Reaction of Ozone with Olefins in the Gas Phase. J. Phys. Chem. 78 (1974) 2318–2320.

106 Cadle, R. D.; Schadt, C.: Kinetics of the Gas Phase Reactions of Olefins with Ozone. J. Am. Chem. Soc. 74 (1952) 6002–6004.

107 Grimsrud, E. P.; Westberg, H. H.; Rasmussen, R. A.: Atmospheric Reactivity of Monoterpene Hydrocarbons, NO_x Photooxidation and Ozonolysis. Int. J. Chem. Kinet. 7 (1975) 183–195.

108 Atkinson, R.; Aschmann, S. M.; Carter, W. P. L.; Pitts, Jr., J. N.: Effects of Ring Strain on Gas-Phase Rate Constants. I. Ozone Reactions with Cycloalkenes. Int. J. Chem. Kinet. 15 (1983) 721–731.

109 Ripperton, L. A.; Jeffries, H. E.: Formation of Aerosols by Reaction of Ozone with Selected Hydrocarbons. Adv. Chem. Ser. 113 (1972) 219–131.

110 Japar, S. M.; Wu, C. H.; Niki, H.: Rate Constants for the Gas Phase Reaction of Ozone with α-Pinene and Terpinolene. Environ. Lett. 7 (1974) 245–249.

111 Atkinson, R.; Winer, A. M.; Pitts, Jr., J. N.: Rate Constants for Gas Phase Reactions of O_3 with the Natural Hydro-carbons Isoprene and α- and β-Pinene. Atmos. Environ. 16 (1982) 1017–1020.

112 Adeniji, S. A.; Kerr, J. A.; Williams, M. R.: Rate Constants for Ozone – Alkene Reactions under Atmospheric Conditions. Int J. Chem. Kinet. 13 (1981) 209–217.

113 Hull, L. A.; Hisatsune, I. C.; Heicklen, J.: The Reaction of O_3 with CCl_2CH_2. Can. J. Chem. 51 (1973) 1504.

114 Heicklein, J.: The Reaction of Ozone with Perfluoroolefins. J. Chem. Phys. 70 (1966) 477–480.

115 Atkinson R.; Aschmann, S. M.; Carter, W. P. L.; Pitts, Jr., J. N.: Rate Constants for the Gas-Phase Reactions of O_3 with Selected Organics at 296 K. Int. J. Chem. Kinet. 14 (1982) 13–18.

116 Atkinson, R.; Aschmann, S. M.; Pitts, Jr., J. N.: Kinetics of the Reactions of Naphthalene and Biphenyl with OH Radicals and with O_3 at 294 ± 1 K. Environ. Sci. Technol. 18 (1984) 110–113.

117 Atkinson, R.; Aschmann, S. M.; Winer, A. M.; Pitts, Jr., J. N.: Rate Constants for the Gas-Phase Reactions of Ozone with a Series of Carbonyls at 296 K. Int. J. Chem. Kinet. 13 (1981) 1133–1142.

118 Atkinson, R.; Aschmann, S. M.; Carter, W. P. L.: Kinetics of the Reactions of Ozone and Hydroxyl Radicals with Furan and Thiophene at 298 ± 2 K. Int. J. Chem. Kinet. 15 (1983) 51–61.

119 Martinez, R. I.; Herron, J. T.: Stopped-Flow Study of the Gas Phase Reactions of Ozone with Organic Sulfides: Thiirane. Chem. Phys. Lett. 72 (1980) 74–76.

120 Kaduk, B. A.; Toby, S.: The Reaction of Ozone with Thiophene in the Gas Phase. Int. J. Chem. Kinet. 9 (1977) 829–840.

121 Daykin, E. P.; Wine, P. H.: A Study of the Reactions of NO_3 Radicals with Organic Sulfides: Reactivity Trend at 298 K. Int. J. Chem. Kinet. 22 (1990) 455–482.

122 Barnes, I.; Bastian, V.; Becker, K. H.; Tong, Zhu.: Kinetics and Products of the Reactions of NO_3 with Monoalkenes, Dialkenes, and Monoterpenes.: J. Phys. Chem. 94 (1990) 2413–2419.

123 Ravishankara, A. R.; Mauldin III, R. L.: Absolute Rate Coefficient for the Reaction of NO_3 with trans-2-Butene. J. Phys. Chem. 89 (1985) 3144–3147.

124 Rahmann, M. M.; Becker, E.; Benter, Th.; Schindler, R. N.: A Gasphase Kinetic Investigation of the System F + HNO_3 and the Determination of Absolute Rate Constants for the Reaction of the NO_3 Radical with CH_3SH, 2-Methylpropene, 1,3-Butadiene and 2,3-Dimethyl-2-Butene. Ber. Bunsenges. Phys. Chem. 92 (1988) 91–100.

125 Canosa-Mas, C.; Smith, S. J.; Toby, S.; Wayne, R. P.: Reactivity of the Nitrate Radical towards Alkynes and some other Molecules. J. Chem. Soc. Faraday Trans. 2, 84 (1988) 247–262.

126 Japar, S. M.; Niki, H.: Gas-Phase Reactions of the Nitrate Radical with Olefins. J. Phys. Chem. 79 (1975) 1629–1632.

127 Benter, Th.; Schindler, R. N.: Absolute Rate Coefficients for the Reaction of NO_3 Radicals with Simple Dienes. Chem. Phys. Lett. 145 (1988) 67–70.

128 Atkinson, R.; Aschmann, S. M.; Winer, A. M.; Pitts, J. N., Jr.: Kinetics of the Gas-Phase Reactions of Nitrate Radicals with a Series of Dialkenes, Cycloalkenes, and Monoterpenes at 295 ± 1 K. Environ. Sci. Technol. 18 (1984) 370-375.

129 Dlugocencky, E. J.; Howard, C. J.: Studies of NO_3 Radical Reactions with Some Atmospheric Organic Compounds at Low Pressures. J. Phys. Chem 93 (1989) 1091–1096.

130 Atkinson, R.; Aschmann, S. M.; Winer, A. M.; Pitts, Jr., J. N.: Kinetics and Atmospheric Implications of the Gas-Phase Reactions of NO_3 Radicals with a Series of Monoterpenes and Related Organics at 294 ± 2 K. Environ. Sci. Technol. 19 (1985) 159–163.

131 Atkinson, R.; Aschmann, S. M.; Carter, W. P. L.: Rate Constants for the Reaction of OH with a Series of Alkenes and Dialkenes at 295 ± 1 K. Int. J. Chem. Kinet. 16 (1984) 1175–1786.

132 Toby, F. S.; Toby, S.: Reaction between Ozone and Allene in the Gas Phase. Int. J. Chem. Kinet. 6 (1974) 417–428.

133 Atkinson, R.; Perry, R. A.; Pitts, Jr., J. N.: Absolute Rate Constants for the Reaction of Hydroxyl Radicals with Allene, 1,3-Butadiene, and 3-Methyl-1-butene over the Temperature Range 299–424 K. J. Chem. Phys. 67 (1977) 3170–3174.

134 Liu, A.; Mulac, W. A.; Jonah, C. D.: Rate Constants for the Gas Phase Reactions of OH Radicals with 1,3-Butadiene and Allene at 1 atm in Ar and over the Temperature Range 305–1173 K. J. Phys. Chem. 92 (1988) 131–134.

135 Bradley, J. N.; Hack, W.; Hoyermann, K.; Wagner, H. G.: Kinetics of the Reaction of Hydroxyl Radicals with Ethylene and with C_3 Hydrocarbon. J. Chem. Soc. Faraday. Trans. 1, 69 (1973) 1889–1898.

136 Rogers, J. D.: Rate Constant Measurement for the Reaction of the Hydroxyl Radical with Cyclohexene, Cyclopentene, and Glutaraldehyde. Environ. Sci. Technol. 23 (1989) 177–181.

137 Becker, K. H.; Schurath, U.; Seitz, H.: Ozone-Olefin Reactions in the Gas Phase 1. Rate Constants and Activation Energies. Int. J. Chem. Kinet. 6 (1974) 725–739.

138 Blume, C. W.; Hisatsune, I. C.; Heicklen, J.: Gas-Phase Ozonolysis of cis – and trans – Dichlorethylene. Int. J. Chem. Kinet. 83 (1976) 235–258.

139 Young, C. J.; Hurley, M. D.; Wallington, T. J.; Mabury, S. A.: Atmospheric Lifetime and Global Warming Potential of a Perfuoropolyether. Environ. Sci. Technol. 40 (2006) 2242–2246.

140 Niki, H.; Maker, P. D.; Savage, C. M.; Breitenbach, L. P.: Fourier Transform Infrared Study of the Gas-Phase Reaction of $^{18}O_3$ with trans-CHCl=CHCl in $^{16}O_2$-Rich Mixtures. Branching Ratio for O-Atom Production via Dissociation of the Primary Criegee Intermediate. J. Phys. Chem. 88 (1984) 766–769.

141 Atkinson, R.; Aschmann, S. M.: Rate Constants for the Reactions of O_3 and OH Radicals with a Series of Alkynes. Int. J. Chem. Kinet. 16 (1984) 259–268.

142 De More, W. B.: Rates and Mechanism of Alkyne Ozonation. Int. J. Chem. Kinet.3 (1971) 161–173.

143 Dillemuth, F. J.; Schubert, C. C.; Skidmore, D. R.: The Reaction of Ozone with Acetylenic Hydrocarbons. Combust. Flame 6 (1962) 211–212.

144 Bufalini, J. J.; Altshuller, A. P.: Kinetics of Vapor-Phase Hydrocarbon-Ozone Reactions. Can. J. Chem. 43 (1965) 2243–2250.

145 Nielsen, I. J.; Sehested, J.; Langer, S.; Ljungstrom, E.; Wangberg, I.: UV Absorption Spectra and Kinetics for Alkyl and Alkyl Peroxy Radicals Originating from di-tert-Butyl Ether. Chem. Phys. Lett. 238 (1995) 359–364.

146 Kamens, R. M.; Gery, M. W.; Jeffries, H. E.; Jackson, M.; Cole, E. I.: Ozone-Isoprene Reactions: Product Formation and Aerosol Potential. Int. J. Chem. Kinet. 14 (1982) 955–975.

147 Hastie, D. R.; Freeman, C. G.; Mc Ewan, M. J.; Schiff, H. I.: The Reactions of Ozone with Methyl and Ethyl Nitrites. Int. J. Chem. Kinet. 8 (1976) 307–313.

148 Tuazon, E. C.; Carter, W. P. L.; Atkinson, R.; Pitts, Jr., J. N.: The Gasphase Reaction of Hydrazine and Ozone: A Nonphotolytic Source of OH Radicals for Measurements of Relative OH Radical Rate Constants. Int. J. Chem. Kinet. 15 (1983) 619–629.

149 Baulch, D. L.; Campbell, I. M.; Saunders, S. M.: The Rate Constant for the Reaction of Hydroxyl Radical with Methyl, n-Propyl, and n-Butyl Nitrites. Int J. Chem. Kinet. 17 (1985) 355–366.

150 Campbell, J. M.; Goddman, K.: Rate Constants for the Reactions of Hydroxyl Radicals with Nitromethane and Methyl Nitrite Vapors at 292 K. Chem. Phys. Lett. 36 (1975) 382–384.

151 Harrison, R. M.; Laxen, D. P. H.: Sink Prozesses for Tetraalkyllead Compounds in the Atmosphere. Environ. Sci. Technol. 12 (1978) 1384–1392.

152 Nielsen, O. J.; Nielsen, T.; Pagsberg, P.: Direct Spectrokinetic Investigation of the Reactivity of OH with Tetraalkyllead Compounds in the Gas Phase. Estimate of Lifetimes of Tetraalkyllead Compounds in Ambient Air. Riso Nat. Lab., DK-4000 Roskilde, Denmark. (1984) Riso-R-480.

153 Niki, H.; Maker, P. D.; Savage, C. M.; Breitenbach, L. P.: A Long-Path Fourier Transform Infrared Study of the Kinetics and Mechanisms for the HO-Radical Initiated Oxidation of Dimethylmercury. J. Phys. Chem. 87 (1983) 4978–4981.

154 Witte, F.; Urbanik, E.; Zetsch, C.: Temperature Dependence of the Rate Constants for the Addition of OH to Benzene and to Some Monosubstituted Aromatics (Aniline, Bromobenzene, and Nitrobenzene) and the Unimolecular Decay of the Adducts. Kinetics into a Quasi-Equilibrium. 2. J. Phys. Chem. 90 (1986) 3251–3259.

155 Rinke, M.; Zetsch, C.: Rate Constants for the Reaction of OH Radicals with Aromatics: Benzene, Phenol, Aniline, and 1,2,4-Trichlorobenzene. Ber. Bunsenges. Phys. Chem. 88 (1984) 55–62.

156 Atkinson, R.; Aschmann, S. M.; Arey, J.; Carter, W. P. L.: Formation of Ring-Retaining Products from the OH Radical-Initiated Reactions of Benzene and Toluene. Int. J. Chem. Kinet. 21 (1989) 801–827.

157 Anderson, L. G.; Stephens, R. D.: Kinetics of the Reaction of Hydroxyl Radical with 2-(Dimethylamino)ethanol from 234–364 K. Int. J. Chem. Kinet. 20 (1988) 103–110.

158 Gaffney, J. S.; Fayer, R.; Senum, G. I.; Lee, J. H.: Measurement of the Reactivity of OH with Methyl Nitrate: Implications for Prediction of Alkyl Nitrate-OH Reaction Rates. Int. J. Chem. Kinet. 18 (1986) 399–407.

159 Atkinson, R.; Aschmann, S. M.: Rate Constants for the Reactions of OH Radical with the Propyl and Butyl Nitrates and 1-Nitrobutane at 298 ± 2 K. Int. J. Chem. Kinet. 21 (1989) 1123–1129.

160 Fritz, B.; Lorenz, K.; Steinert, W.; Zellner, R.: Laboratory Kinetic Investigations of the Tropospheric Oxidation of Selected Industrial Emissions. In: Versino, B.; Ott, H. (Eds.): Physico-Chemical Behaviour of Atmospheric Pollutants. Reidel Publ. Comp., Dordrecht (1982) 192–202.

161 Phillips, L. F.: Rate of Reaction of Hydroxyl (OH) with Hydrogen Cyanide between 298 and 563 K. Aust. J. Chem. 32 (1979) 2571–2577.

162 Harris, G. W.; Kleindienst, T. E.; Pitts, Jr., J. N.: Rate Constants for the Reaction of Hydroxyl Radicals with Acetonitrile, Propionitrile and Acrylonitrile in the Temperature Range 298–424 K. Chem. Phys. Lett. 80 (1981) 479–483.

163 Wallington, T. J.; Atkinson, R.; Winer, A. M.: Rate Constants for the Gas Phase Reaction of OH Radicals with Peroxyacetyl Nitrate (PAN) at 273 and 297 K. Geophys. Res. Lett. 9 (1984) 861–864.

164 Edney, E. O.; Kleindienst, T. E.; Corsre, E. W.: Room Temperature Rate Constants for the Reaction of OH with Selected Chlorinated and Oxygenated Hydrocarbons. Int. J. Chem. Kinet. 18 (1986) 1355–1371.

165 Campbell, J. M.; McLaughlin, D. F.; Handy, B. J.: Rate Constants for the Reactions of Hydroxyl Radicals with Alcohol Vapors at 292 K. Chem. Phys. Lett. 38 (1978) 362–364.

166 Zarbanick, S. S.; Fleming, J. W.; Lin, M. C.: Kinetics of Hydroxyl Radical Reactions with Formaldehyde and 1,3,5-Trioxane between 290 and 600 K. Int. J. Chem. Kinet. 20 (1988) 117–129.

167 Perry, R. A.; Atkinson, R.; Pitts, Jr., J. N.: Rate Constants for the Reaction of OH Radicals with Dimethyl Ether and Vinylmethyl Ether over the Temperature Range 299–427 K. J. Chem. Phys. 67 (1977) 611–614.

168 Campbell, J. M.; Parkinson, P. E.: Rate Constants for Reactions of Hydroxyl Radicals with Ester Vapours at 292 K. Chem. Phys. Lett. 53 (1978) 385–389.

169 Zetzsch, C.; Becker, K. H.: Das luftchemische Verhalten von flüchtigen Organohalogenverbindungen. VDI Bericht 745 Bd. 1, Düsseldorf (1989) 97–127.

170 Dagaut, P.; Wallington, T. J.; Liu, R.; Kurylo, M. J.: The Gas Phase Reactions of Hydroxyl Radicals with a Series of Carboxylic Acids over the Temperature Range 240–440 K. Int. J. Chem. Kinet. 20 (1988) 331–338.

171 Howard, C. J.; Evenson, K. M.: Rate Constants for Reactions of OH with Methane and Fluorine, and Bromine Substituted Methanes at 296 K. J. Chem. Phys. 64 (1976) 197–202.

172 Perry, R. A.; Atkinson, R.; Pitts, Jr., J. N.: Rate Constants for the Reaction of Hydroxyl Radicals with Dichlorofluoromethane and Chloromethane over the Temperature Range 298–423 K, and with Dichloromethane at 298 K. J. Chem. Phys. 64 (1976) 1618–1620.

173 Watson, R.T; Machado, G.; Conoway, B.; Wagner, S.; Davis, D. D.: A Temperature Dependent Kinetics Study of the Reaction of Hydroxyl Radicals with Chlorofluoromethane, Dichlorofluoromethane, Chlorodifluoromethane, 1,1,1-Trichloroethane, 1-Chloro-1,1-Difluoroethane, and 1,1,2-Trichloro-1,2,2-trifluoroethane. J. Phys. Chem. 81 (1977) 256–262.

174 Chekulaev, V. P.; Schevchuk, I. M.: Reactivity of some Polycyclic Hydrocarbons with a Free Hydroxyl Radical. Esti NSV Tead. Akad. Toim., Keem. 30 (1981) 138–140; Chem. Abstr. 95-41889b.

175 Clyne, M. A. A.; Holt, P. M.: Reaction Kinetics Involving Ground X2 and Excited A2u+ Hydroxyl Radicals. Part 2. Rate Constants for Reactions of Hydroxyl X2 with Halogenmethanes and Halogenethanes. J. Chem. Soc. Faraday Trans. 2, 75 (1979) 582–591.

176 Paraskevopoulos, G; Singleton, D. L.; Irwin, R. S.: Rates of OH Radical Reactions. 8. Reactions with CH_2FCl, CHF_2Cl, $CHFCl_2$, CH_3CF_2Cl, CH_3Cl, and C_2H_5Cl at 297 K. J. Phys. Chem. 85 (1981) 561–564.

177 Jeong, K.; Kaufmann, F.: Kinetics of the Reaction of Hydroxyl Radical with Methane and with Nine Cl- and F-Substituted Methanes. 1. Experimental Results, Comparisons, and Applications. J. Phys. Chem. 86 (1982) 1808–1815.

178 Jeong, K.; Hsu, K.; Jeffries, J. B.; Kaufmann, F.: Kinetics of the Reactions of OH with C_2H_2, CH_3CCl_3, $CH_2ClCHCl_2$, $CH_2ClCClF_2$, and CH_2FCF_3. J. Phys. Chem. 88 (1984) 1222–1226.

179 Jeong, K.; Kaufmann, F.: Rates of the Reactions of 1,1,1-Trichloroethane (Methylchloroform) and 1,1,2-Trichloroethane with OH. Geophys. Res. Lett. 6 (1979) 757–759.

180 Harris, S. J.; Kerr, J. A.: Relative Rate Measurements of Some Reactions of Hydroxyl Radicals with Alkanes Studied under Atmospheric Conditions. Int. J. Chem. Kinet. 20 (1988) 939–955.

181 Jolly, G. S.; Paraskevopoulos, G.; Singleton, D. L.: Rates of OH Radical Reactions. XII. The Reactions of OH with c-C_3H_6, c-C_5H_{10}, and c-C_7H_{14}. Correlation of Hydroxyl Rate Constants with Bond Dissociation Energies. Int. J. Chem. Kinet. 17 (1985) 1–10.

182 Wallington, T. J.; Dagaut, P.; Kurylo, M. J.: Correlation between Gas-Phase and Solution-Phase Reactivities of Hydroxyl Radicals toward Saturated Organic Compounds. J. Phys. Chem. 92 (1988) 5024–5028.

183 Bradley, W. R.; Wyatt, S. E.; Wells, J. R.; Henley, M. V.; Graziano, G. M.: The Hydroxyl Radical Reaction Rate Constant and Products of Cyclohexanol. Int. J. Chem. Kinet. 33 (2001) 108–117.

184 Roberts, J. M.; Fajer, R. W.: UV Absorption Cross Sections of Organic Nitrates of Potential Organic Atmospheric Importance and Estimation of Atmospheric Lifetimes. Environ. Sci. Technol. 23 (1989) 945–951.

185 Luke, W. T.; Dickerson, R. R.: Direct Measurements of the Photolysis Rate Coefficient of Ethyl Nitrate. Geophys. Lett.15 (1988) 1181–1184.

186 Luke, W. T.; Dickerson, R. R.; Nunnermacker, L. J.: Direct Measurements of Photolysis Rate Coefficients and Henry's Law Constants of Several Alkyl Nitrates. J. Geophys. Res. Atmos. 94 (D12) (1989) 14,905–14,921.

187 Becker, K. H.; Wirtz, K.: Gas Phase Reactions of Alkyl Nitrates with Hydroxyl Radicals under Tropospheric Conditions in Comparison with Photolysis. J. Atmos. Chem. 9 (1989) 419–433.

188 Mongar, K.; Miller, G. C.: Vapor Phase Photolysis of Trifluralin in an Outdoor Chamber. Chemosphere 17 (1988) 2183–2188.

189 Carlier, P.; Hannachi, H.; Mouvier, G.: The Chemistry of Carbonyl Compounds in the Atmosphere – A Review. Atmos. Environ. 20 (1986) 2079–2099.

190 Cabañas, B.; Baeza, M. T.; Martín, P.; Salgado, S.; Villanueva, F.; Monedero, E.; Díaz De Mera, Y.: Reaction of the NO_3 Radical with some Thiophenes: Kinetic Study and Correlation between Rate Constant and E_{HOMO}. Int. J. Chem. Kinet. 38 (2006) 570–576.

191 Back, R. A.; Cao, J.-R.: The PhotoChemistry of 1,2-Cyclobutanedione in the Gas Phase. J. Photochem. 33 (1986) 161–171.

192 Hee, S. S.; Sutherland, R. G.: Vapor and Liquid Phase Photolysis of the n-Butyl Ester of 2,4-Dichlorophenoxyacetic Acid. Arch. Environ. Contam. Toxicol. 8 (1979) 247–254.

193 Bunce, N. J.; Landers, J. P.; Langshaw, J. A.; Nakal, J. S.: An Assesment of the Importance of Direct Solar Degradation of Some Simple Chlorinated Benzenes and Diphenyls in the Vapor Phase. Environ. Sci. Technol. 23 (1989) 213–218.

194 Lorenz, K.; Zeller, R.: Kinetics of Reactions of OH-Radicals with Benzene, Benzene-d6 and Naphthalene. Ber. Bunsenges. Phys. Chem. 87 (1983) 629–636.

195 Atkinson, R.; Tuazon, E. C.; Wallington, T. J.; Aschmann, S. M.; Arey, J.; Winer, A. M.: Atmospheric Chemistry of Aniline, N,N-Dimethylaniline, Pyridine, 1,3,5-Triazine, and Nitrobenzene. Environ. Sci. Technol. 21 (1987) 64–72.

196 Bignozzi, C. A.; Maldotti, A.; Chiorboli, C.; Bartocci, C.; Carasitti, V.: Kinetics and Mechanisms of Reactions between Aromatic Olefins and Hydroxyl Radicals. Int. J. Chem. Kinet. 13 (1981) 1235–1242.

197 Braslavsky, S.; Heicklein, J.: The Gas-Phase Reactions of Ozone with Formaldehyd. Int. J. Chem. Kinet. 8 (1976) 801–808.

198 Tsalkani, N.; Mellouki, A.; Poulet, G.; Toupance, G.; Le Bras, G.: Rate Constant Measurement for the Reactions of OH and Cl with Peroxyacetyl Nitrate at 298 K. J. Atmos. Chem. 7 (1988) 409–419.

199 Seinfeld, J. H.: Atmospheric Chemistry and Physics of Air Pollution. Wiley, New York 1986.

200 Berndt, T.; Böge, O.: Rate Constants for the Gas-phase Reaction of Hexamethylbenzene with OH Radicals and H Atoms and of 1,3,5-Trimethylbenzene with H Atoms. Int. J. Chem. Kinet. 33 (2001) 124–129.

201 Atkinson, R.; Plum, C. N.; Carter, W. P. L; Winer, A. M.: Kinetics of the Gas-Phase Reactions of NO_3 Radicals with a Series of Alkanes. J. Phys. Chem. 88 (1984) 2361–2364.

202 Atkinson, R.; Aschmann, S.: Rate Constants for the Gas-phase Reactions of the OH Radical with the Cresols and Dimethylphenols at 296 ± 2K. Int. J. Chem. Kinet. 22 (1990) 59–67.

203 Atkinson, R.; Aschmann, S. M.; Long, W. D.; Winer, A. M.: Effects of Ring Strain on Gas-Phase Rate Constants. 3. Nitrogen Trioxide Radical Reactions with Cycloalkenes. Int. J. Chem. Kinet. 17 (1985) 957–966.

204 Atkinson, R.; Aschmann, S. M.; Goodmann, M. A.: Kinetics of the Gas-Phase Reactions of NO_3 Radicals with a Series of Alkynes, Haloalkenes, and α,β-Unsaturated Aldehydes. Int. J. Chem. Kinet. 19 (1987) 299–307.

205 Reissel, A.; Arey, J.; Atkinson, R.: Atmospheric Chemistry of Camphor. Int. J. Kinet. 33 (2001) 56–63.

206 Tuazon, E. C.; Atkinson, R.; Aschmann, S.: Kinetics and Products of the Gas-phase Reactions of the OH Radical and O_3 with Allyl Chloride and Benzyl Chloride at Room Temperature. Int. J. Chem. Kinet. 22 (1990) 981–998.

207 Bethel, H. L.; Atkinson, R.; Arey, J.: Kinetics and Products of the Reactions of Selected Diols with the OH Radical. Int. J. Chem. Kinet. 33 (2001) 310–316.

208 O'rji, L. N.; Stone, D. A.: Relative Rate Constant Measurements for the Gas-phase Reactions of Hydroxyl Radicals with 4-Methyl-2-pentanone, trans-4-Octene, and trans-2-Heptene. Int. J. Chem. Kinet. 24 (1992) 703–710.

209 Papagni, C.; Arey, J.; Atkinson, R.: Rate Constants for the Gas-phase Reactions of OH Radicals with a Series of Unsaturated Alcohols. Int. J. Chem. Kinet. 33 (2001) 142–147.

210 Dagaut, P.; Liu, R.; Wallington, T. J.; Kurylo, M. J.: The Gas Phase Reactivity of Aliphatic Polyethers towards OH Radicals: Measurements and Predictions. Int. J. Chem. Kinet. 21 (1989) 1173–1180.

211 Forester, C. D.; Ham, J. E.; Wells, J. R.: Gas phase Chemistry of Dihydromyrcenol with Ozone and OH Radical: Rate Constants and Products. Int. J. Chem. Kinet. 38 (2006) 451–463.

212 Atkinson, R.; Aschmann, S. M.; Winer, A. M.: Kinetics of the Reactions of NO_3 Radicals with a Series of Aromatic Compounds. Environ. Sci. Technol. 21 (1987) 1123–1126.

213 Atkinson, R.; Aschmann, S.: Kinetics of the Reactions of Naphthalene, 2-Methylnaphthalene, and 2,3-Dimethylnaphthalene with OH Radicals and O_3 at 295 ± 1 K. Int. J. Chem. Kinet. 18 (1986) 569–573.

214 Witter, M.; Berndt, T.; Böge, O.; Stratman, F.; Heintzenberg, J.: Gas-phase Ozonolysis: Rate Coefficients for a Series of Terpenes and Rate Coefficients and OH Yields for 2-Methyl-2-butene and 2,3-Dimethyl-2-butene. Int. J. Chem. Kinet. 34 (2002) 394–403.

215 Akinson, R.; Arey, J.; Tuazon, E. C.; Aschmann, S.: Gas-phase Reactions of 1,4-Benzodioxan, 2,3-Dihydrobenzofuran, and 2,3-Benzofuran with OH Radicals and O_3. Int. J. Chem. Kinet. 24 (1992) 345–358.

216 Klotz, B.; Barnes, I.; Becker, K.-H.: Kinetic Study of the Gas-phase Photolysis and OH Radical Reaction of E,Z- and E,E-2,4-Hexandienal. Int. J. Chem. Kinet. 31 (1999) 689–697.

217 Hellhammer, J.: Untersuchungen von Reaktionen des Radikals Stickstofftrioxid mit Aldehyden und Ketonen. Dissertation, Essen 1988.

218 Atkinson, R.: Gas-Phase Tropospheric Chemistry of Organic Compounds. J. Phys. Chem. Reference Data Monograph No. 2, American Inst. of Physics, New York 1994.

219 Starkey, D. P.; Holbrook, K. A.; Oldershaw, G. A.; Walker, R. W.: Kinetics of the Reactions of Hydroxyl Radicals (OH) and of Chlorine atoms (Cl) with Methylethylether over the Temperature Range 274–345 K. Int. J. Chem. Kinet. 29 (1997) 231–236.

220 Mallard, W. G.; Westley, F.; Herron, J. T.; Hampson, R. F.; Frizzell, D. H.: NIST Chemical Kinetics Database. NIST Standard Reference Database 17, U.S. Department of Commerce, Natl. Inst. of Standards and Technology, Standard Referenced Data, Gaitherburg, MD 20899, 1998.

221 Treacy, J.; El Hag, M.; O'Farrell, D.: Reactions of Ozone with Unsaturated Organic Compounds. Ber. Bunsenges. Phys. Chem. 96 (1992) 422–427.

222 Berges, M. G. M.; Warneck, P.: Product Quantum Yields for the 350 nm Photodecomposition of Pyruvic Acid in Air. Ber. Bunsenges. Phys. Chem. 96 (1992) 413–416.

223 Dobi, S.; Khachatryan, L. A.; Birces, T.: Kinetics of Reactions of Hydroxyl Radicals with a Series of Aliphatic Aldehydes. Ber. Bunsenges. Phys. Chem. 93 (1989) 847–852.

224 Atkinson, R.; Hasegawa, D.; Aschmann, S. M.: Rate Constants for the Gas-phase Reactions of O_3 with a Series of Monoterpenes and Related Compounds at 296 ± 2 K. Int. J. Chem. Kinet. 22 (1990) 871–887.

225 Greene, C. R.; Atkinson, R.: Rate Constants for the Gas-phase Reactions of O3 with a Series of Cycloalkenes and α,β-Unsaturated Ketones at 296 ± 2 K. Int. J. Chem. Kinet. 26 (1994) 37–44.

226 Boyd, A. A.; Canosa-Mas, C. E.; King, A. D.; Wayne, R. P.; Wilson, M. R.: Use of a Stopped-flow Technique to Measure the Rate Constants at Room Temperature for Reactions between the Nitrate Radical and various organic species. J. Chem. Soc. Faraday Trans. 87 (1991) 3.

227 Hynes, A. J.; Wine, P. H.: Kinetics and Mechanism of the Reaction of Hydroxyl Radicals with Acetonitrile under Atmospheric Conditions. J. Phys. Chem. 95 (1991) 1232–1240.

228 Nelson, L.; Shanahan, I.; Sidebottom, H. W.; Treacy, J. J.; Nielson, O. J.: Kinetics and Mechanism for the Oxidation of 1,1,1-Trichloroethane. Int. J. Chem. Kinet. 22 (1990) 577–590.

229 Atkinson, R.; Arey, J.; Aschmann, S. M.: Gas-phase Reactions of Azulene with OH and NO3 Radicals and O3 at 298 ± 2 K. Int. J. Chem. Kinet. 24 (1992) 467–480.

230 Toby, S.; Van den Burgt, L. V.; Toby, F. S.: Kinetics and Chemiluminescence of Ozone-aromatic Reactions in the Gas Phase. J. Phys. Chem. 89 (1985) 1982–1986.

231 Burkholder, J. B.; Wilson, R. R.; Gierczak, T.; Talukdar, R.; McKeen, S. A.; Orlando, J. J.; Vaghjiani, G. L.; Ravishankara, A. R.: Atmospheric Fate of CF_3Br, CF_2Br_2, CF_2ClBr, and CF_2BrCF_2Br. J. Geophys. Res. 96 No. D3 (1991) 5025–5043.

232 Orkin, V. L.; Khamaganov, V. G.: Determination of Rate Constants for Reactions of Some Hydrohaloalkanes with OH Radicals and their Atmospheric Lifetimes. J. Atmos. Chem. 16 (1993) 157–167.

233 Clifford, G. M.; Wenger, J. C.: Rate Coefficients for the Gas-Phase Reaction of Hydroxyl Radicals with the Dimethylbenzaldehydes. Int. J. Chem. Kinet. (2006) 563–569 online: DOI 10.1002/kin.20189.

234 Zetzsch, C.; Stuhl, F.: Rate Constants for OH with Carbonic Acids. Comm. Eur. Communities Rpt. 7624 (1982) 129.

235 Dóbé, S.; Turányi, T.; Iogansen, A. A.; Bérces, T.: Rate Constants of the Reactions of OH Radicals with Cyclopropane and Cyclobutane. Int. J. Chem. Kinet. 24 (1992) 191–198.

236 Greene, C. R.; Atkinson, R.: Rate Constants for the Gas-phase Reactions of O_3 with a Series of Alkenes at 296 ± 2 K. Int. J. Chem. Kinet. 24 (1992) 803–811.

237 Atkinson, R.: Kinetics of the Gas Phase Reactions of a Series of Organosilicon Compounds with OH and NO_3 Radicals and with O_3 at 297 ± 2K. Environ. Sci. Technol. 25 (1991) 863–866.

238 Bagley, A.; Canosa-Mas, C. E.; Little, M. R.; Parr, A. D.; Smith, S. J.; Waygood, S. J.; Wayne, R. P.: Temperature Dependence of Reactions of the Nitrate Radical with Alkanes. J. Chem. Soc. Faraday Trans. 86 (1990) 2109–2114.

239 McIlroy, A.; Tully, F. P.: Kinetic Study of Hydroxyl Reactions with Perfluoropropene and Perfluorobenzene. J. Phys. Chem. 97 (1993) 610–614.

240 Wallington, T. J.; Potts, A. R.; Andino, J. M.; Siegl, W. O.; Zhang, Z.; Kurylo, M. J.; Huie, R. E.: Kinetics of the Reaction of OH Radicals with t-Amyl Methyl Ether Revisited. Int. J. Chem. Kinet. 25 (1993) 265–272.

241 Langer, S.; Ljungstrom, E.; Wangberg, I.; Wallington, T. J.; Hurley, M. D.; Nielsen, O. J.: Atmospheric Chemistry of di-tert-Butyl Ether: Rates and Products of the Reactions with Chlorine Atoms, Hydroxyl Radicals, and Nitrate Radicals. Int. J. Chem. Kinet. 28 (1996) 299–306.

242 Langer, S.; Ljungström, E.; Wängberg, I.: Rates of Reaction between the Nitrate Radical and some Aliphatic Esters. J. Chem. Soc. Faraday Trans. 89 (1993) 425–431.

243 Arnts, R. R.; Seila, R. L.; Bufalini, J. J.: Determination of Room Temperature OH Rate Constants for Acetylene, Ethylene Dichloride, Ethylene Dibromide, p-Dichlorobemnzene and Carbon Disulfide. J. Air Pollut. Control Ass. (JAPCA) 39 (1989) 453–460.

244 Atkinson, R.; Aschmann, S. M.; Arey, J.: Rate Constants for the Gas-phase Reactions of OH and NO_3 Radicals and O_3 with Sabinene and Camphene at 296 ± 2 K. Atmos. Environ. 24A (1990) 2647–2654.

245 Ahn, M.-Y.; Filley, T. R.; Jafvert, C. T.; Nies, L.; Hua, I.; Bezares-Cruz, J.: Photodegradation of Decabromodiphenyl Ether Adsorbed onto Clay Minerals, Metal Oxides, and Sediment. Environ. Sci. Technol. 40 (2006) 215–220.

246 Ravishankara, A. R.; Solomon, S.; Turnipseed, A. A.: Atmospheric Lifetimes of Long-Lived Halogenated Species. Science 259 (1993) 194–199.

247 Taylor, P. H.; Jiang, Z.; Dellinger, B.: Determination of the Gas-phase Reactivity of Hydroxyl with Chlorinated Methanes at High Temperatures: Effects of Laser/Thermal Photochemistry. Int. J. Chem. Kinet. 25 (1993) 9–23.

248 Hurley, M. D.; Misner, J. A.; Ball, J.; Wallington, T. J.; Ellis, D. A.; Martin, J. W.; Mabury, S. A.; Sulbaek Andersen, M. P.: Atmospheric Chemistry of $CF_3CH_2CH_2OH$: Kinetics, Mechanism and Products of Cl Atom and OH Radical Initiated Oxidation in the Presence and Absence of NO_x. J. Phys. Chem. A 109 (2005) 9816–9826.

249 Shorees, B.; Atkinson, R.; Arey, J.: Kinetics of the Gas-phase Reactions of β-Phellandrene with OH and NO_3 Radicals and O_3 at 297 ± 2 K. Int. J. Chem. Kinet. 23 (1991) 897–906.

250 Tuazon, E. C.; Arey, J.; Atkinson, R.; Aschmann, S. M.: Gas-phase Reactions of 2-Vinylpyridine and Styrene with OH and NO_3 Radicals and O_3. Environ. Sci. Technol. 27 (1993) 1832–1841.

251 Atkinson, R.; Aschmann, S. M.: OH Radical Reaction Rate Constants for Polycyclic Alkanes: Effects of Ring Strain and Consequences for Estimation Methods. Int. J. Chem. Kinet. 24 (1992) 983–989.

252 Brown, A. C.; Canosa-Mas, C. E.; Wayne, R. P.: A Kinetic Study of the Reactions of OH with CH_3I and CF_3I. Atmos. Environ. 24A (1990) 361–367.

253 Nielsen, O. J.; O'Farrell, Treacy, J. J.; Sidebottom, H. W.: Rate Constants for the Gas-phase Reactions of Hydroxyl Radicals with Tetramethyllead and Tetraethyllead. Environ. Sci. Technol. 25 (1991) 1098–1103.

254 Qiu, L. X.; Shi, S. H.; Xing, S. B.; Chen, X. G.: Rate Constants for the Reactions of Hydroxyl with Five Halogen-substituted Ethanes from 292 to 366 K. J. Phys. Chem. 96 (1992) 685–689.

255 Markert, F.; Nielsen, O. J.: Rate Constants for the Reaction of OH Radicals with 1-Chloroalkanes at 295 K. Chem. Phys. Lett. 189 (1992) 171–174.

256 Nelsen, Jr., D. D.; Zahniser, M. S.; Kolb, C. E.: Chemical Kinetics of the Reactions of the Hydroxyl Radical with Several Hydrochlorofluoropropanes. J. Phys. Chem. 96 (1992) 249–253.

257 Abbatt, J. P. D.; Anderson, J. G.: High-pressure Discharge Flow Kinetics and Frontier Orbital Mechanistic Analysis for Hydroxyl + 1,1-Dichloroethene, cis-1,2-Dichloroethene, trans-1,2 Dichloroethene, 1-Chloro-1,2,2 trifluoroethene, and 1,1-Dichloro-2,2-difluoroethene → Products. J. Phys. Chem. 95 (1991) 2382–2390.

258 Brown, A. C.; Canosa-Mas, C. E.; Parr, A. D.; Rothwell, K.; Wayne, R. P.: Tropospheric Lifetimes of Three Compounds for Possible Replacement of CFC and Halons. Nature, 347 (1990) 541–543.

259 Atkinson, R.; Aschmann, S. M.: Atmospheric Chemistry of the Monoterpene Reaction Products Nopinone, Camphenilone, and 4-Acetyl-1-Methylcyclohexene. J. Atmos. Chem. 16 (1993) 337–348.

260 Dagaut, P.; Liu, R.; Wallington, T. J.; Kurylo, M. J.: Kinetic Measurements of the Gas-phase Reactions of Hydroxyl Radicals with Hydroxy Ethers, Hydroxy Ketones, and Keto Ethers. J. Phys. Chem. 93 (1989) 7838–7840.

261 Brown, A. C.; Canosa-Mas, C. E.; Parr, A. D.; Wayne, R. P.: Laboratory Studies of Some Halogenated Ethanes and Ethers: Measurements of Rates of Reaction with OH and of Infrared Absorption Cross-sections. Atmos. Environ. 24A (1990) 2499–2511.

262 Corchnoy, S. B.; Atkinson, R.: Kinetics of the Gas-phase Reactions of Hydroxyl and Nitrogen Oxide (NO_3) Radicals with 2-Carene, 1,8-Cineole, p-Cymene, and Terpinolene. Environ. Sci. Technol. 24 (1990) 1497-1502.

263 Bierbach, A.; Barnes, I.; Becker, K. H.: Rate Coefficients for the Gas-phase Reactions of Hydroxyl Radicals with Furan, 2-Methylfuran and 2,5-Dimethylfuran at 300 ± 2 K. Atmos. Environ. 26A (1992) 813–817.

264 Smith, D. F.; Kleindienst, T. E.; Hudgens, E. E.; McIver, C. D.; Bufalini, J. J.: Kinetics and Mechanism of the Atmospheric Oxidation of Ethyl Tertiary Butyl Ether. Int. J. Chem. Kinet. 24 (1992) 199–215.

265 Nielsen, O. J.; Sidebottom, H. W.; Nelson, L.; Rattigan, O.; Treacy, J. J.; O'Farrell, D. J.: Rate Constants for the Reactions of OH Radicals and Cl atoms with Diethyl Sulfide, Di-n-propyl Sulfide, and Di-n-butyl Sulphide. Int. J. Chem. Kinet. 22 (1990) 603–612.

266 Japar, S. M.; Wallington, T. J.; Andino, J. M.; Ball, J. C.: Atmospheric Chemistry of Gaseous Diethyl Sulphate. Environ. Sci. Technol. 24 (1990) 894–897.

267 Nielsen, O. J.; Sidebottom, H. W.; Donlon, M.; Treacy, J.: An Absolute- and Relative-rate Study of the Gas-phase Reaction of OH Radicals and Cl atoms with n-Alkyl Nitrates. Chem. Phys. Lett. 178 (1991) 163–170.

268 Martinez, R. D.; Buitrago, A. A.; Howell, N. W.; Hearn, C. H.; Joens, J. A.: The Near U.V. Absorption Spectra of Several Aliphatic Aldehydes and Ketones at 300 K. Atmos. Environ. 26A (1992) 785–792.

269 Zhu, T.; Barnes, I.; Becker, K. H.: Relative-rate Study of the Gas-Phase Reaction of Hydroxy Radicals with Difunctional Organic Nitrates at 298 K and Atmospheric Pressure. J. Atmos. Chem. 13 (1991) 301–311.

270 Nielson, O. J.; Sidebottom, H. W.; Donlon, M.; Treacy, J.: Rate Constants for the Gas-phase Reactions of OH Radicals and Cl atoms with n-Alkyl Nitrites at Atmospheric Pressure and 298 K. Int. J. Chem. Kinet. 23 (1991) 1095–1109.

271 Liu, R.; Huie, R. E.; Kurylo, M. J.: The Gas Phase Reactions of Hydroxyl Radicals with a Series of Nitroalkanes over the Temperature Range 240–400 K. Chem. Phys. Lett. 167 (1990) 519–523.

272 Wängberg, I.; Ljungström, E.; Hjorth, J.; Ottobrini, G.: FTIR Studies of Reactions between the Nitrate Radical and Chlorinated Butanes. J. Phys. Chem. 94 (1990) 8036–8040.

273 Aird, R. W. S.; Canosa-Mas, C. E.; Cook, D. J.; Marston, G.; Monks, P. S.; Wayne, R. P.; Ljungström, E.: Kinetics of the Reactions of the Nitrate Radical with a Series of Halogenobutenes. A Study of the Effect of Substituents on the Rate of Addition of NO_3 to Alkenes. J. Chem. Soc. Faraday Trans. 88 (1992) 1093–1099.

274 Benter, Th.; Becker, E.; Wille, U.; Schindler, R. N.; Canosa-Maas, C. E.; Smith, S. J.; Waygood, S. J.; Wayne, R. P.: Nitrate Radical Reactions: Interactions with Alkynes. J. Chem. Soc. Faraday Trans. 87 (1991) 2141–2145.

275 Batha, A.; Simonaitis, R.; Heicklen, J.: Reactions of Ozone with Olefins: Ethylene, Allene, 1,3-Butadiene, and trans-1,3-Pentadiene. Int. J. Chem. Kinet. 16 (1984) 1227–1246.

276 Mellouki, A.; Talukdar, R. K.; Schmoltner, A.-M.; Gierczak, T.; Mills, M. J.; Solomon, S.; Ravishankara, A. R.: Atmospheric Lifetimes and Ozone Depletion Potentials of Methyl Bromide (CH_3Br) and Bibromomethane (CH_2Br_2). Geophys. Res. Lett. 19 (1992) 2059–2062.

277 Jiang, Z.; Taylor, P. H.; Dellinger, B.: Laser Photolysis/Laser-induced Fluorescence Studies of the Reaction of Hydroxyl with 1,1,1,2- and 1,1,2,2-Tetrachloroethane over an Extended Temperature Range. J. Phys. Chem. 97 (1993) 5050–5053.

278 Nielsen, O. J.; Jorgensen, O.; Donlon, M.; Sidebottom, H. W.; O'Farell, D. J.; Treacy, J.: Rate Constants for Gas-phase Reactions of OH Radicals with Nitroethene, 3-Nitropropene and 1-Nitrocyclohexene at 298 K and 1 atm. Chem. Phys. Lett. 168 (1990) 319.

279 Donaghy, T.; Shanahan, I.; Hande, M.; Fitzpatrick, S.: Rate Constants and Atmospheric Lifetimes for the Reactions of OH Radicals and Cl Atoms with Haloalkanes. Int. J. Chem. Kinet. 25 (1993) 273–284.

280 Nelson, Jr., D. D.; Zahniser, M. S.; Kolb, C. E.: OH Reaction Kinetics and Atmospheric Lifetimes of CF_3CFHCF_3 and CF_3CH_2Br. Geophys. Res. Lett. 20 (1993) 197–200.

281 Zhang, Z.; Saini, R. D.; Kurylo, M. J.; Huie, R. E.: Rate Constants for the Reactions of the Hydroxyl Radical with $CHF_2CF_2CF_2CHF_2$ and $CF_3CHFCHFCF_2CF_3$. Chem. Phys. Lett. 200 (1992) 230–234.

282 Williams, D. C.; O'Rji, L. N.; Stone, D. A.: Kinetics of the Reactions of OH Radicals with Selected Acetates and other Esters under Simulated Atmospheric Conditions. Int. J. Chem. Kinet. 25 (1993) 539–548.

283 McLoughlin, P.; Kane, R.; Shanahan, I.: A Relative Rate Study of the Reaction of Chlorine Atoms (Cl) and Hydroxyl Radicals (OH) with a Series of Ethers. Int. J. Chem. Kinet. 25 (1993) 137–149.

284 Ellermann, T.; Nielsen, O. J.; Skow, H.: Absolute Rate Constants for the Reaction of NO_3 Radicals with a Series of Dienes at 295 K. Chem. Phys. Lett. 200 (1992) 224–229.

285 Atkinson, R.; Baulch, D. L.; Cox, R. A.; Hampson, R. F.; Kerr, J. A.; Rossi, M. J.; Troe, J.: Evaluated Kinetic, Photochemical and Heterogeneous Data for Atmospheric Chemistry: Supplement V, IUPAC Subcommittee on Gas Kinetic Data Evaluation for Atmospheric Chemistry. J. Phys. Chem. Ref. Data 26 (1997) 521–1011.

286 Nielsen, O. J.; Sidebottom, H. W.; O'Farrell, D. J.; Donlon, M.; Treacy, J.: Rate Constants for the Gas-phase Reactions of OH Radicals and Cl Atoms with $CH_3CH_2NO_2$, $CH_3CH_2CH_2NO_2$, $CH_3CH_2CH_2CH_2NO_2$, and $CH_3CH_2CH_2CH_2CH_2NO_2$. Chem. Phys. Lett. 156 (1990) 312–318.

287 Kwok, E. S. C.; Atkinson, R.; Arey, J.: Gas-Phase Atmospheric Chemistry of Selected Thiocarbamates. Environ. Sci. Technol. 26 (1992) 1798–1807.

288 Zhang, Z.; Saini, R. D.; Kurylo, M. J.; Huie, R. E.: Rate Constants for Reactions of the Hydroxyl Radical with Several Partially Fluorinated Ethers. J. Phys. Chem. 96 (1992) 9301–9304.

289 Talukdar, R. K.; Herndon, S. C.; Burkholder, J. B.; Roberts, J. M.; Ravishankara, A. R.: Atmospheric Fate of Several Alkyl Nitrates. Part 1. Rate Coefficients of the Reactions of Alkyl Nitrates with Isotopically Labelled Hydroxyl Radicals. J. Chem. Soc. Faraday Trans. 93 (1997) 2787–2796.

290 Atkinson, R.: Estimation of OH Radical Reaction Rate Constants and Atmospheric Lifetimes for Polychlorobiphenyls, Dibenzo-p-dioxins, and Dibenzofurans. Environ. Sci. Technol. 21 (1987) 305–307.

291 Scollard, D. J.; Treacy, J. J.; Sidebottom, H. W.; Balestra-Garcia, C.; Laverdet, G.; LeBras, G.; MacLeod, H.; Teton, S.: Rate Constants for the Reactions of Hydroxyl Radicals and Chlorine Atoms with Halogenated Aldehydes. J. Phys. Chem. 97 (1993) 4683–4688.

292 Aschmann, S. M.; Tuazon, E. C.; Atkinson, R.: Atmospheric Chemistry of Dimethyl Phosphonate, Dimethyl Methylphosphonate, and Dimethyl Ethylphosphonate. J. Phys. Chem. A 109 (2005) 11828–11836.

293 Anderson, P. N.; Hites, R. A.: System to Measure Relative Rate Constants of Semi-volatile Organic Compounds with Hydroxyl Radicals. Environ. Sci. Techol. 30 (1996) 301–306.

294 Anderson, P. N.; Hites, R. A.: OH Radical Reactions: The Major Removal Pathway for Polychlorinated Biphenyls from the Atmos. Environ. Sci. Technol. 30 (1996) 1756–1763.

295 Grosjean, D.; Williams, E. L., II.: Environmental Persistence of Organic Compounds Estimated from Structure-reactivity and Linear Free-energy Relationships. Unsaturated Aliphatics. Atmos. Environ. Part A 26 (1992) 1395–1405.

296 Bierbach, A.; Barnes, I.; Becker, K. H.: Product and Kinetic Study of the OH-initiated Gas-phase Oxidation of 2-Methylfuran and Furanaldehydes at 300 K. Atmos. Environ. 29 (1995) 2651–2660.

297 Brubaker, Jr., W. W.; Hites, R. A.: OH Reaction Kinetics of Polycyclic Aromatic Hydrocarbons and Polychlorinated Dibenzo-p-dioxins and Dibenzofurans. J. Phys. Chem. A 102 (1998) 915–921.

298 Kwok, E. S. C.; Atkinson, R.; Arey, J.: Kinetics of the Gas-phase Reactions of Indan, Indene, Fluorene, and 9,10-Dihydroanthracene with OH Radicals, NO_3 Radicals, and O_3. Int. J. Chem. Kinet. 29 (1997) 299–308.

299 LeCalve, S.; Le Bras, G.; Mellouki, A.: Temperature Dependence for the Rate Coefficients of the Reactions of the OH Radical with a Series of Formates. J. Phys. Chem. A 101 (1997) 5489–5493.

300 Aschmann, S. M.; Long, W. D.; Atkinson, R.: Temperature-dependent Rate Constants for the Gas-Phase Reactions of OH Radicals with 1,3,5-Trimethylbenzene, Triethyl Phosphate, and a Series of Alkylphosphonates. J. Phys. Chem. A 110 (2006) 7393–7400.

301 DeMore, W. B.; Sander, S. P.; Golden, D. M.; Hampson, R. F.; Kurylo, M. J.; Howard, C. J.; Ravishankara, A. R.; Kolb, C. E.; Molina, M. J.: Chemical Kinetics and Photochemical Data for use in Stratospheric Modeling. Evaluation Number 12. JPL Publication 97-4, (1997) 1–266.

302 Koch, R.; Palm, W. U.; Zetzsch, C.: First Rate Constants for Reactions of OH Radicals with Amides. Int. J. Chem. Kinet. 29 (1997) 81–87.

303 Bierbach, A.; Barnes, I.; Becker, K. H.; Wiesen, E.: Atmospheric Chemistry of Unsaturated Carbonyls: Butenedial, 4-Oxo-2-pentenal, 3-Hexene-2,5-dione, Maleic Anhydride, 3H-Furan-2-one, and 5-Methyl-3H-furan-2-one. Environ. Sci. Technol. 28 (1994) 715–729.

304 Markgraf, S. J.; Wells, J. R.: The Hydroxyl Radical Reaction Rate Constants and Atmospheric Reaction Products of Three Siloxanes. Int. J. Chem. Kinet. 29 (1994) 445–451.

305 Teton, S.; Mellouki, A.; LeBras, G.; Sidebottom, H.: Rate Constants for Reactions of OH Radicals with a Series of Asymmetrical Ethers and tert-Butyl Alcohol. Int. J. Chem. Kinet. 28 (1996) 291–297.

306 Le Calve, S.; Hitier, D.; Le Bras, G.; Mellouki, A.: Kinetic Studies of OH Reactions with a Series of Ketones. J. Phys. Chem. A 102 (1998) 4579–4584.

307 Atkinson, R.; Tuazon, E. C.; Arey, J.; Aschmann, S. M.: Atmospheric and Indoor Chemistry of Gas-phase Indole, Quinoline, and Isoquinoline. Atmos. Environ. 29 (1995) 3423–3432.

308 Rudich, Y.; Talukdar, R.; Burkholder, J. B.; Ravishankara, A. R.: Reaction of Methylbutenol with Hydroxyl Radical: Mechanism and Atmospheric Implications. J. Phys. Chem. 99 (1995) 12188–12194.

309 Jourdain, J.-L.; Laveredet, G.; Le Bras, G.; Combourieu, J.: Étude cinétique par RPE et spectrometrie de masse des réactions élémentaires du trichlorure de phosphore avec les atomes H et O et les radicaux OH. J. Chim. Phys. 77 (1980) 809.

310 Chew, A. A.; Atkinson, R.: OH Radical Formation Yields from the Gas-phase Reactions of O_3 with Alkenes and Monoterpenes. J. Geophys. Res. 101 (1996) 2864–2865.

311 Wiesen, E.; Barnes, I.; Becker, K. H.: Study of the OH-Initiated Degradation of the Aromatic Photooxidation Product 3,4-Dihydroxy-3-hexene-2,5-dione. Environ. Sci. Technol. 29 (1995) 1380–1386.

312 Jourdain, J.-L.; Laverdet, G.; Le Bras, G.; Combourieu, J.: Étude cinétique par RPE et spectrometrie de masse des réactions élémentaires du trichlorure de bore avec les atomes H et O et les radicaux OH. J. Chim. Phys. 78 (1981) 253.

313 Atkinson, R.; Arey, J.; Aschmann, S. M.; Corchnoy, S. B.; Shu, Y.: Rate Constants for the Gas-phase Reactions of cis-3-Hexen-1-ol, cis-3-Hexenylacetate, trans-2-Hexenal, and Linalool with OH and NO_3 Radicals and O_3 at 296 ± 2 K and OH Radical Formation yields from O_3 Reactions. Int. J. Chem. Kinet. 27 (1995) 941–955.

314 Kwok, E. S. C.; Atkinson, R.; Arey, J.: Rate Constants for the Gas-Phase Reactions of the OH Radical with Dichlorobiphenyls, 1-Chlorodibenzo-p-dioxin. 1,2-Dimethoxybenzene, and Diphenyl Ether: Estimation of OH Radical Reaction Rate Constants for PCBs, PCDDs, and PCDFs. Environ. Sci. Technol. 29 (1995) 1591–1598.

315 Orkin, V. L.; Huie, R. E.; Kurylo, M. J.: Rate Constants for the Reactions of OH with HFC-245cb ($CH_3CF_2CF_3$) and some Fluoroalkenes (CH_2CHCF_3, CH_2CFCF_3, CF_2CFCF_3, and CF_2CF_2). J. Phys. Chem. A 101 (1997) 9118–9124.

316 Saunders, S. M.; Baulch, D. L.; Cooke, K. M.; Pilling, M. J.; Smurthwaite, P. I.: Kinetics and Mechanisms of the Reactions of OH with some Oxygenated Compounds of Importance in Tropospheric Chemistry. Int. J. Chem. Kinet. 26 (1994) 113–130.

317 Porter, E.; Wenger, J.; Treacy, J.; Sidebottom, H.; Mellouki, A.; Teton, S.; LeBras, G.: Kinetic Studies on the Reactions of Hydroxyl Radicals with Diethers and Hydroxyethers. J. Phys. Chem. A 101 (1997) 5770–5775.

318 Merkgraf, S.; Wells, J. R.: The Hydroxyl Radical Reaction Rate Constants and Atmospheric Reaction Products of Three Siloxanes. Int. J. Chem. Kinet. 29 (1997) 445–451.

319 Chiorboli, C.; Piazza, R.; Tosato, M. L.; Carassiti, V.: Atmospheric Chemistry: Rate Constants of Gas-phase Reactions between Haloalkanes of Environmental Interest and Hydroxyl Radicals. Coord. Chem. Rev. 125 (1993) 241–250.

320 Gal, G.; Bar-Ziv, E.: The Kinetics of the B_2H_6/O (^3P) System at Room Temperature. Int. J. Chem. Kinet. 27 (1995) 235–259.

321 Aschmann, S. M.; Atkinson, R.: Rate Constants for the Gas-phase Reactions of Selected Dibasic Esters with the OH Radical. Int. J. Chem. Kinet. 30 (1998) 471–474.

322 Le Calve, S.; Le Bras, G.; Mellouki, A.: Kinetic Studies of OH Reactions with iso-Propyl, sec-Butyl, and tert-Butyl Acetate. Int. J. Chem. Kinet. 29 (1997) 683–688.

323 Koch, R.; Kruger, H.-U.; Elend, M.; Palm, W.-U.; Zetzsch, C.: Rate Constants for the Gas-phase Reaction of OH with Amines: tert-Butyl Amine, 2,2,2-Trifluoroethyl Amine, and 1,4-Diazabicyclo[2.2.2]octane. Int. J. Chem. Kinet. 28 (1996) 807–815.

[324 Atkinson, R.; Tuazon, E. C.; Kwok, E. C. S.; Arey, J.; Aschmann, S. M.; Bridler, I.: Kinetics and Product of Gas-phase Reactions of $(CH_3)_4Si$, $(CH_3)_3SiCH_2OH$, $(CH_3)_3SiOSi(CH_3)_3$ and $(CD_3)_3SiOSi(CD_3)_3$ with Cl atoms and OH Radicals. J. Chem. Soc. Faraday Trans. 91 (1995) 3033–3039.

325 Kwok, E. C. S.; Aschmann, S. M.; Atkinson, R.: Rate Constants for the Gas-phase Reactions of the OH Radical with Selected Carbamates and Lactates. Environ. Sci. Technol. 30 (1996) 329–334.

326 Donahue, N. M.; Anderson, J. G.; Demerjian, K. L.: New Rate Constants for the OH alkane Reactions from 300 to 400 K: An Assessment of Accuracy. J. Phys. Chem. A 102 (1998) 3121–3126.

327 Stemmler, K.; Kinnison, D. J.; Kerr, J. A.: Room Temperature Rate Coefficients for the Reaction of OH Radicals with some Monoethylene Glycol Monoalkyl Ethers. J. Phys. Chem. 100 (1996) 2114–2116.

328 Martinez, E.; Cabañas, B.; Aranda, A.; Martin, P.; Wayne, R. P.: Kinetic Study of the Reactions of NO_3 with 3-Chloropropene, 3-Bromopropene and 3-Iodopropene using LIF Detection. J. Chem. Soc. Faraday Trans. 92 (1996) 4385–4389.

329 Klotz, B.; Barnes, K. H.; Golding, B. T.: Atmospheric Chemistry of Benzeneoxide/Oxepin. J. Chem. Soc. Faraday Trans. 93 (1997) 1507–1516.

330 a) Stemmler, K.; Mengon, W.; Kerr, J. A.: Hydroxyl-Radical-initiated Oxidation of Isobutyl Isopropyl Ether under Laboratory Conditions Related to the Troposphere. J. Chem. Soc. Faraday Trans. 93 (1997) 2865–287; b) Atkinson, R.; Arey, J.; Zielinska, B.; Aschmann, S. M.: Kinetics and Nitro-products of the Gas-phase OH and NO_3 Radical-initiated Reactions of Naphthalene-d10, fluoranthene-d10, and Pyrene. Int J. Chem. Kinet. 22 (1990) 999–1014.

332 Shu, Y.; Atkinson, R.: Atmospheric Lifetimes and Fates of a Series of Sesquiterpenes. J. Geophys. Res. 100 (1995) 7275–7281.

333 Coeur-Tourneur, C.; Henry, F.; Janquin, M.-A.; Brutier, L.: Gas-Phase Reaction of Hydroxyl Radicals with m-, o- and p-Cresol. Int. J. Chem. Kinet. (2006) 553–562 online: DOI 10.1002/kin.20186.

334 Grosjean, E.; Grosjean, D.: Rate Constants for the Gas-phase Reaction of Ozone with 1,2-Disubstituted Alkenes. Int. J. Chem. Kinet. 28 (1996) 461–466.

335 Kramp, F.; Paulson, S. E.: On the Uncertainties in the Rate Coefficients for OH Reactions with Hydrocarbons, and Rate Coefficients of the 1,3,5-Trimethylbenzene and m-Xylene Reactions with OH Radicals in the Gas Phase. J. Phys. Chem. A: 102 (1998) 2685–2690.

336 Bilde, M.; Mogelberg, T. E.; Sehested, J.; Nielsen, O. J.; Wallington, T. J.; Hurley, M. D.; Japar, S. M., Dill, M.; Orkin, V. L.; Buckley, T. J.; Huie, R. E.; Kurylo, M. J.: Atmospheric Chemistry of Dimethyl Carbonate: Reaction with OH Radicals, UV Spectra of $CH_3OC(O)OCH_2$ and $CH_3OC(O)OCH_2O_2$ Radicals, Reactions of $CH_3OC(O)OCH_2O_2$ with NO and NO_2, and Fate of $CH_3OC(O)OCH_2O$ Radicals. J. Phys. Chem. A: 101 (1997) 3514–3525.

337 Atkinson, R.; Aschmann, S. M.: Alkoxy Radical Isomerization Products from the Gas-phase OH Radical-initiated Reactions of 2,4-Dimethyl-2-penthanol and 3,5-Dimethyl-3-hexanol. Environ. Sci. Technol. 29 (1995) 528–536.

338 Tse, C. W.; Flagan, R. C.; Seinfeld, J. H.: Rate Constants of the Gas-phase Reaction of the Hydroxyl Radical with a Series of Dimethylbenzaldehydes and Trimethylphenols at Atmospheric Pressure. Int. J. Chem. Kinet. 29 (1997) 523–525.

339 Bilde, M.; Wallington, T. J.; Ferronato, C.; Orlando, J. J.; Tyndall, G. S.; Estupinan, E.; Haberkorn, S.: Atmospheric Chemistry of CH_2BrCl, $CHBrCl_2$, $CHBr_2Cl$, $CF_3CHBrCl$, and CBr_2Cl_2. J. Phys. Chem. A: 102 (1998) 1976–1986.
340 Glasius, M.; Calogirou, A.; Jensen, N. R.; Hjorth, J.; Nielsen, C. J.: Kinetic Study of Gas-phase Reactions of Pinonaldehyde and Structurally Related Compounds. Int. J. Chem. Kinet. 29 (1997) 527–531.
341 Baxley, J. S.; Henley, M. V.; Wells, J. R.: The Hydroxyl Radical Reaction Rate Constant and Products of Ethyl 3-ethoxypropionate. Int. J. Chem. Kinet. 29 (1997) 637–644.
342 Volman, D. H.: The Vapor-Phase Photo Decomposition of Hydrogen Peroxide. J. Chem. Phys. 17 (1949) 947–950.
343 Kwok, E. C. S.; Arey, J.; Atkinson, R.: Gas-phase Atmospheric Chemistry of Dibenzo-p-dioxin and Dibenzofuran. Environ. Sci. Technol. 28 (1994) 528–533.
344 Klöpffer, W.; Daniel, B.: Reaktionskonstanten zum abiotischen Abbau von organischen Chemikalien. Umweltbundesamt (Datenbank ABIOTIKx), Berlin (Hrsg.) UBA Texte, Berlin (1991), 51/91.
345 Espinoza-Garcia, J.; Corchado, J. C.; Sana, M.: Theoretical Thermochemistry and Kinetics of some Hydrogen Abstraction Reactions on Nitrogen. J. Chim. Phys. 90 (1993) 1181–1200.
346 El Boudali, A.; Le Calvé, S.; Le Bras, G.; Mellouki, A.: Kinetic Studies of OH Reactions with a Series of Acetates. J. Phys. Chem. 100 (1996) 12364–12368.
347 LeCalve, S.; LeBras, G.; Mellouki, A.: Kinetic Studies of OH Reactions with a Series of Methyl Esters. J. Phys. Chem. A: 101 (1997) 9137–9141.
348 Villenave, E.; Orkin, V. L.; Huie, R. E.; Kurylo, M. J.: Rate Constant for the Reaction of OH Radicals with Dichloromethane. J. Phys. Chem. A: 101 (1997) 8513–8517.
349 Ferrari, C.; Roche, A.; Jacob, V.; Foster, P.; Baussand, P.: Kinetics of the Reaction of OH Radicals with a Series of Esters under Simulated Conditions at 295 K. Int. J. Chem. Kinet. 28 (1996) 609–614.
350 Vaghjiani, G. L.: Discharge Flow-tube Studies of $O(^3P)+N_2H_4$ Reaction: the Rate Coefficient Values over the Temperature Range 252–423 K and the OH(X2II) Product Yield at 298 K. J. Chem. Phys. 104 (1996) 547954–89.
351 Langer, S.; Ljungström, E.: Rates of Reaction between the Nitrate Radical and some Aliphatic Ethers. Int. J. Chem. Kinet. 26 (1994) 367–380.
352 Wallington, T. J.; Atkinson, R.; Winer, A. M.; Pitts, Jr., J. N.: Absolute Rate Constants for the Gas-phase Reactions of the Nitrogen Trioxide Radical with Methanethiol, Dimethyl Sulfide, Dimethyl Disulfide, Hydrogen Sulfide, Sulfur Dioxide, and Dimethyl Ether over the Temperature Range 280–350 K. J. Phys. Chem. 90 (1986) 5393–5396.
353 Aschmann, S.; Atkinson, R.: Rate Constants for the Reaction of the NO_3 Radical with Alkanes at 296 ± 2K. Atmos. Environ. 29 (1995) 2311–2316.
354 Martinez, E.; Cabanas, B.; Aranda, A.; Albaladeja, J.; Wayne, R. P.: Absolute Rate Coefficients for the Reaction of NO_3 with Pent-1-ene and Hex-1-ene at T = 298 to 433 K Determined using LIF Detection. J. Chem. Soc. Faraday Trans. 93 (1997) 2943–2047.
355 Berndt, T.; Böge, O.; Kind, I.; Rolle, W.: Reaction of NO_3 Radicals with 1,3-Cyclohexadiene, α-Terpinene, and α-Phellandrene: Kinetics and Products. Ber. Bunsenges. Phys. Chem. 100 (1996) 462–469.
356 D'Anna, B.; Nielsen, C. J.: Kinetic Study of the Vapour-phase Reaction between Aliphatic Aldehydes and the Nitrate Radical. J. Chem. Soc. Faraday Trans. 93 (1997) 3479–3483.
357 Chew, A. A.; Atkinson, R.; Aschmann, S. M.: Kinetics of the Gas-phase Reactions of NO_3 Radicals with a Series of Alcohols, Glycol Ethers, Ethers and Chloroalkenes. J. Chem. Soc. Faraday Trans. 94 (1998) 1083–1089.
358 Rudich, Y.; Talukdar, R. K.; Fox, R. W.; Ravishankara, A. R.: Rate Coefficients for Reactions of NO_3 with a few Olefins and Oxygenated Olefins. J. Phys. Chem. 100 (1996) 5374–5381.
359 Hallquist, M.; Langer, S.; Ljungstrom, E.; Wangberg, I.: Rates of Reaction between the Nitrate Radical and some Unsaturated Alcohols. Int. J. Chem. Kinet. 28 (1996) 467–474.
360 Kind, I.; Berndt, T.; Boge, O.: Gas-phase Rate Constants for the Reaction of NO_3 Radicals with a Series of Cyclic Alkenes, 2-Ethyl-1-butene and 2,3-Dimethyl-1,3-butadiene. Chem. Phys. Lett. 288 (1998) 111–118.

361 Ljungstrom, E.; Wangberg, I.; Langer, S.: Absolute Rate Coefficients for the Reaction between Nitrate Radicals and some Cyclic Alkenes. J. Chem. Soc. Faraday Trans. 89 (1993) 2977–2982.
362 Noremsaune, I. M. W.; Langer, S.; Ljungstrom, E.; Nielsen, C. J.: Rate Coefficients and Arrhenius Parameters for the Reactions between Chloroethenes and the Nitrate Radical. J. Chem. Soc. Faraday Trans. 93 (1997) 525–531.
363 Marston, G.; Monks, P. S.; Canosa-Mas, C. E.; Wayne, R. P.: Correlations between Rate Parameters and Calculated Molecular Properties in the Reactions of the Nitrate Radical with Alkenes. J. Chem. Soc. Faraday Trans. 89 (1993) 3899–3905.
364 Urbanski, S. P.; Stickel, R. E.; Wine, P. H.: Mechanistic and Kinetic Study of the Gas-phase Reaction of Hydroxyl Radical with Dimethyl Sulfoxide. J. Phys. Chem. A 102 (1998) 10.522–10.529.
365 Kind, I.; Böge, O.; Rolle, W.: Gas-phase Rate Constants for the Reaction of NO_3 with Furan and Methyl-sustituted Furans. Chem. Phys. Lett. 256 (1996) 679–683.
366 Noremsaune, I. M. W.; Hjorth, J.; Nielsen, C. J.: FTIR Studies of Reactions between the Nitrate Radical and Haloethenes. J. Atmos. Chem. 21 (1995) 223–250.
367 Galan, B. C.; Marston, G.; Wayne, R. P.: Arrhenius Parameters for the Reaction of the Nitrate Radical with 1,1-Dichloroethene and 1,2-Dichloroethene. J. Chem. Soc. Faraday Trans. 91 (1995) 1185–1189.
368 Seland, J. G.; Noremsaune, I. M. W.; Nielsen, C. J.: FTIR Studies of the NO_3 initiated Degradation of but-2-yne: Mechanism and Rate Constant Determination. J. Chem. Soc. Faraday Trans. 92 (1996) 3459–3465.
369 Shallcross, D. E.; Biggs, P.; Canosa-Mas, C. E.; Clemishaw, K. C.; Harrison, M. G.; Alanon, M. R. L.; Pyle, J. A.; Vipond, A.; Wayne, R. P.: Rate Constants for the Reaction between OH and CH_3ONO_2 and $C_2H_5ONO_2$ over a Range of Pressure and Temperature. J. Chem. Soc. Faraday Trans. 93 (1997) 2807–2811.
370 Rudich, Y.; Talukdar, R. K.; Fox, R. W.; Ravishankara, A. R.: Rate Coefficients for Reactions of NO_3 with a few Olefins and Oxygenated Olefins. J. Phys. Chem. 100 (1996) 5374–5381.
371 Canosa-Mas, C. E.; Monks, P. S.; Wayne, R. P.: Temperature Dependence of the Reaction of the Nitrate Radical with But-1-ene. J. Chem. Soc. Faraday Trans. 1, 88 (1992) 11–14.
372 Burrows, J. P.; Tyndall, G. S.; Moortgat, G. K.: Absorption Spectrum of NO_3 and Kinetics of the Reaction of NO_3 with NO_2, Cl, and Several Stable Atmospheric Species at 298 K. J. Phys. Chem. 89 (1985) 4848–4856.
373 Sommar, J.; Hallquist, M.; Ljungstrom, E.: Rate of Reaction between the Nitrate Radical and Dimethylmercury in the Gas Phase. Chem. Phys. Lett. 257 (1996) 434–438.
374 Boyd, A. A.; Marston, G.; Wayne, R. P.: Kinetic Studies of the Reaction between NO_3 and OClO at T = 300 K and P = 2–8 Torr. J. Phys. Chem. 100 (1996) 130–137.
375 Tuazon, E. C.; Atkinson, R.; Aschmann, S. M.; Arey, J.: Kinetics and Products of the Gas-phase Reactions of O_3 with Amines and Related Compounds. Res. Chem. Intermed. 20 (1994) 303–320.
376 Grosjean, E.; Grosjean, D.: Rate Constants for the Gas-phase Reaction of Ozone with Unsaturated Oxygenates. Int. J. Chem. Kinet. 30 (1998) 21–29.
377 Shu, Y.; Atkinson, R.: Rate Constants for the Gas-phase Reactions of O_3 with a Series of Terpenes and OH Radical Formation from O_3 Reactions with Sequiterpenes at 296 ± 2 K. Int. J. Chem. Kinet. 26 (1994) 1193–1205.
378 Grosjean, E.; Grosjean, D.: Rate Constants for the Gas-phase Reaction of Ozone with 1,1-Disubstituted Alkenes. Int. J. Chem. Kinet. 28 (1996) 911–918.
379 Grosjean, E.; Grosjean, D.: Rate Constants for the Gas-phase Reaction of C5–C10 Alkenes with Ozone. Int. J. Chem. Kinet. 27 (1995) 1945–1054.
380 Treacy, J.; Curley, W.; Wenger, J.; Sidebottom, H.: Determination of Arrhenius Parameters for the Reactions of Ozone with Cycloalkenes. J. Chem. Soc. Faraday Trans. 93 (1997) 2877–2881.
381 Wangberg, I.; Barnes, I.; Becker, K. H.: Atmospheric Chemistry of Bifunctional Cycloalkyl Nitrates. Chem. Phys. Lett. 261 (1996) 138–144.

382 Nelson, D. D.; Zahniser, M. S.; Kolb, C. E.; Magid, H.: OH Reaction Kinetics and Atmospheric Lifetime Estimates for Several Hydrofluorocarbons. J. Phys. Chem. 99 (1995) 16301–16306.

383 Grosjean, E.; Grosjean, D.: Rate Constants for the Gas-phase Reactions of Ozone with Unsaturated Aliphatic Alcohols. Int. J. Chem. Kinet. 26 (1994) 1185–1191.

384 Grosjean, E.; Grosjean, D.: A Kinetic and Product Study of the Gas-phase Reaction of Ozone with Vinylcyclohexane and Methylene Cyclohexane. Int. J. Chem. Kinet. 29 (1997) 855–860.

385 Grosjean, E.; Grosjean, D.; Seinfeld, J. H.: Gas-phase Reaction of Ozone with trans-2-Hexenal, trans-2-Hexenyl Acetate, Ethylvinyl Ketone, and 6-Methyl-5-hepten-21-one. Int. J. Chem. Kinet. 28 (1996) 373–382.

386 Mellouki, A.M Talukdar, R. K.; Howard, C. J.: Kinetics of the Reaction of HBr with O_3 and HO_2: the Yield of HBr from HO_2 + BrO. J. Geophys. Res. 99 (1994) 2294–2295.

387 Grosjean, D.; Grosjean, E.; Williams, E. L., II: Rate Constants for the Gas-phase Reactions of Ozone with Unsaturated Alcohols, Esters, and Carbonyls. Int. J. Chem. Kinet. 25 (1993) 783–794.

388 Alvarado, A.; Atkinson, R.; Arey, J.: Kinetics of the Gas-phase Reactions of NO_3 Radicals and O_3 with 3-Methylfuran and the OH Radical Yield from the O_3 Reaction. Int. J. Chem. Kinet. 28 (1996) 905–909.

389 Ninomiya, Y.; Kawasaki, M.; Guschin, A.; Molina, L. T.; Molina, M. I.; Wallington, T. J.: Atmospheric Chemistry of n-$C_3F_7OCH_3$: Reaction with OH Radicals and Cl Atoms and Atmospheric Fate of n-$C_3F_7OCH_2O(\cdot)$ Radicals. Environ. Sci. Technol. 34 (2000) 2973–2978.

390 Dagaut, P.; Liu, R.; Wallington, T. J.; Kurylo, M. J.: Flash Photolysis Resonance Fluorescence Investigation of the Gas-Phase Reactions of Hydroxyl Radicals with Cyclic Ethers. J. Phys. Chem. 94 (1990) 1881–1883.

391 Leu, M. T.; Hatakeyama, S.; Hsu, K. J.: Rate Constants for the Reactions between Atmospheric Reservoir Species. 1. Hydrogen Chloride. J. Phys. Chem. 93 (1989) 5778–5784.

392 National Institute of Standards and Technology (NIST) Chemical Kinetics Database. Standard Reference Database 17, Version 7.0 (Web), Release 1.3. A Compilation of Kinetics Data on Gas-phase Reactions(1999). Internet: http://kinetics.nist.gov/index.php

393 National Institute of Standards and Technology (NIST) Chemical Kinetics Database, Standard Reference Database 17, Version 7.0 (Web), Release 1.3. A Compilation of Kinetics Data on Gas-phase Reactions (2000). Internet: http://kinetics.nist.gov/index.php

394 Acerboni, G.; Jensen, N. R.; Rindone, B.; Hjorth, J.: Kinetics and Products Formation of the Gas-phase Reactions of Tetrafluoroethylene with OH and NO_3. Chem. Phys. Lett. 309 (1999) 364–368.

395 Martinez, E.; Cabanas, B.; Aranda, A.; Martin, P.; Salgado, S.: Absolute Rate Coefficients for the Gas-Phase Reactions of NO_3 Radical with a Series of Monoterpenes at T = 298 to 433 K. J. Atmos. Chem. 33 (1999) 265–282.

396 Grosjeans, E.; Grosjeans, D.: The Reaction of Unsaturated Aliphatic Oxygenates with Ozone. J. Atmos. Chem. 32 (1999) 205–232.

397 Aschmann, S. M.; Atkinson, R.: Atmospheric Chemistry of 1-Methyl-2-pyrrolidone. Atmos. Environ. 33 (1999) 591–599.

398 Becker, K. H.; Dinis, C. M. F.; Geiger, H.; Wiesen, P.: The Reactions of OH Radicals with Di-i-Propoxymethane and Di-sec-Butoxymethane: Kinetic Measurements and Structure Activity Relationship. Phys. Chem. Chem. Phys. 1 (1999) 4721–4726.

399 Canosa-Mas, C. E.; Carr, S.; King, M. D.; Shallcross, D. E.; Thompson, K. C.; Wayne, R. P.: A Kinetic Study of the Reactions of NO_3 with Methyl Vinyl Ketone, Methacrolein, Acrolein, Methyl Acrylate, and Methyl Methacrylate. Phys. Chem. Chem. Phys. 1 (1999) 4195–4202.

400 DeMore, W. B.; Bayers, K. D.: Rate Constants for the Reactions of Hydroxyl Radical with Several Alkanes, Cycloalkanes, and Dimethyl Ether. J. Phys. Chem. A 103 (1999) 2649–2654.

401 Zetzsch, C.: Photooxidativer Abbau mittelflüchtiger, partikelgebundener Stoffe. Gesellschaft Deutscher Chemiker (Hrsg.): Stofftransport und Transformation in der Atmosphäre (10. BUA Koll. 25.11.2003), GDCH Monographie Bd. 28, Frankfurt/M (2004) 95–118.

402 Behnke, W.; Nolting, F.; Zetzsch, C.: The Atmospheric Fate of di-(2-Ethylhexyl-) Phthalate, Adsorbed on Various Metal Oxide Model Aerosols and on Coal Fly Ash. J. Aerosol Sci. 18 (1987) 849–852.

403 Behnke, W.; Holländer, W.; Koch, W.; Nolting, F.; Zetzsch, C.: A Smog Chamber for Studies of the Photochemical Degradation of Chemicals in the Presence of Aerosols. Atmos. Environ. 22 (1988) 1113–1120.

404 Palm, W.-U.; Elend, M.; Krueger, H.-U.; Zetzsch, C.: OH Radical Reactivity of Airborne Terbuthylazine Adsorbed on Inert Aerosols. Environ. Sci. Technol. 31 (1997) 3389–3396.

405 Klöpffer, W.; Kohl, E.-G.: Bimolecular OH-Rate Constants of Organic Compounds in Solution. Part 2. Measurements in 1,2,2-Trichlorotrifluoroethane Using Hydrogen Peroxide as OH-Source. Ecotox. Environ. Safety 26 (1993) 346–356.

406 Dilling, W. L.; Gonsior, S. J.; Boggs, G. V.; Mendoza, C. G.: Organic Photochemistry 20. Relative Rate Measurements for Reactions of Organic Compounds with Hydroxyl Radicals in 1,1,2-Trichhlorotrifluoroethane Solution – A New Method for Estimating Gas Phase Rate Constants for Reactions of Hydroxyl Radicals with Organic Compounds Difficult to Study in the Gas Phase. Environ. Sci. Technol. 22 (1988) 1447–1453.

407 Orkin, V. L.; Villenave, E.; Huie, R.; Kurylo, M. J.: Atmospheric Lifetimes and Global Warming Potentials of Hydrofluoroethers: Reactivity toward OH, UV Spectra, and IR Absorption Sections. J. Phys. Chem. A 103 (1999) 9770–9779.

408 Vesine, E.; Bossoutrot, V.; Mellouki, A.; LeBras, G.; Wenger, J.; Sidebottom, H.: Kinetic and Mechanistic Study of OH- and Cl-Initiated Oxidation of Two Unsaturated HFCs: $C_4H_9CH=CH_2$ and $C_6F_{13}CH=CH_2$. J. Phys. Chem. A 104 (2000) 8512–8520.

409 Aschmann, S. M.; Arey, J.; Atkinson, R.: Atmospheric Chemistry of Selected Hydroxycarbonyls. J. Phys. Chem. A 104 (2000) 3998–4003.

410 Tokuhashi, K.; Takahashi, A.; Kaise, M.; Kondo, S.; Sekiya, A.; Yamashita, S.; Ito, H.: Rate Constants for the Reactions of OH Radicals with $CH_3OCF_2CHF_2$, $CHF_2OCH_2CF_2CHF_2$, $CHF_2OCH_2CF_2CF_3$, and $CF_3CH_2OCF_2CHF_2$ over the Temperature Range 250–430 K. J. Phys. Chem. A 104 (2000) 1165–1170.

411 Mashino, M.; Kawasaki, M.; Wallington, T. J.; Hurley, M. D.: Atmospheric Degradation of $CF_3OCF=CF_2$: Kinetics and Mechanism of its Reaction with OH Radical and Cl Atoms. J. Phys. Chem. A 104 (2000) 2925–2930.

412 Li, Z.; Tao, Z.; Naik, V.; Good, D. A.; Hansen, J. C.; Jeong, G.-R.; Francisco, J. S.; Jain, A. K.; Wuebbles, D. J.: Global Warming Potential Assessment for $CF_3OCF=CF_2$. J. Geophys. Res. 105 (2000) 4019–4029.

413 Pagagni, C.; Arey, J.; Atkinson, R.: Rate Constants for the Gas-Phase Reactions of a Series of C_3-C_6 Aldehydes with OH and NO_3 Radicals. Int. J. Chem. Kinet. 32 (2000) 79–84.

414 D'Anna, B.; Andresen, O.; Gefen, Z.; Nielsen, C. J.: Kinetic Study of OH and NO_3 Radical Reactions with 14 Aliphatic Aldehydes. Phys. Chem. Chem. Phys. 3 (2001) 3057–3063.

415 Cabanas, B.; Martín, P.; Salgado, S.; Ballesteros, B.; Martínez, E.: An Experimental Study on the Temperature Dependence for the Gas-Phase Reactions of NO_3 Radical with a Series of Aliphatic Aldehydes. J. Atmos. Chem. 40 (2001) 23–39.

416 Atkinson, R.; Baulch, D. L.; Cox, R. A.; Crowley, J. N.; Hampson, R. F.; Kerr, J. A.; Rossi, M. J.; Troe, J.: Summary of Evaluated Kinetic and Photochemical Data for Atmospheric Chemistry. IUPAC Subcommittee on Gas Kinetic Data Evaluation for Atmospheric Chemistry; web Version internet (December 2001).

417 Klotz, B.; Barnes, I.; Golding, B. T.; Becker, K.-H.: Atmospheric Chemistry of Toluene-1,2-oxide/2-methyloxepin. Phys. Chem. Chem. Phys. 2 (2000) 227–235.

418 Wyatt, S. E.; Baxley, J. S.; Wells, J. R.: The Hydroxyl Radical Reaction Rate Constant and Products of Methyl Isobutyrate. Int. J. Chem. Kinet. 31 (1999) 551–557.

419 Chen, L.; Fukuda, K.; Takenaka, N.; Bandow, H.; Maeda, Y.: Kinetics of the Gas-Phase Reaction of $CF_3CF_2CH_2OH$ with OH Radicals and its Atmospheric Lifetime. Int. J. Chem. Kinet. 32 (2000) 73–78.

420 Xing, S. B.; Shi, S.-H.; Qiu, L.-X.: Kinetics Studies of Reactions of OH Radicals with Four Haloethanes. Part I. Experiment and BEBO Calculation. Int. J. Chem. Kinet. 24 (1992) 1–10.

421 Thevenet, R.; Mellouki, A.; Le Bras, G.: Kinetics of OH and Cl Reactions with a Series of Aldehydes. Int. J. Chem. Kinet. 32 (2000) 676–685.

422 Yamada, T.; Fang, T. D.; Taylor, P. H.; Berry, R. J.: Kinetics and Thermochemistry of the OH Radical Reaction with CF_3CCl_2H and CF_3CFClH. J. Phys. Chem. A 104 (2000) 5013–5022.

423 Magneron, I.; Bossoutrot, V.; Mellouki, A.; Laverdet, G.; Le Bras, G.: The OH-initiated Oxidation of Hexylene Glycol and Diacetone. Environ. Sci. Technol. 37 (2003) 4170–4181.

424 Atkinson, R.; Tuazon, E. C.; Aschmann, S. M.: Atmospheric Chemistry of 2-Pentanone and 2-Heptanone. Environ. Sci. Technol. 34 (2000) 623–631.

425 Olariu, R. I.; Barnes, I.; Becker, K. H.; Klotz, B.: Rate Coefficients for the Gas-Phase Reactions of OH Radicals with Selected Dihydroxybenzenes and Benzoquinones. Int. J. Chem. Kinet. 32 (2000) 696–702.

426 Johnson, D.; Rickard, A. R.; McGill, C. D.; Marston, G.: The Influence of Orbital Asymmetry on the Kinetics of the Gas-Phase Reactions of Ozone with Unsaturated Compounds. Phys. Chem. Chem. Phys. 2 (2000) 323–328.

427 Kramp. F.; Paulson, S. E.: The Gas Phase Reaction of Ozone with 1,3-Butadiene: Formation and Yield of some Toxic Products. Atmos. Environ. 34 (2000) 35–43.

428 Mashino, M.; Ninomiya, Y.; Kawasaki, M.; Wallington, T. J.; Hurley, M. D.: Atmospheric Chemistry of $CF_3CF=CF_2$: Kinetics and Mechanism of its Reactions with OH Radicals, Cl Atoms, and Ozone. J. Phys. Chem. A 104 (2000) 7255–7260.

429 Atkinson, R.; Baulch, D. L.; Cox, R. A.; Crowley, J. N.; Hampson, R. F.; Hynes, R. G.; Jenkin, M. E.; Rossi, M. J.; Troe, J.: Evaluated Kinetic and Photochemical Data for Atmospheric Chemistry. Volume 1- Gas Phase Reactions of O_x, HO_x, NO_x and SO_x Species. Atmos. Chem. Phys. 4 (2004) 1461–1738.

430 Cabanas, B.; Salgado, S.; Martin, B.; Baeza, M. T.; Martínez, E.: Night-time Atmospheric Loss Process for Unsaturated Aldehydes: Reaction with NO_3 Radicals. J. Phys. Chem. A 105 (2001) 4440–4445.

431 Cabanas, B.; Salgado, S.; Martin, P.; Baeza, M. T.; Albaladejo, J.; Martinez, E.: Gas-phase Rate Coefficients and Activation Energies for the Reaction of NO_3 Radicals with Selected Branched Aliphatic Aldehydes. Phys. Chem. Chem. Phys. 5 (2003) 112–116.

432 Thuner, L. P.; Barnes, I.; Maurer, T.; Sauer, C. G.; Becker, K. H.: Kinetic Study of the Reaction of OH with a Series of Acetals at 298 ± 4 K. Int. J. Chem. Kinet. 31 (1999) 797–803.

433 Heathfield, A. E.; Anastasi, C.; Pagsberg, P.; McCullouch, A.: Atmospheric Lifetimes of Selected Fluorinated Ether Compounds. Atmos. Environ. 32 (1998) 711–717.

434 Chen, L.; Kutsuna, S.; Tokuhashi, K.; Sekiya, A.; Takeuchi, K.; Ibusuki, T.: Kinetics for the Gas-Phase Reactions of OH Radicals with the Hydrofluoroethers $CH_2FCF_2OCHF_2$, $CHF_2CF_2OCH_2CF_3$, $CF_3CHFCF_2OCH_2CF_3$, and $CF_3CHFCF_2OCH_2CF_2CHF_2$ at 268–308 K. Int. J. Chem. Kinet. 35 (2003) 239–245.

435 Tokuhashi, K.; Takahashi, A.; Kaise, M.; Kondo, S.; Sekiya, A.; Yamashita, S.; Ito, H.: Rate Constants for the Reactions of OH Radicals with $CH_3OCF_2CF_3$, $CH_3OCF_2CF_2CF_3$, and $CH_3OCF(CF_3)_2$. Int. J. Chem. Kinet. 31 (1999) 846–853.

436 Suh, I.; Lei, W.; Zhang, R.: Experimental and Theoretical Studies of Isoprene Reaction with NO_3. J. Phys. Chem. A 105 (2001) 6471–6478.

437 Tokuhashi, K.; Takahashi, A.; Kaise, M.; Kondo, S.: Rate Constants for the Reactions of OH Radicals with CH_3OCF_2CHFCl, CHF_2OCF_2CHFCl, $CHF_2OCHClCF_3$, and $CH_3CH_2OCF_2CHF_2$. J. Geophys. Res. 104 (1999) 18681–18688.

438 Khamaganov, V. G.; Hites, R. A.: Rate Constants for the Gas-Phase Reactions of Ozone with Isoprene, alpha- and beta-Pinene, and Limonene as a Function of Temperature. J. Phys. Chem. A 105 (2001) 815–822.

439 Neeb, P.; Moortgat, G. K.: Formation of OH Radicals in the Gas-Phase Reaction of Propene, Isobutene and Isoprene with O_3: Yields and Mechanistic Implications. J. Phys. Chem. A 103 (1999) 9003–9012.

440 Lewin, A. G.; Johnson, D.; Price, D. W.; Marston, G.: Aspects of the Kinetics and Mechanism of the Gas-Phase Reactions of Ozone with Conjugated Dienes. Phys. Chem. Chem. Phys. 3 (2001) 1253–1261.

441 Tobias, H. J.; Ziemann, P. J.: Kinetics of the Gas-Phase Reactions of Alcohols, Aldehydes, Carboxylic Acids, and Water with the C_{13} Stabilized Criegee Intermediate Formed from Ozonolysis of 1-Tetradecene. J. Phys. Chem. A 105 (2001) 6129–6135.

442 Martínez, E.; Cabanas, B.; Aranda, A.; Martin, P.; Salgado, S.: A Temperature Dependence Study of the Gas-phase Reaction of Nitrate Radical with 3-Fluoro followed by Laser Induced Fluorescence Detection. Int. J. Chem. Kinet. 29 (1997) 927–932.

443 Chowdhury, P. K.; Upadhyaya, H. P.; Naik, P. D.; Mittal, J. P.: ArF Laser Photodissociation Dynamics of Hydroxyacetone: LIF Observation of OH and its Reaction Rate with the Parent. Chem. Phys. Lett. 351 (2002) 201–207.

444 Li, Z.; Jeong, G.-R.; Hansen, J. C.; Good, D. A.; Francisco, J. S.: Rate Constants for the Reactions of $CF_3OCHFCF_3$ with OH and Cl. Chem. Phys. Lett. 320 (2000), 70–76.

445 Mellouki, A.; Mu, Y. J.: On the Atmospheric Degradation of Pyruvic Acid in the Gas Phase. J. Photochem. Photobio. A Chem. 157 (2003) 295–300.

446 Goumri, A.; Yuan, J.; Hommel, E. L.; Marshall, P.: Kinetic Studies of the Reactions of Hydroxyl Radicals with the Methyl-substituted Silanes $SiH_n(CH_3)(4-n)$ (n = 0 to 4) at Room Temperature. Chem. Phys. Lett. 375 (2003) 179–156.

447 Tuazon, E. C.; Aschmann, S. M.; Nguyen, M. V.; Atkinson, R.: H-Atom Abstraction from Selected C–H Bonds in 2,3-Dimethylpentanal, 1,4-Cyclohexadiene, and 1,3,5-Cycloheptatriene. Int. J. Chem. Kinet. 35 (2003), 415–426.

448 Moriarty, J.; Sidebottom, H.; Wenger, J.; Mellouki, A., LeBras, G.: Kinetic Studies on the Reactions of Hydroxyl Radicals with Cyclic Ethers and Aliphatic Diethers. J. Phys. Chem. A 107 (2003) 1499–1505.

449 Kukui, A.; Borissenko, D.; Laverdet, G.; Le Bras, G.: Gas-Phase Reactions of OH Radicals with Dimethyl Sulfoxide and Methane Sulfinic Acid Using Turbulent Flow Reactor and Chemical Ionization Mass Spectrometry. J. Phys. Chem. A 107 (2003) 5732–5742.

450 Wirtz, K.: Determination of Photolysis Frequencies and Quantum Yields for Small Carbonyl Compounds using the EUROPHORE Chamber. Presentation on the Combined US/German Environmental Chamber Workshop Riverside (full manuscript available via internet) EU Project ENV4-CT97-0419, 1999.

451 Tanigushi, N.; Wallington, T. J.; Hurley, M. D.; Guschin, A. G.; Molina, L. T.; Molina, M. J.: Atmospheric Chemistry of $C_2F_5C(O)CF(CF_3)_2$. J. Phys. Chem. A 107 (2003) 2674–2679.

452 Taylor, P. H.; Yamada, T.; Neuforth, A.: Kinetics of OH Radical Reactions with Dibenzo-p-dioxin and Selected Chlorinated Dibenzo-p-dioxins. Chemosphere 58 (2005), 243–252.

453 Moortgat, G. K.: Important Photochemical Processes in the Atmosphere. Pure Appl. Chem. 73 (2001) 487–490.

454 Brubaker, Jr., W. W.; Hites, R. A.: OH Reaction Kinetics of Gas-Phase alpha-and gamma-Hexachlorocyclohexane and Hexachlorobenzene. Environ. Sci. Technol. 32 (1998) 766–769.

455 Prinn, R. G.; Weiss, R. F.; Miller, B. R.; Huang, J.; Alyea, F. N.; Cunnold, D. M.; Frazer, P. J.; Hartley, H. E.; Simmonds, P. G.: Atmospheric Trends and Lifetime of CH_3CCl_3 and Global OH Concentrations. Science 269 (1995) 187–192.

456 Palm, W.-U.; Elend, M.; Krüger, H.-U.; Zetzsch, C.: Atmospheric Degradation of a Semivolatile Aerosol-Borne Pesticide: Reaction of OH with Pyrifenox (an Oxime-Ether), Adsorbed on SiO_2. Chemosphere 38 (1999) 1241–1252.

457 Atkinson, R.; Guicheret, R.; Hites, R. A.; Palm, W.-U.; Seiber, J. N.; De Voogt, P.: Transformations of Pesticides in the Atmosphere: A State of the Art. Water Air Soil Pollut. 115 (1999) 219–243.

458 Rühl, E.: Messung von Reaktionsgeschwindigkeitskonstanten zum Abbau von langlebigen, partikelgebundenen Substanzen durch indirekte Photooxidation. Bericht an das Umweltbundesamt, FKZ 202 67 434. Berlin 2004.

459 Alvarez, R. A.; Moore, C. B.: Quantum Yield for Production of CH_3NC in the Photolysis of CH_3NCS. Science 263 (1994) 205–207.

460 Hebert, V. R.; Geddes, J. D.; Mendosa, J.; Miller, G. C.: Gas-phase Photolysis of Phorate, a Phosphorothioate Insecticide. Chemosphere 36 (1998) 2057–2066.
461 Carter, W. P. L.; Luo, D.; Malkina, I. L.: Investigation of the atmospheric Reactions of Chloropicrin. Atmos. Environ. 31 (1997) 1425–1439.
462 Brubaker, Jr., W. W.; Hites, R. A.: Polychlorinated Dibenzo-p-dioxins and Dibenzofurans: Gas-Phase Hydroxyl Radical Reactions and Related Atmospheric Removal. Environ. Sci. Technol. 31 (1997) 1805–1810.
463 Geiger, H.; Barnes, I.; Becker, K. H.; Birger, B.; Brauers, T.; Donner, B.; Dorn, H.-P.; Elend, M.; Freitas Dinis, C. M.; Grossmann, D.; Hass, H.; Hein, H.; Hoffmann, A.; Hoppe, L.; Hülsemann, F.; Kley, D.; Klotz, B.; Libuda, H. G.; Maurer, T.; Mihelcic, D.; Moortgat, G. K.; Olariu, R.; Neeb, B.; Poppe, D.; Ruppert, L.; Sauer, C. G.; Shestakov, O.; Somnitz, H.; Stockwell, W. R.; Thüner, L. P.; Wahner, A.; Wiesen, P.; Zabel, F.; Zellner, R.; Zetzsch, C.: Chemical Mechanism Development: Laboratory Studies and Model Applications. J. Atmos. Chem. 42 (2002) 323–357.
464 Barnes, I.; Donner, B.: Wirkung reformulierter und alternativer Kraftstoffe auf die Bildung von Photooxidantien in urbaner Luft. In: Becker, K. H. (Ed.) TFS-LT3 Annual Report 1999, University of Wuppertal, Germany, 2000.
465 Sauer, C. G.; Barnes, I.; Becker, K. H.; Geiger, H.; Wallington, T. J.; Christensen, L. K.; Platz, J.; Nielsen, O. J.: Atmospheric Chemistry of 1,3-Dioxolane: Kinetic, Mechanistic, and Modeling Study of OH Radical Initiated Oxidation. J. Phys. Chem. A 103 (1999) 5959–5966.
466 Hass, H.: Untersuchungen zum atmosphärischen Abbau von alternativen Kraftstoffen und Kraftstoffzusätzen. In: Becker, K. H. (Ed.) TFS-LT3 Annual Report 1999, University of Wuppertal, Germany, 2000.
467 Maurer, T.; Hass, H.; Barnes, I.; Maurer, Becker, K. H.: Kinetic and Product Study of the Atmospheric Photooxidation of 1,4-Dioxane and Its Main Reaction Product Ethylene Glycol Diformate. J. Phys. Chem. A 103 (1999) 5032–5039.
468 Becker, K. H.; Dinis, C.; Geiger, H.; Wiesen, P.: Kinetics and Reaction of OH with di-n-Butoxymethane (DBM) in the Range 298–710 K. Chem. Phys. Lett. 300 (1999) 460–464.
469 Platz, J.; Christensen, L. K.; Sehested, J.; Nielsen, O. J.; Wallington, T.J-; Sauer, C.; Barnes, I.; Becker, K. H.; Vogt, R.: Atmospheric Chemistry of 1,3,5-Trioxane. J. Phys. Chem. A 102 (1998) 4829–4838.
470 Plagens, Heike: Untersuchungen zum atmosphärenchemischen Abbau langkettiger Aldehyde. PhD Thesis, University of Wuppertal, Germany 2001.
471 Liu, X.; Jeffries, H. E.; Sexton, K. G.: Atmospheric Photochemical Degradation of 1,4-Unsaturated Dicarbonyls. Environ. Sci. Technol. 33 (1999) 4212–4220.
472 Wayne, R. P.; Barnes, I.; Biggs, P.; Burrows, J. P.; Canosa-Mas, C. E.; Hjorth, J.; Le Bras, G.; Moortgat, G. K.; Perner, D.; Poulet, G.; Restelli, G.; Sidebottom, H.: The Nitrate Radicals: Physics, Chemistry, and the Atmosphere. Atmos. Environ. Part A 25 (1991) 1–203.
473 Picquet-Varrault, B.; Doussin, J.-F.; Durant-Joilbois, R.; Pirali, O.; Carlier, P.: Kinetic and Mechanistic Study of the Atmospheric Oxidation by OH Radicals of Allyl Acetate. Environ. Sci. Technol. 36 (2002) 4081–4086.
474 Le Calvi, S.: PhD Thesis, University of Orléans, France, 1998.
475 Phousongphouang, P. T.; Arey, J.: Rate Constants for Gas-Phase Reactions of a Series of Alkylnaphthalenes with the OH Radical. Environ. Sci. Technol. 36 (2002) 1947–1952.
476 Krüger, H.-U.; Gavrilov, R.; Liu, Q.; Zetzsch, C.: Entwicklung eines Persistenz-Messverfahrens für den troposphärischen Abbau von mittelflüchtigen Pflanzenschutzmitteln durch OH-Radikale. Bericht an das Umweltbundesamt, FKZ 201 67 424/02, Berlin 2005.
477 Reisen, F.; Arey, J.: Reactions of Hydroxyl Radicals and Ozone with Acenaphthene and Acenaphthylene. Environ. Sci. Technol. 36 (2002) 4302–4311.
478 Carl, S. A.; Crowley, J. N.: 298 K Rate Coefficients for the Reaction of OH with i-C_3H_7I, n-C_3H_7I and C_3H_8: Atmos. Chem. Phys. Discuss. 1 (2001) 23–41.
479 Smith, A. M.; Rigler, E.; Kwok, E. S. C.; Atkinson, R.: Kinetics and Products of the Gas-Phase Reactions of 6-Methyl-5-hepten-2-one and trans-Cinnamaldehyde with NO_3 Radicals and O_3 at 296 ± 2 K. Environ. Sci. Technol. 30 (1996) 1781–1785.

480 Tuazon, E. C.; Aschmann, S. M.; Atkinson, R.: Atmospheric Degradation of Volatile Methyl-Silicon Compounds. Environ. Sci. Technol. 34 (2000) 1970–1976.
481 Aschmann, S. M.; Atkinson, R.: Kinetics of the Gas-phase Reactions of the OH Radical with Selected Glycol Ethers, Glycols, and Alcohols. Int. J. Chem. Kinet. 30 (1998) 533–540.
482 Aschmann, S. M.; Arey, J.; Atkinson, R.; Simonich, S. L.: Atmospheric Lifetimes and Fates of Selected Fragrance Materials and Volatile Model Compounds. Environ. Sci. Technol. 35 (2001) 3595–3600.
483 Gierczak, T.; Burkholder, J. R.; Talukdar, R. K.; Mellouki, A.; Baronc, S. R.; Ravishankara, A. R.: Atmospheric Fate of Methyl Vinyl Ketone and Methacrolein. J. Photochem. Photobiol. A: Chem. 110 (1997) 1–10.
484 Roehl, C. M.; Burkholder, J. B.; Moortgat, G. K.; Ravishankara, A. R.; Crutzen, P. J.: Temperature Dependence of UV Absorption Cross Sections and Atmospheric Implications of Several Alkyl Iodides. J. Geophys. Res. 102 D11 (1997) 12 819–12 829.
485 Fantechi, G.; Jensen, N. R.; Hjorth, J.; Peeters, J.: Determination of the Rate Constants for the Gas-phase Reactions of Methyl Butenol with OH Radicals, Ozone, NO_3 Radicals, and Cl Atoms. Int. J. Chem. Kinet. 30 (1998) 589–594.
486 Cavalli, F.; Barnes, I.; Becker, K.-H.: FT-IR Kinetic and Product Study of the OH Radical and Cl-Atom-initiated Oxidation of Dibasic Esters. Int. J. Chem. Kinet. 33 (2001) 431–439.
487 Harry, C.; Arey, J.; Atkinson, R.: Rate Constants for the Reactions of OH Radicals and Cl Atoms with Di-n-Propyl Ether and Di-n-Butyl Ether and their Deuterated Analogs. Int. J. Chem. Kinet. 31 (1999) 425–431.
488 Noda, J.; Holm, C.; Nyman, G.; Langer, S.; Ljungström, E.: Kinetics of the Gas-Phase Reaction of n-C_6-C_{10} Aldehydes with the Nitrate Radical. Int. J. Chem. Kinet. 35 (2003) 120–129.
489 Wu, H.; Mu, X.; Zhang, X.; Jiang, G.: Relative Rate Constants for the Reactions of Hydroxyl Radicals and Chlorine Atoms with a Series of Aliphatic Alcohols. Int. J. Chem. Kinet. 35 (2003) 81–87.
490 Davis, D. D.; Prusazcyk, J.; Dwyer, M.; Kim, P.: A Stop-Flow Time-of-Flight Mass Spectrometry Kinetics Study. Reaction of Ozone with Nitrogen Dioxide and Sulfur Dioxide. J. Phys. Chem. 78 (1974) 1775–1779.
491 Tomas, A.; Olariu, R. I.; Barnes, I.; Becker, K. H.: Kinetics of the Reaction of O_3 with selected Benzenediols. Int. J. Chem. Kinet. 35 (2003) 223–230.
492 Olariu, R. I.; Bejan, I.; Barnes, I.; Klotz, B.; Becker, K. H.; Wirtz, K.: Rate Coefficients for the Gas-Phase Reaction of NO_3 Radicals with Selected Dihydroxybenzenes. Int. J. Chem. Kinet. 36 (2004) 577–583.
493 Imamura, T.; Iida, Y.; Obi, K.; Nagatani, I.; Nakagawa, K.; Patroescu-Klotz, I.; Hatakeyama, S.: Rate Coefficients for the Gas-Phase Reactions of OH Radicals with Methylbutenols at 298 K. Int. J. Chem. Kinet. 36 (2004) 379–385.
494 Chen, L.; Tokuhaski, S.; Kutsuna, S.; Sekiya, A.: Rate Constants for the Gas-Phase Reaction of $CF_3CF_2CF_2CF_2CHF_2$ with OH Radicals at 250–430 K. Int. J. Chem. Kinet. 36 (2004) 26–33.
495 Chen, L.; Kutsuna, S.; Tokuhashi, K., Sekiya, A.: Kinetics of the Gas-Phase Reaction of $CF_3OC(O)H$ with OH Radicals at 242–328 K. Int. J. Chem. Kinet. 36 (2004) 337–344.
496 Cabanas, B.; Martín, P.; Salgado, S.; Baeza, M. T.; Albaladejo, J. Martínez, E.: Kinetic Study of the Gas-Phase Reaction of the Nitrate Radical with Methyl-Substituted Thiophenes. Int. J. Chem. Kinet. 35 (2003) 286–293.
497 Colomb, A.; Jacob, V.; Kaluzny, P.; Baussand, P.: Kinetic Investigation of Gas-phase Reactions between the OH-Radical and o-, m-, p-Ethyltoluene and n-Nonane in Air. Int. J. Chem. Kinet. 36 (2004) 367–378.
498 Wells, J. R.: The Hydroxyl Radical Reaction Rate Constant and Products of 3,5-Dimethyl-1-hexyn-3-ol. Int. J. Chem. Kinet. 36 (2004) 534–544.
499 Stabel, J. R.; Johnson, M. S.; Langer, S.: Rate Coefficients for the Gas-Phase Reaction of Isoprene with NO_3 and NO_2. Int. J. Chem. Kinet. 37 (2005) 57–65.
500 Cusik, R. D.; Atkinson, R.: Rate Constants for the Gas-Phase Reactions of O_3 with a Series of Cycloalkenes at 296 ± 2 K. Int. J. Kinet. 37 (2005) 183–190.

501 Aschmann, S. M.; Arey, J.; Atkinson, R.: Products and Mechanism of the Reaction of OH Radicals with 2,3,4-Trimethylpentane in the Presence of NO. Environ. Sci. Technol. 38 (2004) 5038–5045.
502 Oyaro, N.; Sellevag, S. G.; Nielsen, C. J.: Study of the OH and Cl-Initiated Oxidation, IR Absorption Cross-Section, Radiative Forcing, and Global warming Potential of four C_4-Hydrofluoroethers. Environ. Sci. Technol. 38 (2004) 5567–5576.
503 Wenger, J. C.; Le Calvé, S.; Sidebottom, H. W.; Wirtz, K.; Reviejo, M. M.; Franklin, J. A.: Photolysis of Chloral under Atmospheric Conditions. Environ. Sci. Technol. 38 (2004) 831–837.
504 Esteve, W.; Budzinski, H.; Vilenave, E.: Relative Rate Constants for the Heterogeneous Reactions of OH, NO_2 and NO Radicals with Polycyclic Aromatic Hydrocarbons Adsorbed on Carbonaceous Particles. Part 1: PAH Adsorbed on 1–2 µm Calibrated Graphite Particles. Atmos. Environ. 38 (2004) 6063–6072.
505 Sellevag, S. R.; Nielsen, C. J.; Sovde, O. A.; Myhre, G.; Sundet, J. K.; Stordal, F.; Isaksen, I. S. A.: Atmospheric Gas-phase Degradation and Global Warming Potentials of 2-Fluoroethanol, 2,2-Difluoroethanol, and 2,2,2-Trifluoroethanol. Atmos. Environ. 38 (2004) 6725–6735.
506 Davis, M. E.; Stevens, P. S.: Measurements of the Kinetics of the OH-initiated Oxidation of alpha-Pinene: Radical Propagation in the OH+alpha-pinene+O_2+NO Reaction System. Atmos. Environ. 39 (2005) 1765–1774.
507 Goto, M.; Inoue, Y.; Kawasaki, M.; Guschin, A. G.; Molina, L. T.; Molina, M. J.; Wallington, T. J.; Hurley, M. D.: Atmospheric Chemistry of HFE-7500 [n-C_3F_7CF(OC_2H_5)CF(CF_3)$_2$]: Reaction with OH Radicals and Cl Atoms and Atmospheric Fate of n-C_3F_7CF(OCHO·)CF(CF_3)$_2$ and n-C_3F_7CF(OCH_2CH_2O·)CF(CF_3)$_2$ Radicals. Environ. Sci. Technol. 36 (2002) 2395–2402.
508 Cavalli, F.; Glasius, M.; Hjorth, J.; Rindone, B.; Jensen, N. R.: Atmospheric Lifetimes, Infrared Spectra and Degradation Products of a Series of Hydrofluoroethers. Atmos. Environ. 32 (1998) 3767–3773.
509 Aschmann, S. M.; Martin, P.; Tuazon, E. C.; Arey, J.; Atkinson, R.: Kinetic and Product Studies of the Reactions of Selected Glycol Ethers with OH Radicals. Environ. Sci. Technol. 35 (2001) 4080–4088.
510 Veillerot, M.; Foster, P.; Guillermo, R.; Galloo, J. C.: Gas-phase Reaction of n-Butyl Acetate with the Hydroxyl Radical under Simulated Tropospheric Conditions: Relative Rate Constants and Product Study. Int. J. Chem. Kinet. 28 (1996) 235–243.
511 Sun, F.; Zhu, T.; Shang, J.; Han, L.: Gas-phase Reaction of Dichlorphos, Carbaryl, Chlordimeform, and 2,4-D Butylester with OH Radicals. Int. J. Chem. Kinet. 37 (2005) 755–762.
512 Picquet, B.; Heroux, S.; Chebbi, A.; Doussin, J.-F.; Durant-Jolibois, R.; Monod, A.; Loirat, H.; Carlier, P.: Kinetics of the Reactions of OH Radicals with some Oxygenated Volatile Organic Compounds under Simulated Atmospheric Conditions. Int. J. Chem. Kinet. 30 (1998) 839–847.
513 Taketani, F.; Nakayama, T.; Takahashi, K.; Matsumi, Y.; Hurley, M. D.; Wallington, T. J.; Toft, A.; Sulbaek Andersen, M. P.: Atmospheric Chemistry of CH_3CHF_2 (HFC-152a): Kinetics, Mechanisms, and Products of Cl Atom- and OH Radical-Initiated Oxidation in the Presence and Absence of NO_x. J. Phys. Chem. 109 (2005) 9061–9069.
514 Wallington, T. J.; Hurley, M. D.; Nielsen, O. J.; Sulbaek Andersen, M. P.: Atmospheric Chemistry of CF_3CFHCF_2OCF_3 and CF_3CFHCF_2OCF_2H: Reaction with Cl Atoms and OH Radicals, Degradation Mechanism, and Global Warming Potentials. J. Phys. Chem. A 108 (2004) 11333–11338.
515 Sulbaek Andersen, M. P.; Nielsen, O. J.; Wallington, T. J.; Hurley, M. D.; DeMore, W. B.: Atmospheric Chemistry of CF_3OCF_2CF_2H and CF_3OC(CF3)$_2$H: Reaction with Cl Atoms and OH Radicals, Degradation Mechanism, Global Warming Potentials, and Empirical Relationship between k(OH) and k(Cl) for Organic Compounds. J. Phys. Chem. A 109 (2005) 3926–3934.

516 Christensen, L. K.; Sehested, J.; Nielsen, O. J.; Bilde, M.; Wallington, T. J.; Guschin, A.; Molina, L. T.; Molina, M. J.: Atmospheric Chemistry of HFE-7200 ($C_4F_9OC_2H_5$): Reaction with OH Radicals and Fate of $C_4F_9OCH_2CH_2O(\cdot)$ and $C_4F_9OCHO(\cdot)CH_3$ Radicals. J. Phys. Chem. A 102 (1998) 4839–4845.

517 D'Eon, J. C.; Hurley, M. D.; Wallington, T. J.; Mabury, S. A.: Atmospheric Chemistry of N-methyl Perfluorobutane Sulfonamidoethanol, $C_4F_9SO_2N(CH_3)CH_2CH_2OH$: Kinetics and Mechanism of Reaction with OH. Environ. Sci. Technol. 40 (2006) 1862–1868.

518 Baker, J.; Arey, J.; Atkinson, R.: Formation and Reaction of Hydroxycarbonyls from the Reaction of OH Radicals with 1,3-Butadiene and Isoprene. Environ. Sci. Technol. 39 (2005) 4091–4099.

519 Bolzacchini, E.; Bruschi, M.; Hjorth, J.; Meinardi, S.; Orlandi, M.; Rindone, B.; Rosenbohm, E.: Gas-Phase Reaction of Phenol with NO_3. Environ. Sci. Technol. 35 (2001) 1791–1797.

520 Martin, J. W.; Ellis, D. A.; Mabury, S. A.; Hurley, M. D.; Wallington, T. J.: Atmospheric Chemistry of Perfluoroalkanesulfonamides: Kinetic and Product Studies of the OH Radical and Cl Atom Initiated Oxidation of N-Ethyl Perfluorobutanesulfonamide. Environ. Sci. Technol. 40 (2006) 864–872.

521 Wells, J. R.: Gas-Phase Chemistry of α-Terpineol with Ozone and OH Radical: Rate Constants and Products. Environ. Sci Technol. 39 (2005) 6937–6943.

522 Fruekilde, P.; Hjorth, J.; Jensen, N. R.; Kotzias, D.; Larsen, B.: Ozonolysis at Vegetation Surfaces: A Source of Acetone, 4-Oxopentanal, 6-Methyl-5-hepten-2-on, and Geranyl Acetone in the Troposphere. Atmos. Environ. 32 (1998) 1893–1902.

523 Falbe-Hansen, H.; Sørensen, S.; Jensen, N. R.; Pedersen, T.; Hjorth, J.: Atmospheric Gas-phase Reactions of Dimethylsulphoxide and Dimethylsulphone with OH and NO_3 Radicals, Cl Atoms and Ozone. Atmos. Environ. 34 (2000) 1543–1551.

524 Albaladejo, J.; Ballesteros, B.; Jiménez, E.; Martín, P.; Martínez, E.: A PLP-LIF Kinetic Study of the Atmospheric Reactivity of a Series of C_4-C_7 Saturated and Unsaturated Aliphatic Aldehydes with OH. Atmos. Environ. 36 (2002) 3231–3239.

525 Feigenbrugel, V.; Le Person, A.; Le Calvé, S.; Mellouki, A.; Muñoz, A.; Wirtz, K.: Atmospheric Fate of Dichlorvos: Photolysis and OH-Initiated Oxidation Studies. Environ. Sci. Technol. 40 (2006) 850–857.

526 Martínez, E.; Cabañas, B.; Aranda, A.; Martín, P.: Kinetics of the Reactions of NO_3 Radicals with Selected Monoterpenes: A Temperature Dependence Study. Environ. Sci. Technol. 32 (1998) 3730–3734.

527 Albaladejo, J.; Ballesteros, B.; Jiménez, E.; Díaz, Y.; Martínez, E.: Gas-phase OH Radical-initiated Oxidation of the 3-Halopropenes Studied by PLP-LIF in the Temperature Range 228–388 K. Atmos. Environ. 37 (2003) 2919–2926.

528 Ham, J. E.; Proper, S. P.; Wells, J. R.: Gas-phase Chemistry of Citronellol with Ozone and OH Radical: Rate Constants and Products. Atmos. Environ. 40 (2006) 726–735.

529 Carrasco, M.; Doussin, J.-F.; Picquet-Varrault, B.; Carlier, P.: Tropospheric Degradation of 2-Hydroxy-2-methylpropanal, a Photo-oxidation Product of 2-Methyl-3-buten-2-ol: Kinetic and Mechanistic Study of its Photolysis and its Reaction with OH Radicals. Atmos. Environ. 40 (2006) 2011–2019.

530 Teruel, M. A.; Lane, S. I.; Mellouki, A.; Solignac, G.; Le Bras, G.: OH Reaction Rate Constants and UV Absorption Cross-sections of Unsaturated Esters. Atmos. Environ. 40 (2006) 3764–3772.

531 Klotz, B. G.; Bierbach, A.; Barnes, I.; Becker, K.-H.: Kinetic and Mechanistic Study of the Atmospheric Chemistry of Mucoaldehydes. Environ. Sci. Technol. 29 (1995) 2322–2332.

532 Martin, P.; Tuazon, E. C.; Atkinson, R.; Maugham, A. D.: Atmospheric Gas-Phase Reactions of Selected Phosphorus-Containing Compounds. J. Phys. Chem. A 106 (2002) 1542–1550.

533 Aschmann, S. M.; Tuazon, E. C.; Atkinson, R.: Atmospheric Chemistry of Diethyl Methylphosphonate, Diethyl Ethylphosphonate, and Triethyl Phosphate. J. Phys. Chem. A 109 (2005) 2282–2291.

534 Lee, W.; Stevens, P. S.; Hites, R. A.: Rate Constants for the Gas-Phase Reactions of Methylphenanthrenes with OH as a Function of Temperature. J. Phys. Chem. A 107 (2003) 6603–6608.

535 DeMore, W. B.; Wilson, Jr., E. W.: Rate Constant and Temperature Dependence for the Reaction of Hydroxyl Radicals with 2-Fluoropropane (HFC-281ea). J. Phys. Chem. A 103 (1999) 573–576.

536 Langer, S.; Ljungstrom, E.: Reaction of the Nitrate Radical with some Potential Automotive Fuel Additives. A Kinetic and Mechanistic Study. J. Phys. Chem. 98 (1994) 5906–5912.

Appendix: CAS Register

CAS Number	Chemical Name
50-00-0	Formaldehyde
50-29-3	4,4'-DDT
50-32-8	Benzo[a]pyrene
51-79-6	Urethane
56-23-5	Methane, tetrachloro-
56-38-2	Parathion
56-55-3	Benz[a]anthracene
57-06-7	1-Propene, isothiocyanato
57-14-7	Hydrazine, 1,1-dimethyl-
57-55-6	1,2-Propanediol
58-89-9	Lindane
60-29-7	Diethylether
60-34-4	Hydrazine, methyl-
62-53-3	Aniline
62-73-7	Dichlorvos
62-75-9	Methanamine, N-methyl-N-nitroso-
63-25-2	Carbaryl
64-17-5	Ethanol
64-18-6	Formic acid
64-19-7	Acetic acid
64-67-5	Sulfuric acid, diethyl ester
66-25-1	Hexanal
67-56-1	Methanol
67-63-0	2-Propanol
67-64-1	Acetone
67-66-3	Chloroform
67-68-5	Dimethylsulfoxide
67-71-0	Dimethyl sulfone
71-23-8	1-Propanol
71-36-3	1-Butanol

Atmospheric Degradation of Organic Substances. W. Klöpffer and B. O. Wagner
Copyright © 2007 WILEY-VCH Verlag GmbH & Co. KGaA, Weinheim
ISBN: 978-3-527-31606-9

CAS Number	Chemical Name
71-41-0	1-Pentanol
71-43-2	Benzene
71-55-6	Ethane, 1,1,1-trichloro-
74-82-8	Methane
74-83-9	Methane, bromo-
74-84-0	Ethane
74-85-1	Ethene
74-86-2	Ethyne
74-87-3	Methane, chloro-
74-88-4	Methane, iodo-
74-89-5	Methanamine
74-90-8	Hydrocyanic acid
74-93-1	Methanethiol
74-95-3	Methane, dibromo-
74-96-4	Ethane, bromo-
74-97-5	Methane, bromochloro-
74-98-6	Propane
74-99-7	1-Propyne
75-00-3	Ethane, chloro-
75-01-4	Ethene, chloro-
75-02-5	Ethene, fluoro-
75-03-6	Ethane, iodo-
75-04-7	Ethanamine
75-05-8	Acetonitrile
75-07-0	Acetaldehyde
75-08-1	Ethanethiol
75-09-2	Methane, dichloro-
75-10-5	Methane, difluoro-
75-11-6	Methane, diiodo-
75-15-0	Carbon disulfide
75-17-2	Formaldehyde, oxime
75-18-3	Methane, thiobis-
75-19-4	Cyclopropane
75-21-8	Oxirane
75-25-2	Methane, tribromo-
75-26-3	Propane, 2-bromo-
75-27-4	Methane, bromodichloro-
75-28-5	Propane, 2-methyl-
75-29-6	Propane, 2-chloro-
75-30-9	Propane, 2-iodo-
75-33-2	2-Propanethiol

CAS Number	Chemical Name
75-34-3	Ethane, 1,1-dichloro-
75-35-4	Ethene, 1,1-dichloro-
75-36-5	Acetyl chloride
75-37-6	Ethane, 1,1-difluoro-
75-38-7	Ethene, 1,1-difluoro-
75-43-4	Methane, dichlorofluoro-
75-44-5	Phosgene
75-45-6	Methane, chlorodifluoro-
75-46-7	Methane, trifluoro-
75-50-3	Methanamine, N,N-dimethyl-
75-52-5	Methane, nitro-
75-56-9	Oxirane, methyl-
75-61-6	Methane, dibromodifluoro-
75-63-8	Methane, bromotrifluoro-
75-64-9	2-Propanamine, 2-methyl-
75-65-0	2-Propanol, 2-methyl-
75-66-1	2-Propanethiol, 2-methyl-
75-68-3	Ethane, 1-chloro-1,1-difluoro-
75-69-4	Methane, trichlorofluoro-
75-71-8	Methane, dichlorodifluoro-
75-72-9	Methane, chlorotrifluoro-
75-73-0	Methane, tetrafluoro-
75-74-1	Lead, tetramethyl-
75-76-3	Silane, tetramethyl-
75-81-0	Ethane, 1,2-dibromo-1,1-dichloro-
75-82-1	Ethane, 1,2-dibromo-1,1-difluoro-
75-83-2	Butane, 2,2-dimethyl-
75-84-3	1-Propanol, 2,2-dimethyl-
75-87-6	Acetaldehyde, trichloro-
75-88-7	Ethane, 2-chloro-1,1,1-trifluoro-
75-89-8	Ethanol, 2,2,2-trifluoro-
75-90-1	Acetaldehyde, trifluoro-
75-91-2	Hydroperoxide, 1,1-dimethylethyl-
75-95-6	Ethane, pentabromo-
75-97-8	2-Butanone, 3,3-dimethyl-
76-01-7	Ethane, pentachloro-
76-05-1	Acetic acid, trifluoro-
76-06-2	Methane, trichloronitro-
76-13-1	Ethane, 1,1,2-trichloro-1,2,2-trifluoro-
76-14-2	Ethane, 1,2-dichloro-1,1,2,2-tetrafluoro-
76-22-2	Camphor

CAS Number	Chemical Name
76-49-3	Bornyl acetate
77-76-9	Propane, 2,2-dimethoxy-
77-78-1	Sulfuric acid, dimethyl ester
78-00-2	Lead, tetraethyl-
78-38-6	Phosphonic acid, ethyl-, diethyl ester
78-40-0	Phosphoric acid, triethyl ester
78-70-6	Linalool
78-78-4	Butane, 2-methyl-
78-79-5	1,3-Butadiene, 2-methyl-
78-84-2	Propanal, 2-methyl-
78-85-3	2-Propenal, 2-methyl-
78-87-5	Propane, 1,2-dichloro-
78-88-6	1-Propene, 2,3-dichloro-
78-92-2	2-Butanol
78-93-3	2-Butanone
78-94-4	3-Buten-2-one
78-98-8	Propanal, 2-oxo-
79-00-5	Ethane, 1,1,2-trichloro-
79-01-6	Ethene, trichloro-
79-02-7	Acetaldehyde, dichloro-
79-09-4	Propanoic acid
79-10-7	2-Propenoic acid
79-16-3	Acetamide, N-methyl-
79-20-9	Acetic acid, methyl ester
79-24-3	Ethane, nitro-
79-27-6	Ethane, 1,1,2,2-tetrabromo-
79-29-8	Butane, 2,3-dimethyl-
79-31-2	Propanoic acid, 2-methyl-
79-34-5	Ethane, 1,1,2,2-tetrachloro-
79-35-6	Ethene, 1,1-dichloro-2,2-difluoro-
79-41-4	2-Propenoic acid, 2-methyl-
79-46-9	Propane, 2-nitro-
79-92-5	Camphene
80-62-6	2-Propenoic acid, 2-methyl-, methyl ester
83-32-9	Acenaphthylene, 1,2-dihydro-
85-01-8	Phenanthrene
86-57-7	Naphthalene, 1-nitro-
86-73-7	9H-Fluorene
87-44-5	Caryophyllene
87-85-4	Benzene, hexamethyl-
88-72-2	Benzene, 1-methyl-2-nitro-

CAS Number	Chemical Name
88-75-5	Phenol, 2-nitro-
90-12-0	Naphthalene, 1-methyl-
91-16-7	Benzene, 1,2-dimethoxy-
91-20-3	Naphthalene
91-22-5	Quinoline
91-57-6	Naphthalene, 2-methyl-
92-52-4	1,1'-Biphenyl
94-80-4	2,4-D-butyl ester
95-13-6	1H-Indene
95-47-6	Benzene, 1,2-dimethyl-
95-48-7	Phenol, 2-methyl-
95-50-1	Benzene, 1,2-dichloro-
95-63-6	Benzene, 1,2,4-trimethyl-
95-65-8	Phenol, 3,4-dimethyl-
95-80-7	1,3-Benzenediamine, 4-methyl-
95-87-4	Phenol, 2,5-dimethyl-
96-12-8	Propane, 1,2-dibromo-3-chloro-
96-14-0	Pentane, 3-methyl-
96-17-3	Butanal, 2-methyl-
96-22-0	3-Pentanone
96-33-3	2-Propenoic acid, methyl ester
96-41-3	Cyclopentanol
96-43-5	Thiophene, 2-chloro-
96-49-1	1,3-Dioxolane-2-one
97-96-1	Butanal, 2-ethyl-
98-01-1	2-Furancarboxaldehyde
98-06-6	Benzene, (1,1-dimethylethyl)-
98-08-8	Benzene, (trifluoromethyl)-
98-55-5	α-Terpineol
98-56-6	Benzene, 1-chloro-4-(trifluoromethyl)-
98-82-8	Benzene, (1-methylethyl)-
98-83-9	Benzene, (1-methylethenyl)-
98-86-2	Ethanone, 1-phenyl-
98-95-3	Benzene, nitro-
99-08-1	Benzene, 1-methyl-3-nitro-
99-85-4	γ-Terpinene
99-86-5	α-Terpinene
99-87-6	Benzene, 1-methyl-4-(1-methylethyl)-
100-41-4	Benzene, ethyl-
100-42-5	Styrene
100-44-7	Benzene, (chloromethyl)-

CAS Number	Chemical Name
100-47-0	Benzonitrile
100-51-6	Benzenemethanol
100-52-7	Benzaldehyde
100-66-3	Benzene, methoxy-
100-69-6	Pyridine, 2-ethenyl-
101-77-9	Benzenamine, 4,4'-methylenebis-
101-84-8	Benzene, 1,1'-oxybis-
103-65-1	Benzene, propyl-
103-72-0	Benzene, isothiocyanato-
105-37-3	Propanoic acid, ethyl ester
105-46-4	Acetic acid, 1-methylpropyl ester
105-54-4	Butanoic acid, ethyl ester
105-58-8	Carbonic acid, diethyl ester
105-66-8	Butanoic acid, propyl ester
105-67-9	Phenol, 2,4-dimethyl-
106-22-9	Citronellol
106-36-5	Propanoic acid, propyl ester
106-42-3	Benzene, 1,4-dimethyl-
106-44-5	Phenol, 4-methyl-
106-46-7	Benzene, 1,4-dichloro-
106-47-8	Benzenamine, 4-chloro-
106-51-4	2,5-Cyclohexadiene-1,4-dione
106-65-0	Butanedioic acid, dimethyl ester
106-88-7	Oxirane, ethyl-
106-89-8	Oxirane, (chloromethyl)-
106-93-4	Ethane, 1,2-dibromo-
106-94-5	Propane, 1-bromo-
106-95-6	1-Propene, 3-bromo-
106-97-8	Butane
106-98-9	1-Butene
106-99-0	1,3-Butadiene
107-00-6	1-Butyne
107-02-8	2-Propenal
107-03-9	1-Propanethiol
107-04-0	Ethane, 1-bromo-2-chloro-
107-05-1	1-Propene, 3-chloro-
107-06-2	Ethane, 1,2-dichloro-
107-07-3	Ethanol, 2-chloro-
107-08-4	Propane, 1-iodo-
107-12-0	Propanenitrile
107-13-1	2-Propenenitrile

CAS Number	Chemical Name
107-18-6	2-Propen-1-ol
107-20-0	Acetaldehyde, chloro-
107-21-1	Ethylene glycol
107-22-2	Ethanedial
107-25-5	Ethene, methoxy-
107-29-9	Acetaldehyde, oxime
107-31-3	Formic acid, methyl ester
107-40-4	2-Pentene, 2,4,4-trimethyl-
107-41-5	2,4-Pentanediol, 2-methyl-
107-46-0	Disiloxane, hexamethyl-
107-51-7	Trisiloxane, octamethyl-
107-54-0	1-Hexyn-3-ol, 3,5-dimethyl-
107-83-5	Pentane, 2-methyl-
107-87-9	2-Pentanone
107-88-0	1,3-Butanediol
107-92-6	Butanoic acid
107-98-2	2-Propanol, 1-methoxy-
108-01-0	Ethanol, 2-(dimethylamino)-
108-03-2	Propane, 1-nitro-
108-05-4	Acetic acid, ethenyl ester
108-08-7	Pentane, 2,4-dimethyl-
108-10-1	2-Pentanone, 4-methyl-
108-20-3	Propane, 2,2'-oxybis-
108-21-4	Acetic acid, 1-methylethyl ester
108-22-5	1-Propen-2-ol, acetate
108-31-6	2,5-Furandione
108-38-3	Benzene, 1,3-dimethyl-
108-39-4	Phenol, 3-methyl-
108-67-8	Benzene, 1,3,5-trimethyl-
108-68-9	Phenol, 3,5-dimethyl-
108-83-8	4-Heptanone, 2,6-dimethyl-
108-86-1	Benzene, bromo-
108-87-2	Cyclohexane, methyl-
108-88-3	Benzene, methyl-
108-90-7	Benzene, chloro-
108-93-0	Cyclohexanol
108-94-1	Cyclohexanone
108-95-2	Phenol
108-98-5	Benzenethiol
109-21-7	Butanoic acid, butyl ester
109-59-1	Ethylene glycol isopropyl ether

CAS Number	Chemical Name
109-60-4	Acetic acid, propyl ester
109-65-9	Butane, 1-bromo-
109-66-0	Pentane
109-67-1	1-Pentene
109-68-2	2-Pentene
109-69-3	Butane, 1-chloro-
109-79-5	1-Butanethiol
109-86-4	Ethylene glycol methyl ether
109-87-5	Methane, dimethoxy-
109-92-2	Ethene, ethoxy-
109-94-4	Formic acid, ethyl ester
109-95-5	Nitrous acid, ethyl ester
109-97-7	1H-Pyrrole
109-99-9	Furan, tetrahydro-
110-00-9	Furan
110-01-0	Thiophene, tetrahydro-
110-02-1	Thiophene
110-06-5	Disulfide, bis(1,1-dimethylethyl)-
110-12-3	2-Hexanone, 5-methyl-
110-13-4	2,5-Hexanedione
110-19-0	Acetic acid, 2-methylpropyl ester
110-43-0	2-Heptanone
110-53-2	Pentane, 1-bromo-
110-54-3	Hexane
110-62-3	Pentanal
110-71-4	Ethane, 1,2-dimethoxy-
110-74-7	Formic acid, propyl ester
110-80-5	Ethylene glycol ethyl ether
110-82-7	Cyclohexane
110-83-8	Cyclohexene
110-88-3	1,3,5-Trioxane
110-93-0	5-Hepten-2-one, 6-methyl-
111-13-7	2-Octanone
111-15-9	Ethylene glycol ethyl ether acetate
111-25-1	Hexane, 1-bromo-
111-27-3	1-Hexanol
111-30-8	Pentanedial
111-35-3	1-Propanol, 3-ethoxy-
111-43-3	Propane, 1,1'-oxybis-
111-46-6	Diethylene glycol
111-47-7	Propane, 1,1'-thiobis-

CAS Number	Chemical Name
111-65-9	Octane
111-66-0	1-Octene
111-70-6	1-Heptanol
111-71-7	Heptanal
111-76-2	Ethylene glycol butyl ether
111-78-4	1,5-Cyclooctadiene
111-84-2	Nonane
111-87-5	1-Octanol
111-90-0	Diethylene glycol ethyl ether
111-96-6	Diethylene glycol dimethyl ether
112-31-2	Decanal
112-34-5	Diethylene glycol butyl ether
112-36-7	Diethylene glycol diethyl ether
112-40-3	Dodecane
115-07-1	1-Propene
115-10-6	Methane, oxybis-
115-11-7	1-Propene, 2-methyl-
115-18-4	3-Buten-2-ol, 2-methyl-
115-20-8	Ethanol, 2,2,2-trichloro-
115-22-0	2-Butanone, 3-hydroxy-3-methyl-
115-32-2	Dicofol
116-09-6	2-Propanone, 1-hydroxy-
116-14-3	Ethene, tetrafluoro-
116-15-4	1-Propene, 1,1,2,3,3,3-hexafluoro-
117-81-7	1,2-Benzenedicarboxylic acid, bis(2-ethylhexyl) ester
118-74-1	Benzene, hexachloro-
119-61-9	Benzophenone
119-64-2	Naphthalene, 1,2,3,4-tetrahydro-
119-65-3	Isoquinoline
120-12-7	Anthracene
120-72-9	1H-Indole
120-80-9	1,2-Benzenediol
120-82-1	Benzene, 1,2,4-trichloro-
120-83-2	Phenol, 2,4-dichloro-
120-92-3	Cyclopentanone
121-45-9	Phosphorous acid, trimethyl ester
121-46-0	2,5-Norbornadiene
121-69-7	Benzenamine, N,N-dimethyl-
123-15-9	Pentanal, 2-methyl-
123-35-3	1,6-Octadiene, 7-methyl-3-methylene-
123-38-6	Propanal

CAS Number	Chemical Name
123-42-2	2-Pentanone, 4-hydroxy-4-methyl-
123-51-3	1-Butanol, 3-methyl-
123-54-6	2,4-Pentanedione
123-72-8	Butanal
123-73-9	2-Butenal, (E)-
123-86-4	Acetic acid, butyl ester
123-91-1	1,4-Dioxane
124-13-0	Octanal
124-18-5	Decane
124-19-6	Nonanal
124-40-3	Methanamine, N-methyl-
124-48-1	Methane, dibromochloro-
124-68-5	1-Propanol, 2-amino-2-methyl-
124-72-1	Ethane, 1-bromo-1,2,2,2-tetrafluoro-
124-73-2	Ethane, 1,2-dibromo-1,1,2,2-tetrafluoro-
126-84-1	Propane, 2,2-diethoxy-
127-17-3	Propanoic acid, 2-oxo-
127-18-4	Ethene, tetrachloro-
127-19-5	Acetamide, N,N-dimethyl-
129-00-0	Pyrene
130-15-4	1,4-Naphthalenedione
132-64-9	Dibenzofuran
138-86-3	Limonene
140-88-5	2-Propenoic acid, ethyl ester
141-46-8	Acetaldehyde, hydroxy-
141-62-8	Tetrasiloxane, decamethyl-
141-78-6	Acetic acid, ethyl ester
142-28-9	Propane, 1,3-dichloro-
142-29-0	Cyclopentene
142-68-7	2H-Pyran, tetrahydro-
142-82-5	Heptane
142-83-6	2,4-Hexadienal, (E,E)-
142-96-1	Butane, 1,1'-oxybis-
149-73-5	Methane, trimethoxy-
150-78-7	Benzene, 1,4-dimethoxy-
151-10-0	Benzene, 1,3-dimethoxy-
151-56-4	Aziridine
151-67-7	Ethane, 1-bromo-1-chloro-2,2,2-trifluoro-
151-67-7	Ethane, 2-bromo-2-chloro-1,1,1-trifluoro-
152-18-1	Phosphorothioic acid, O,O,O-trimethyl ester
152-20-5	Phosphorothioic acid, O,O,S-trimethyl ester

CAS Number	Chemical Name
156-59-2	Ethene, 1,2-dichloro-, (Z)-
156-60-5	Ethene, 1,2-dichloro-, (E)-
191-24-2	Benzo[ghi]perylene
192-97-2	Benzo[e]pyrene
198-55-0	Perylene
206-44-0	Fluoranthene
207-08-9	Benzo[k]fluoranthene
208-96-8	Acenaphthylene
218-01-9	Chrysene
262-12-4	Dibenzo[b,e][1,4]dioxin
271-89-6	Benzofuran
275-51-4	Azulene
278-06-8	Tetracyclo[3.2.0.02,7.04,6]heptane
279-23-2	Norbornane
280-33-1	Bicyclo[2.2.2]octane
281-23-2	Tricyclo[3.3.1.13,7]decane
287-23-0	Cyclobutane
287-92-3	Cyclopentane
288-32-4	1H-Imidazole
288-42-6	Oxazole
288-47-1	Thiazole
288-88-0	1H-1,2,4-Triazole
290-87-9	1,3,5-Triazine
291-64-5	Cycloheptane
291-70-3	Oxepin
292-64-8	Cyclooctane
298-02-2	Phorate
302-01-2	Hydrazine
306-83-2	Ethane, 2,2-dichloro-1,1,1-trifluoro-
309-00-2	Aldrin
319-84-6	α-Hexachlorocyclohexane
333-36-8	Ethane, 1,1'-oxybis[2,2,2,-trifluoro-
334-88-3	Methane, diazo-
352-93-2	Ethane, 1,1'-thiobis-
353-36-6	Ethane, fluoro-
353-59-3	Methane, bromochlorodifluoro-
354-04-1	Ethane, 1,2-dibromo-1,1,2-trifluoro-
354-06-3	Ethane, 1-bromo-2-chloro-1,1,2-trifluoro-
354-11-0	Ethane, 1,1,1,2-tetrachloro-2-fluoro-
354-12-1	Ethane, 1,1,1-trichloro-2,2-difluoro-
354-14-3	Ethane, 1,1,2,2-tetrachloro-1-fluoro-

CAS Number	Chemical Name
354-15-4	Ethane, 1,2-difluoro-1,2,2-trichloro-
354-21-2	Ethane, 1,2,2-trichloro-1,1-difluoro-
354-23-4	Ethane, 1,2-dichloro-1,1,2-trifluoro-
354-33-6	Ethane, pentafluoro-
355-37-3	Hexane, 1,1,1,2,2,3,3,4,4,5,5,6,6-tridecafluoro-
359-11-5	Ethene, trifluoro-
359-13-7	Ethanol, 2,2-difluoro-
359-15-9	Methane, difluoromethoxy-
359-19-3	Ethane, 1,1-dibromo-2,2-difluoro-
359-29-5	Ethene, trichlorofluoro-
359-35-3	Ethane, 1,1,2,2-tetrafluoro-
360-89-4	2-Butene, 1,1,1,2,3,4,4,4-octafluoro-
371-62-0	Ethanol, 2-fluoro-
375-03-1	Propane, 1,1,1,2,2,3,3-heptafluoro-3-methoxy-
377-36-6	Butane, 1,1,2,2,3,3,4,4-octafluoro-
378-16-5	Propane, 1,1,1,2,2-pentafluoro-3-methoxy-
382-34-3	Propane, 1,1,1,2,3,3-hexafluoro-3-methoxy
392-56-3	Benzene, hexafluoro-
406-58-6	Butane, 1,1,1,3,3-pentafluoro-
406-78-0	Ethane, 1,1,2,2-tetrafluoro-1-(2,2,2-trifluoroethoxy)-
407-59-0	Butane, 1,1,1,4,4,4-hexafluoro-
420-04-2	Cyanamide
420-12-2	Thiirane
420-26-8	Propane, 2-fluoro-
420-46-2	Ethane, 1,1,1-trifluoro-
421-04-5	Ethane, 1-chloro-1,1,2-trifluoro-
421-06-7	Ethane, 2-bromo-1,1,1-trifluoro-
421-07-8	Propane, 1,1,1-trifluoro-
421-14-7	Methane, trifluoromethoxy-
421-50-1	2-Propanone, 1,1,1-trifluoro-
422-05-9	1-Propanol, 2,2,3,3,3-pentafluoro-
422-56-0	Propane, 3,3-dichloro-1,1,1,2,2-pentafluoro-
425-82-1	Oxetane, hexafluoro-
425-87-6	Ethane, 2-chloro-1,1,2-trifluoro-1-methoxy-
425-88-7	Ethane, 1,1,2,2-tetrafluoro-1-methoxy-
426-05-6	Nitric acid, 1,1-dimethylethyl ester
430-57-9	Ethane, 1,2-dichloro-1-fluoro-
430-66-0	Ethane, 1,1,2-trifluoro-
430-69-3	Acetaldehyde, difluoro-
431-03-8	2,3-Butanedione
431-06-1	Ethane, 1,2-dichloro-1,2-difluoro-

CAS Number	Chemical Name
431-07-2	Ethane, 1-chloro-1,2,2-trifluoro-
431-31-2	Propane, 1,1,1,2,3-pentafluoro-
431-47-0	Acetic acid, trifluoro-, methyl ester
431-63-0	Propane, 1,1,1,2,3,3-hexafluoro-
431-89-0	Propane, 1,1,1,2,3,3,3-heptafluoro-
452-86-8	1,2-Benzenediol, 4-methyl-
460-12-8	1,3-Butadiyne
460-19-5	Ethanedinitrile
460-40-2	Propanal, 3,3,3-trifluoro-
460-43-5	Ethane, 1,1,1-trifluoro-2-methoxy-
460-73-1	Propane, 1,1,1,3,3-pentafluoro-
462-06-6	Benzene, fluoro-
462-95-3	Diethoxymethane
463-04-7	Nitrous acid, pentyl ester
463-49-0	1,2-Propadiene
463-51-4	Ethenone
463-58-1	Carbon oxide sulfide (COS)
463-82-1	Propane, 2,2-dimethyl-
464-06-2	Butane, 2,2,3-trimethyl-
469-61-4	α-Cedrene
470-82-6	1,8-Cineol
475-20-7	(+)-Longifolene
488-17-5	1,2-Benzenediol, 3-methyl-
493-01-6	Naphthalene, decahydro-, cis-
493-02-7	Naphthalene, decahydro-, trans-
493-05-0	1H-2-Benzopyran, 3,4-dihydro-
493-09-4	1,4-Benzodioxin, 2,3-dihydro-
496-11-7	1H-Indene, 2,3-dihydro-
496-16-2	Benzofuran, 2,3-dihydro-
498-60-2	3-Furancarboxaldehyde
498-66-8	2-Norbornene
503-17-3	2-Butyne
503-30-0	Oxetane
503-60-6	2-Butene, 1-chloro-3-methyl-
505-22-6	1,3-Dioxane
505-57-7	2-Hexenal
507-20-0	Propane, 2-chloro-2-methyl-
507-55-1	Propane, 1,3-dichloro-1,1,2,2,3-pentafluoro-
508-32-7	Tricyclo[2.2.1.02,6]heptane, 1,7,7-trimethyl-
512-51-6	Ethane, 2-ethoxy-1,1,2,2-tetrafluoro-
512-56-1	Phosphoric acid, trimethyl ester

CAS Number	Chemical Name
513-35-9	2-Butene, 2-methyl-
513-37-1	1-Propene, 1-chloro-2-methyl-
513-44-0	1-Propanethiol, 2-methyl-
513-53-1	2-Butanethiol
513-81-5	1,3-Butadiene, 2,3-dimethyl-
513-85-9	2,3-Butanediol
513-86-0	2-Butanone, 3-hydroxy-
526-73-8	Benzene, 1,2,3-trimethyl-
526-75-0	Phenol, 2,3-dimethyl-
532-55-8	Benzoyl isothiocyanate
534-15-6	Ethane, 1,1-dimethoxy-
534-22-5	Furan, 2-methyl-
540-54-5	Propane, 1-chloro-
540-67-0	Ethane, methoxy-
540-80-7	Nitrous acid, 1,1-dimethylethyl ester
540-84-1	Pentane, 2,2,4-trimethyl-
540-88-5	Acetic acid, 1,1-dimethylethyl ester
541-02-6	Cyclopentasiloxane, decamethyl-
541-05-9	Cyclotrisiloxane, hexamethyl-
541-73-1	Benzene, 1,3-dichloro-
542-52-9	Carbonic acid, dibutyl ester
542-56-3	Nitrous acid, 2-methylpropyl ester
542-75-6	1-Propene, 1,3-dichloro-
542-85-8	Ethane, isothiocyanato-
543-29-3	Nitric acid, 2-methylpropyl ester
543-59-9	Pentane, 1-chloro-
543-67-9	Nitrous acid, propyl ester
543-87-3	1-Butanol, 3-methyl-, nitrate
544-10-5	Hexane, 1-chloro-
544-16-1	Nitrous acid, butyl ester
544-25-2	1,3,5-Cycloheptatriene
544-40-1	Butane, 1,1'-thiobis-
544-76-3	Hexadecane
547-63-7	Propanoic acid, 2-methyl-, methyl ester
554-12-1	Propanoic acid, methyl ester
554-14-3	Thiophene, 2-methyl-
554-61-0	2-Carene
555-10-2	β-Phellandrene
556-56-9	1-Propene, 3-iodo-
556-61-6	Methane, isothiocyanato-
556-67-2	Cyclotetrasiloxane, octamethyl-

CAS Number	Chemical Name
556-82-1	2-Buten-1-ol, 3-methyl-
557-91-5	Ethane, 1,1-dibromo-
558-37-2	1-Butene, 3,3-dimethyl-
563-45-1	1-Butene, 3-methyl-
563-46-2	1-Butene, 2-methyl-
563-47-3	1-Propene, 3-chloro-2-methyl-
563-52-0	1-Butene, 3-chloro-
563-58-6	1-Propene, 1,1-dichloro-
563-78-0	1-Butene, 2,3-dimethyl-
563-79-1	2-Butene, 2,3-dimethyl-
563-80-4	2-Butanone, 3-methyl-
565-75-3	Pentane, 2,3,4-trimethyl-
565-80-0	3-Pentanone, 2,4-dimethyl-
569-41-5	Naphthalene, 1,8-dimethyl-
571-58-4	Naphthalene, 1,4-dimethyl-
571-61-9	Naphthalene, 1,5-dimethyl-
573-98-8	Naphthalene, 1,2-dimethyl-
575-37-1	Naphthalene, 1,7-dimethyl-
575-41-7	Naphthalene, 1,3-dimethyl-
575-43-9	Naphthalene, 1,6-dimethyl-
576-24-9	Phenol, 2,3-dichloro-
576-26-1	Phenol, 2,6-dimethyl-
581-40-8	Naphthalene, 2,3-dimethyl-
581-42-0	Naphthalene, 2,6-dimethyl-
581-89-5	Naphthalene, 2-nitro-
582-16-1	Naphthalene, 2,7-dimethyl-
584-02-1	3-Pentanol
584-03-2	1,2-Butanediol
584-84-9	Benzene, 2,4-diisocyanato-1-methyl-
586-62-9	Terpinolene
589-38-8	3-Hexanone
590-14-7	1-Propene, 1-bromo-
590-18-1	2-Butene, (Z)-
590-19-2	1,2-Butadiene
590-35-2	Pentane, 2,2-dimethyl-
590-73-8	Hexane, 2,2-dimethyl-
590-86-3	Butanal, 3-methyl-
590-90-9	2-Butanone, 4-hydroxy-
591-12-8	2(3H)-Furanone, 5-methyl-
591-47-9	Cyclohexene, 4-methyl-
591-48-0	Cyclohexene, 3-methyl-

CAS Number	Chemical Name
591-49-1	Cyclohexene, 1-methyl-
591-50-4	Benzene, iodo-
591-78-6	2-Hexanone
591-87-7	1-Propen-3-ol, acetate
591-93-5	1,4-Pentadiene
591-95-7	1,2-Pentadiene
591-97-9	2-Butene, 1-chloro-
592-41-6	1-Hexene
592-42-7	1,5-Hexadiene
592-46-1	2,4-Hexadiene
592-47-2	3-Hexen (Z)
592-48-3	cis,trans-1,3-Hexadiene
592-57-4	1,3-Cyclohexadiene
592-76-7	1-Heptene
592-77-8	2-Heptene
592-84-7	Formic acid, butyl ester
592-90-5	Oxepane
593-53-3	Methane, fluoro-
593-60-2	Ethene, bromo-
593-70-4	Methane, chlorofluoro-
593-71-5	Methane, chloroiodo-
593-74-8	Mercury, dimethyl-
593-79-3	Methane, selenobis-
594-56-9	1-Butene, 2,3,3-trimethyl-
594-82-1	Butane, 2,2,3,3-tetramethyl-
597-07-9	Phosphoramidic acid, dimethyl-, dimethyl ester
598-25-4	1,2-Butadiene, 3-methyl-
598-26-5	Dimethylketene
598-32-3	3-Buten-2-ol
598-58-3	Nitric acid, methyl ester
598-75-4	2-Butanol, 3-methyl-
611-14-3	Benzene, 1-ethyl-2-methyl-
613-31-0	Anthracene, 9,10-dihydro-
614-69-7	Benzene, 1-isothiocyanato-2-methyl
616-12-6	2-Pentene, 3-methyl-, (E)-
616-25-1	1-Penten-3-ol
616-38-6	Carbonic acid, dimethyl ester
616-44-4	Thiophene, 3-methyl-
617-50-5	Propanoic acid, 2-methyl-, 1-methylethyl ester
620-02-0	2-Furancarboxaldehyde, 5-methyl-
620-14-4	Benzene, 1-ethyl-3-methyl-

CAS Number	Chemical Name
622-96-8	Benzene, 1-ethyl-4-methyl-
623-42-7	Butanoic acid, methyl ester
623-43-8	2-Butenoic acid, methyl ester, (E)-
623-96-1	Carbonic acid, dipropyl ester
624-24-8	Pentanoic acid, methyl ester
624-64-6	2-Butene, (E)-
624-72-6	Ethane, 1,2-difluoro-
624-89-5	Ethane, (methylthio)-
624-91-9	Nitrous acid, methyl ester
624-92-0	Disulfide, dimethyl-
625-27-4	2-Pentene, 2-methyl-
625-33-2	3-Penten-2-one
625-46-7	1-Propene, 3-nitro-
625-55-8	Formic acid, 1-methylethyl ester
625-58-1	Nitric acid, ethyl ester
625-86-5	Furan, 2,5-dimethyl-
626-93-7	2-Hexanol
626-96-0	Pentanal, 4-oxo-
627-05-4	Butane, 1-nitro-
627-13-4	Nitric acid, propyl ester
627-19-0	1-Pentyne
627-20-3	2-Pentene, (Z)-
627-27-0	3-Buten-1-ol
627-42-9	Ethane, 1-chloro-2-methoxy-
627-58-7	1,5-Hexadiene, 2,5-dimethyl-
627-93-0	1,6-Hexanedioic acid, dimethyl ester
628-05-7	Pentane, 1-nitro-
628-28-4	Butane, 1-methoxy-
628-41-1	1,4-Cyclohexadiene
628-55-7	Propane, 1,1'-oxybis[2-methyl-
628-63-7	Acetic acid, pentyl ester
628-80-8	Pentane, 1-methoxy-
628-81-9	Butane, 1-ethoxy-
628-82-0	Ethanol, 2-methoxy-, formate
628-92-2	Cycloheptene
629-14-1	Ethylene glycol diethyl ether
629-15-2	1,2-Ethanediol, diformate
629-20-9	1,3,5,7-Cyclooctatetraene
629-50-5	Tridecane
629-59-4	Tetradecane
629-62-9	Pentadecane

CAS Number	Chemical Name
630-16-0	Ethane, 1,1,1,2-tetrabromo-
630-19-3	Propanal, 2,2-dimethyl-
630-20-6	Ethane, 1,1,1,2-tetrachloro-
637-92-3	Propane, 2-ethoxy-2-methyl-
638-02-8	Thiophene, 2,5-dimethyl-
646-04-8	2-Pentene, (E)-
646-06-0	1,3-Dioxolane
674-76-0	2-Pentene, 4-methyl-, (E)-
676-97-1	Phosphonic dichloride, methyl-
677-21-4	1-Propene, 3,3,3-trifluoro-
677-56-5	Propane, 1,1,1,2,2,3-hexafluoro-
679-86-7	Propane, 1,1,2,2,3-pentafluoro-
683-08-9	Phosphonic acid, methyl-, diethyl ester
690-08-4	2-Pentene, 4,4-dimethyl-, (E)-
690-39-1	Propane, 1,1,1,3,3,3-hexafluoro-
690-93-7	3-Hexene, 2,2-dimethyl-, (E)-
691-37-2	1-Pentene, 4-methyl-
692-70-6	3-Hexene, 2,5-dimethyl-, (E)-
693-02-7	1-Hexyne
693-54-9	2-Decanone
693-65-2	Pentane, 1,1'-oxybis-
693-89-0	Cyclopentene, 1-methyl-
694-72-4	Bicyclo[3.3.0]octane
695-12-5	Cyclohexane, ethenyl-
697-82-5	Phenol, 2,3,5-trimethyl-
753-90-2	Ethanamine, 2,2,2-trifluoro-
754-12-1	1-Propene, 2,3,3,3-tetrafluoro-
756-13-8	3-Pentanone, 1,1,1,2,2,4,5,5,5-nonafluoro-4-(trifluoromethyl)-
756-79-6	Phosphonic acid, methyl-, dimethyl ester
758-96-3	Propanamide, N,N-dimethyl-
759-94-4	Eptam
760-20-3	1-Pentene, 3-methyl-
760-21-4	Pentane, 3-methylene-
762-49-2	Ethane, 1-bromo-2-fluoro-
762-75-4	Formic acid, 1,1-dimethylethyl ester
763-29-1	1-Pentene, 2-methyl-
763-30-4	1,4-Pentadiene, 2-methyl-
763-32-6	3-Buten-1-ol, 3-methyl-
763-69-9	Propanoic acid, 3-ethoxy-, ethyl ester
764-13-6	2,4-Hexadiene, 2,5-dimethyl-
764-35-2	2-Hexyne

CAS Number	Chemical Name
764-41-0	2-Butene, 1,4-dichloro-
768-49-0	Benzene, (2-methyl-1-propenyl)-
811-96-1	Acetaldehyde, chlorodifluoro-
811-97-2	Ethane, 1,1,1,2-tetrafluoro-
818-92-8	1-Propene, 3-fluoro-
820-69-9	3-Hexene-2,5-dione, (E)-
821-07-8	1,3,5-Hexatriene, (E)-
821-55-6	2-Nonanone
823-40-5	1,3-Benzenediamine, 2-methyl-
832-69-9	Phenanthrene, 1-methyl-
832-71-3	Phenanthrene, 3-methyl-
868-85-9	Phosphonic acid, dimethyl ester
872-05-9	1-Decene
872-50-4	2-Pyrrolidinone, 1-methyl-
872-55-9	Thiophene, 2-ethyl-
873-66-5	Benzene, 1-propenyl-, (E)-
881-03-8	Naphthalene, 2-methyl-1-nitro-
883-20-5	Phenanthrene, 9-methyl-
920-66-1	2-Propanol, 1,1,1,3,3,3-hexafluoro-
922-62-3	2-Pentene, 3-methyl-, (Z)-
924-52-7	Nitric acid, 1-methylpropyl ester
924-52-7	Nitrous acid, 1-methylpropyl ester
926-42-1	1-Propanol, 2,2-dimethyl-, nitrate
926-56-7	1,3-Pentadiene, 4-methyl-
928-45-0	Nitric acid, butyl ester
928-96-1	3-Hexen-1-ol, (Z)-
929-22-6	4-Heptenal, (E)-
930-22-3	Oxirane, ethenyl-
930-27-8	Furan, 3-methyl-
930-29-0	Cyclopentene, 1-chloro-
930-68-7	2-Cyclohexen-1-one
931-64-6	Bicyclo[2.2.2]oct-2-ene
931-87-3	Cyclooctene, (Z)-
933-11-9	Cyclooctene, 1-methyl-
939-27-5	Naphthalene, 2-ethyl-
992-94-9	Silane, methyl-
993-07-7	Silane, trimethyl-
993-95-3	Propane, 1-(2,2,2-trifluoroethoxy)-1,1,2,3,3,3-hexafluoro-
994-05-8	Butane, 2-methoxy-2-methyl-
1001-26-9	2-Propenoic acid, 3-ethoxy-, ethyl ester
1002-16-0	Nitric acid, pentyl ester

CAS Number	Chemical Name
1066-40-6	Silanol, trimethyl-
1066-42-8	Silanediol, dimethyl-
1067-20-5	Pentane, 3,3-diethyl-
1069-53-0	Hexane, 2,3,5-trimethyl-
1111-74-6	Silane, dimethyl-
1118-00-9	Propane, 1-methoxy-2,2-dimethyl-
1118-58-7	1,3-Pentadiene, 2-methyl-
1119-16-0	Pentanal, 4-methyl-
1119-40-0	1,5-Pentanedioic acid, dimethyl ester
1120-21-4	Undecane
1120-36-1	1-Tetradecene
1120-56-5	Cyclobutane, methylene-
1120-62-3	Cyclopentene, 3-methyl-
1120-97-4	1,3-Dioxane, 4-methyl-
1123-56-4	Benzaldehyde, 2,6-dimethyl
1127-76-0	Naphthalene, 1-ethyl-
1134-23-2	Cycloate
1163-19-5	Decabromodiphenylether
1187-58-2	Propanamide, N-methyl-
1187-93-5	Ethene, trifluoro-(trifluoromethoxy)-
1191-95-3	Cyclobutanone
1192-37-6	Cyclohexane, methylene-
1222-05-5	Galoxolide
1453-25-4	Cycloheptene, 1-methyl-
1493-02-3	Formylfluoride
1493-11-4	Methanol, trifluoro-
1511-62-2	Methane, bromodifluoro-
1515-79-3	1,3-Hexadiene, 5,5-dimethyl-
1528-30-9	Cyclopentane, methylene-
1551-27-5	Thiophene, 2-propyl-
1574-41-0	1,3-Pentadiene, (Z)-
1576-85-8	4-Penten-1-ol, acetate
1576-87-0	2-Pentenal, (E)-
1576-95-0	2-Penten-1-ol, (Z)-
1582-09-8	Trifluralin
1615-75-4	Ethane, 1-chloro-1-fluoro-
1630-77-9	Ethene, 1,2-difluoro-, (Z)-
1630-78-0	Ethene, 1,2-difluoro-, (E)-
1634-04-4	Propane, 2-methoxy-2-methyl-
1649-08-7	Ethane, 1,2-dichloro-1,1-difluoro-
1674-10-8	Cyclohexene, 1,2-dimethyl-

CAS Number	Chemical Name
1691-17-4	Methane, oxybis[difluoro-
1708-29-8	Furan, 2,5-dihydro-
1712-64-7	Nitric acid, 1-methylethyl ester
1717-00-6	Ethane, 1,1-dichloro-1-fluoro-
1746-01-6	Dibenzo[b,e][1,4]dioxin, 2,3,7,8-tetrachloro-
1814-88-6	Propane, 1,1,1,2,2-pentafluoro-
1825-31-6	Naphthalene, 1,4-dichloro-
1871-57-4	1-Propene, 3-chloro-2-(chloromethyl)-
1885-48-9	Ethane, 2-(difluoromethoxy)-1,1,1-trifluoro-
1912-24-9	Atrazine
2004-70-8	1,3-Pentadiene, (E)-
2050-67-1	1,1'-Biphenyl, 3,3'-dichloro-
2050-68-2	1,1'-Biphenyl, 4,4'-dichloro-
2051-60-7	1,1'-Biphenyl, 2-chloro-
2051-61-8	1,1'-Biphenyl, 3-chloro-
2051-62-9	1,1'-Biphenyl, 4-chloro-
2108-66-9	Nitric acid, cyclohexyl ester
2211-70-3	1-Butene, 2-chloro-
2216-34-4	Octane, 4-methyl-
2235-12-3	1,3,5-Hexatriene
2240-88-2	1-Propanol, 3,3,3-trifluoro-
2278-22-0	Peroxide, acetyl nitro-
2284-20-0	Benzene, 1-isothiocyanato-4-methoxy-
2314-97-8	Methane, trifluoroiodo-
2356-61-8	Ethane, 1,1,2,2-tetrafluoro-2-(trifluoromethoxy)-
2356-62-9	Ethane, 1,1,1,2-tetrafluoro-2-(trifluoromethoxy)-
2370-63-0	2-Propenoic acid, 2-methyl-, 2-ethoxyethyl ester
2403-24-9	Cyclohexanone, 2-hydroxy-, nitrate
2416-94-6	Phenol, 2,3,6-trimethyl-
2436-90-0	1,6-Octadiene, 3,7-dimethyl-
2437-79-8	1,1'-Biphenyl, 2,2',4,4'-tetrachloro-
2497-18-9	2-Hexen-1-ol, acetate, (E)-
2517-43-3	1-Butanol, 3-methoxy-
2524-03-0	Phosphorochloridothioic acid, O,O-dimethyl ester
2531-84-2	Phenanthrene, 2-methyl-
2562-37-0	Cyclohexene, 1-nitro-
2568-89-0	Bis-isopropoxymethane
2568-90-3	Butane, 1,1'-[methylenebis(oxy)]bis-
2568-91-4	Bis-(2-methylpropoxy)-methane
2568-92-5	Bis-(1-methylpropoxy)-methane
2612-46-6	1,3,5-Hexatriene, (Z)-

CAS Number	Chemical Name
2704-78-1	Pinonaldehyde
2757-23-5	Formylchloride
2807-30-9	Ethylene glycol propyl ether
2837-89-0	Ethane, 2-chloro-1,1,1,2-tetrafluoro-
2953-29-9	Phosphorodithioic acid, O,O,S-trimethylester
2987-16-8	Butanal, 3,3-dimethyl-
3013-02-3	Carbamothioic acid, dimethyl-, S-methyl ester
3031-73-0	Hydroperoxide, methyl-
3073-66-3	Cyclohexane, 1,1,3-trimethyl-
3208-16-0	Furan, 2-ethyl-
3219-63-4	Methanol, (trimethylsilyl)-
3221-61-2	Octane, 2-methyl-
3268-87-9	Dibenzo[b,e][1,4]dioxin, octachloro-
3296-50-2	trans-Bicyclo[4.3.0]nonane
3338-55-4	1,3,6-Octatriene, 3,7-dimethyl-, (Z)-
3387-41-5	Sabinene
3393-64-4	2-Butanone, 4-hydroxy-3-methyl-
3638-35-5	Cyclopropane, (1-methylethyl)-
3638-64-0	Ethene, nitro-
3675-14-7	Butenedial, (E)-
3681-71-8	3-Hexen-1-ol, acetate, (Z)-
3710-43-8	Furan, 2,4-dimethyl-
3710-84-7	Ethanamine, N-ethyl-N-hydroxy-
3806-59-5	1,3-Cyclooctadiene, (Z,Z)-
3822-68-2	Methane, (difluoromethoxy)trifluoro-
3856-25-5	Copaene
4049-81-4	1,5-Hexadiene, 2-methyl-
4054-38-0	1,3-Cycloheptadiene
4125-18-2	1,3-Cyclopentadiene, 5,5-dimethyl-
4164-28-7	Methanamine, N-methyl-N-nitro-
4221-98-1	(−)-α-Phellandrene
4382-75-6	Methanol, methoxy-, formate
4461-41-0	2-Butene, 2-chloro-
4461-42-1	1-Butene, 1-chloro-
4549-74-0	1,3-Pentadiene, 3-methyl-
4551-51-3	cis-Bicyclo[4.3.0]nonane
4984-85-4	3-Hexanone, 4-hydroxy-
5073-63-2	Ethane, 2-chloro-1,1,2-trifluoro-1-(fluoromethoxy)-
5077-67-8	2-Butanone, 1-hydroxy-
5131-66-8	2-Propanol, 1-butoxy-
5162-44-7	1-Butene, 4-bromo-

CAS Number	Chemical Name
5194-50-3	2,4-Hexadiene, (E,Z)-
5194-51-4	2,4-Hexadiene, (E,E)-
5409-83-6	Dibenzofuran, 2,8-dichloro-
5502-88-5	Cyclohexene, 1-methyl-4-(1-methylethyl)-
5729-47-5	2-Pentenal, 4-oxo-
5779-93-1	Benzaldehyde, 2,3-dimethyl-
5779-94-2	Benzaldehyde, 2,5-dimethyl-
5779-95-3	Benzaldehyde, 3,5-dimethyl-
5878-19-3	2-Propanone, 1-methoxy-
5915-41-3	Terbuthylazine
5973-71-7	Benzaldehyde, 3,4-dimethyl-
5989-27-5	d-Limonene
6004-38-2	4,7-Methano-1H-indene, octahydro-
6004-44-0	Methylketene
6032-29-7	2-Pentanol
6065-93-6	1-Propene, 1,1-dichloro-2-methyl-
6090-09-1	4-Acetyl-1-methylcyclohexene
6117-91-5	2-Buten-1-ol
6142-73-0	Cyclopropane, methylene-
6163-66-2	Di-tert-butyl ether
6163-75-3	Phosphonic acid, ethyl-, dimethyl ester
6164-98-3	Chlordimeform
6415-12-9	Hydrazine, tetramethyl-
6423-43-4	1,2-Propanediol, dinitrate
6423-45-6	2,3-Butanediol, dinitrate
6482-24-2	Ethane, 1-bromo-2-methoxy-
6629-91-0	Formaldehyde hydrazone
6642-30-4	Carbamic acid, methyl-, methyl ester
6728-26-3	2-Hexenal, (E)-
6745-71-7	2-Propanone, 1-(nitrooxy)-
6753-98-6	1,4,8-Cycloundecatriene, 2,6,6,9-tetramethyl-, (E,E,E)-
7012-37-5	1,1'-Biphenyl, 2,4,4'-trichloro-
7125-99-7	Propane, 1,1-dichloro-1,2,2-trifluoro-
7319-00-8	1,4-Hexadiene, (E)-
7433-78-5	5-Decene, (Z)-
7446-09-5	Sulfur dioxide
7642-09-3	3-Hexene, (Z)-
7642-15-1	4-Octene, (Z)-
7647-01-0	Hydrochloric acid
7664-41-7	Ammonia
7697-37-2	Nitric acid

CAS Number	Chemical Name
7719-12-2	Phosphorous trichloride
7722-84-1	Hydrogen peroxide (H_2O_2)
7726-95-6	Bromine
7778-85-0	Propane, 1,2-dimethoxy-
7783-06-4	Hydrogen sulfide (H_2S)
7785-26-4	(–)-α-Pinene
7803-51-2	Phosphine
7803-62-5	Silane
10025-87-3	Phosphoric trichloride
10035-10-6	Hydrobromic acid
10049-04-4	Chlorine oxide (ClO_2)
10061-01-5	1-Propene, 1,3-dichloro-, (Z)-
10061-02-6	1-Propene, 1,3-dichloro-, (E)-
10153-61-4	3-Hexene-2,5-dione, 3,4-dihydroxy-
10281-53-5	(–)-.trans.-Pinane
10294-34-5	Borane, trichloro-
10574-37-5	2-Pentene, 2,3-dimethyl-
10599-58-3	Furan, tetramethyl-
13029-08-8	1,1′-Biphenyl, 2,2′-dichloro-
13152-05-1	Cyclooctene, 3-methyl-
13171-18-1	Propane, 1,1,1,3,3,3-hexafluoro-2-methoxy-
13179-96-9	Butane, 1,4-dimethoxy-
13211-15-9	Camphenilone
13269-52-8	3-Hexene, (E)-
13294-71-8	2-Butene, 2-bromo-
13466-78-9	3-Carene
13602-13-6	2-Butene, 1,2-dichloro-
13838-16-9	Ethane, 2-chloro-1-(difluoromethoxy)-1,1,2-trifluoro-
13877-91-3	1,3,6-Octatriene, 3,7-dimethyl-
14315-97-0	Propane, 1,1,3-trimethoxy-
14371-10-9	2-Propenal, 3-phenyl-, (E)-
14686-13-6	2-Heptene, (E)-
14850-23-8	4-Octene, (E)-
14920-89-9	Furan, 2,3-dimethyl-
15764-16-6	Benzaldehyde, 2,4-dimethyl-
15862-07-4	1,1′-Biphenyl, 2,4,5-trichloro-
15877-57-3	Pentanal, 3-methyl-
16479-77-9	Oxepin, 2-methyl-
16606-02-3	1,1′-Biphenyl, 2,4′,5-trichloro-
17081-21-9	Propane, 1,3-dimethoxy-
17249-80-8	Thiophene, 3-chloro-

CAS Number	Chemical Name
17321-47-0	Phosphoramidothioic acid, O,O-dimethyl ester
17559-81-8	3-Hexene-2,5-dione, (Z)-
17696-73-0	Methanesulfinic acid
18172-67-3	(–)-β-Pinene
18259-05-7	1,1'-Biphenyl, 2,3,4,5,6-pentachloro-
18409-46-6	2,4-Hexadienedial, (E.E)-
18479-58-8	7-Octen-2-ol, 2,6-dimethyl-
18829-55-5	2-Heptenal, (E)-
19287-45-7	Diborane(6)
19430-93-4	1-Hexene, 3,3,4,4,5,5,6,6,6-nonafluoro-
19931-40-9	2-Butene-1,4-diol, dinitrate
20280-41-1	1,2-Butanediol, dinitrate
20334-52-5	Ethylketene
20818-81-9	Propanal, 2-hydroxy-2-methyl-
20825-71-2	2(3H)-Furanone
21981-48-6	2-Pentanol, nitrate
21981-49-7	2-Hexanol, nitrate
22037-73-6	1-Butene, 3-bromo-
22052-84-2	Propane, 1,1,1,2,3,3,3-heptafluoro-2-methoxy-
22410-44-2	Ethane, pentafluoromethoxy-
22608-53-3	Phosphorodithioic acid O,S,S-trimethylester
22615-23-3	1-Butene-3,4-diol, dinitrate
24270-66-4	Propane, 1,1,2,3,3-pentafluoro-
24903-95-5	Nopinone
25291-17-2	1-Octene, 3,3,4,4,5,5,6,6,7,7,8,8,8-tridecafluoro-
26675-46-7	Ethane, 2-chloro-2-(difluoromethoxy)-1,1,1-trifluoro-
26981-93-1	Methyl diazene
28167-51-3	Phosphoramidothioic acid, dimethyl-, O,O-dimethyl ester
28523-86-6	Propane, 1,1,1,3,3,3-hexafluoro-2-propyl-2-(fluoromethoxy)-
29343-64-4	2-Butenal, 4-hydroxy-
29446-15-9	Dibenzo[b,e][1,4]dioxin, 2,3-dichloro-
30746-58-8	Dibenzo[b,e][1,4]dioxin, 1,2,3,4-tetrachloro-
31464-99-0	Phosphoramidothioic acid, methyl-, O,O-dimethyl ester
32388-55-9	Acetyl cedrene
32665-20-6	Methanol, ethoxy-, formate
32749-94-3	Pentanal, 2,3-dimethyl-
33284-50-3	1,1'-Biphenyl, 2,4-dichloro-
33689-28-0	1,2-Cyclobutanedione
33857-26-0	Dibenzo[b,e][1,4]dioxin, 2,7-dichloro-
34454-97-2	1-Butanesulfonamide, 1,1,2,2,3,3,4,4,4-nonafluoro-N-(2-hydoxyethyl)-N-methyl-

CAS Number	Chemical Name
34846-90-7	2-Propenoic acid, 3-methoxy-, methyl ester
34862-07-2	Ethane, 1,1-dichloro-2-methoxy-
34883-41-5	1,1'-Biphenyl, 3,5-dichloro-
34883-43-7	1,1'-Biphenyl, 2,4'-dichloro-
35042-99-0	Propane, 3-(difluoromethoxy)-1,1,2,2-tetrafluoro-
37680-68-5	1,1'-Biphenyl, 2',3,5-trichloro-
38379-99-6	1,1'-Biphenyl, 2,2',3,5',6-pentachloro-
38380-03-9	1,1'-Biphenyl, 2,3,3',4',6-pentachloro-
38444-86-9	1,1'-Biphenyl, 2',3,4-trichloro-
38444-87-0	1,1'-Biphenyl, 3,3',5-trichloro-
38444-88-1	1,1'-Biphenyl, 3,4',5-trichloro-
38964-22-6	Dibenzo[b,e][1,4]dioxin, 2,8-dichloro-
39227-53-7	Dibenzo[b,e][1,4]dioxin, 1-chloro-
39227-54-8	Dibenzo[b,e][1,4]dioxin, 2-chloro-
40630-67-9	1-Butanesulfonamide, 1,1,2,2,3,3,4,4,4-nonafluoro-N-ethyl-
41464-39-5	1,1'-Biphenyl, 2,2',3,5'-tetrachloro-
42125-48-4	1,2-Propanediol, 2-methyl-, 1-acetate
50396-87-7	4-Hexen-3-one, (4E)-
51731-17-0	3-Buten-2-one, 4-methoxy-, (E)-
52144-69-1	Benzene, pentafluoropropyl-
53042-85-6	2,4-Hexadienedial, (E,Z)-
53398-76-8	2,4-Hexadienal, (2E,4Z)-
54396-97-3	Propanoic acid, 2-methyl-, 2-ethoxyethyl ester
54464-57-2	Ethanone, 1-(1,2,3,4,5,6,7,8-octahydro-2,3,8,8-tetramethyl-2-naphthalenyl)-
56860-81-2	Propane; 1,1,1,2,2-pentafluoro-3-difluoromethoxy-
56860-85-6	Propane, 1,1,1,2,3,3-hexafluoro-(3-difluoromethoxy)-
59643-69-5	2-Hexene, 3,4-diethyl-, (Z)-
60598-17-6	Propane, 1,1,2,2-tetrafluoro-3-methoxy-
63034-44-6	Acetaldehyde, dichlorofluoro-
64712-27-2	Propane, 1,1-dichloro-1,3,3,3-tetrafluoro-
65064-78-0	Propane, 1,1,1,2,3,3-hexafluoro-3-(2,2,3,3-tetrafluoropropoxy)-
69948-24-9	Ethane, 1-difluoromethoxy-1,1,2-trifluoro-
73602-63-8	Ethane, 2-chloro-1,1,1-trifluoro-2-(fluoromethoxy)-
78448-33-6	Propane, 2-methyl-1-(1-methylethoxy)-
78522-47-1	Methane, bis(difluoromethoxy)difluoro-
81729-06-8	Ethane, 2-chloro-1,1-difluoro-1-(1-fluoroethoxy)-
82944-59-0	3-Pentanol, nitrate
82944-60-3	3-Hexanol, nitrate
82944-61-4	3-Heptanol, nitrate
82944-62-5	3-Octanol, nitrate

CAS Number	Chemical Name
85358-65-2	Methanol, trifluoro-, formate
86777-83-5	Oxirane, 2-ethenyl-2-methyl-
88283-41-4	Pyrifenox
95576-25-3	Hexane, 1,1,1,2,2,5,5,6,6,6-decafluoro-
102526-10-3	Pentane, 1,1,1,3,3,5,5,5-octafluoro-
111823-35-9	1-Butene, 2-isopropyl-3-methyl-
121020-77-7	2,4-Hexadienedial, 2-methyl, (2E,4E)-
123024-70-4	2-Pentanol, 3-methyl-, nitrate
123041-25-8	2-Butanol, 3-methyl-, nitrate
127191-96-2	1-Butanol, 2-methyl-, nitrate
133764-33-7	2-Pentanol, 2-methyl-, nitrate
138495-42-8	Pentane, 1,1,1,2,3,4,4,5,5,5-decafluoro-
138689-24-4	Acetaldehyde, chlorofluoro-
138779-12-1	2-Butanone, 1-(nitrooxy)-
144109-03-5	Oxetane, 2,2,3,4,4-pentafluoro-
161791-33-9	Butane, 1,1,1,2,2,4-hexafluoro-
162401-05-0	Propane, 1,1,1,3,3,3-hexafluoro-2-(trifluoromethoxy)-
163702-05-4	Butane, 1-ethoxy-1,1,2,2,3,3,4,4,4-nonafluoro-
163702-06-5	Propane, 2-(ethoxydifluoromethyl)-1,1,1,2,3,3,3-heptafluoro-
163702-07-6	Butane, 1,1,1,2,2,3,3,4,4-nonafluoro-4-methoxy-
188479-35-8	Nitrous acid, 2-hydroxycyclopentyl ester
188690-78-0	Ethane, 1,2-bis(difluoromethoxy)-1,1,2,2-tetrafluoro-
196881-04-6	2-Propanol, 2-methyl-1-(1-methylethoxy)-, nitrate
273223-83-9	Methanol, butoxy-, formate
297730-93-9	Hexane, 3-ethoxy-1,1,1,2,3,4,4,5,5,6,6,6- dodecafluoro-2-(trifluoromethyl)-
428454-68-6	Propane, 1,1,1,2,3,3-hexafluoro-3-(trifluoromethoxy)-
No CAS No.	Hexane, 1,1,1,2,3-pentafluoro-2-(trifluoromethyl)-3-ethoxy-
No CAS No.	Perfluoropolymethylisopropyl ethers

Subject Index

a
abiotic degradation   21 ff.
– droplet phase   59
– measurement   75
– photo-degradation   2
absorbance   37, 73, 90
absorption   75
absorption coefficient   37, 90
absorption cross section   37, 41
absorption spectrum   89
acetone   39 ff.
acetonitrile   27
acetylene   27
aerosol chamber   76 f.
– disadvantage   77
aerosol particle   76 f.
Agenda 21 (United Nations)   3
alkenes   26
aniline   63
aromatic compounds
– polycyclic aromatic hydrocarbons   25
– reaction with nitrate radicals   31
Arrhenius equation   111
Arrhenius plot   75
Atmospheric Oxidation Programme (AOP)   11, 15
Avogadro constant   37

b
benzene   26 f.
bioaccumulation   4 f.
bioaccumulative chemicals   7

c
carbon monoxide   23
carrier-particle   43
CAS register   209 ff.
chlorine   48 f.
chlorofluorohydrocarbons   39
chromophore   93
Convention on Long-range Transboundary Air Pollution   6

d
degradation
– abiotic, *see* abiotic degradation
– direct   1
– direct photochemical   89 ff.
– heterogeneous, *see* heterogeneous degradation
– in droplets   50
– in the absorbed state   75 ff.
– in the troposphere   31
– indirect   1
– indirect photochemical   60 ff., 82 ff.
– on artificial aerosols   45 ff.
– on fly ash and soot   45
– on solid surfaces   43 ff.
di(2-ethylhexyl)-phthalate (DEHP)   46 f.
dimethyl sulphide (DMS)   32, 86
dinitrogen pentoxide   30 f., 70
distribution coefficient   44
droplet phase   57 ff.
durability   2

Atmospheric Degradation of Organic Substances. W. Klöpffer and B. O. Wagner
Copyright © 2007 WILEY-VCH Verlag GmbH & Co. KGaA, Weinheim
ISBN: 978-3-527-31606-9

e

ecotoxicity 3
electrodynamic trap 78
environmental risk assessment 10
equilibrium constant 29 f.
– dependence on temperature 30
Esbjerg declaration 4
ethane 48

f

Fourier transform infrared spectrometer (FTIR) 14
freon 11 26
freon 12 26
freon 113 47

g

gas chromatographic analysis 14
gas cuvette 71 ff.
gas-phase degradation 75 ff.

h

half-distance 9
half-life 9, 80, 82, 109
– lindane 14
– photochemical 38
Henry coefficient 44, 57
Henry's law 51
heterogeneous degradation 43 ff.
– on artificial aerosols 45 ff.
– on fly ash and soot 45
– on solid surfaces 43 ff.
hexafluorobenzene 27
hydrazine 66
hydrofluorocarbons (HFCs) 5
hydrofluoroethers (HFEs) 5
hydrogen peroxide 23, 51 ff.
– diffusion limit 54
– dissociation 52
– mixed reaction 52
– photochemical reaction 53
– photolysis 52
– reformation 53
– source 56
– UV absorption 53

hydroperoxyl 52 ff.
– average concentration in the aqueous phase 54 f.
hydroxyl radical 11, 76, 107
– averaged tropospheric concentration 24, 82 ff.
– bimolecular reaction 60 ff.
– concentrations 55
– day/night average 55
– globally averaged concentration 83, 85
– measurement 85
– reactions with organic compounds 24 ff.
– sink 22 ff., 66
– source 22 ff., 64 ff., 77

i

irradiance table 91
irradiation 36, 73 f., 76
– time 73 f.

j

Junge formula 43

l

Lambert-Beer's law 90
lifetime 26, 80, 89
– acetonitrile 27
– atmospheric lifetime 78
– benzene 27
– calculation 78 ff.
– di(2-ethylhexyl)-phthalate (DEHP) 47
– direct photo-transformation 93
– freon 11 26
– d-limonene 34 f.
– nitrate radicals 31 f.
– organic oxygenated compounds 58
– ozone 31, 56
– persistent organic pollutants 13
– photochemical lifetime 38, 41, 72, 92
– photolytic lifetime 42

– propene 35
– stratospheric lifetime 90
– total chemical lifetime 80
– tropospheric lifetime 28, 90
d-limonene 31, 34 f.
lindane 14
long-range transport 27, 82
long-range transport potential (LRTP) 2, 5 ff.
– multimedia models 8 ff., 15

m

methanol 25
methyl radicals 41
methylchloroform (MCF) 27, 83 f.
– MCF emission 84
mixing ratio 24, 29 ff., 33, 40, 86 ff.
models
– environmental fate model 8
– Moguntia model 83
monochromatic radiation 71
Montreal protocol on substances that deplete the ozone layer 5
multimedia models 8 ff., 13, 15
– data requirements 10
– long-range transport potential 8 f.
– overall persistence 9

n

naphthalene 26, 31
nitrate 57
nitrate radicals 28 ff., 107
– average concentration in the troposphere 86 f.
– average night-time concentration 86 f.
– bimolecular reaction 68
– globally averaged concentration 86 f.
– reactions with organic compounds 31 f.
– sink 28, 30
– source 28 ff.
– time of activity 28
nitric acid 30 f., 57

nitrite 57
nitrogen dioxide 29
nitrogen monoxide 23
nitrogen oxides 48
nitrogen trioxide 29
nitrous acid 23, 57

o

OECD committee for scientific research 2
OECD environment committee 2
OECD monograph 92
olefins 31, 63
organic substances 30
– CAS register 209 ff.
– degradation in the troposphere 31
– degradation reactions 57
overall persistence (Pov), multimedia models 9 f.
oxalic acid 59
ozone 22 f., 29, 107
– average concentration in the droplet phase 55
– average tropospheric concentration 55, 87 f.
– bimolecular reaction 70 f.
– connection to hydrogen peroxid cycle 56
– globally averaged concentration 87 f.
– Henry coefficient 55
– lifetime 56
– photolysis 64 ff.
– reactions with organic compounds 34 f.
– sink 32 ff., 56
– source 32 f.
– UV/VIS absorption 56

p

partition coefficient 10
– air/water 57
– Henry coefficient 54
– n-octanol/air 75
penetration depth 50

peroxy radicals   40
peroxyacetylnitrate   57
peroxynitric acid   66
persistence   2 ff., 5, 7 ff., 82
– definition   9
– multimedia models   8 f., 15
– quantification   78
persistent organic pollutants (POPs)   5 ff., 22, 93
– Stockholm Convention   7
photo-degradation   45, 107 ff., 114 ff.
– in the troposphere   22
– indirect   11
– organic substances   2, 11
– semi-volatile substances   13
– significance   1 ff., 10
photo-irradiation process   77
photochemical degradation
– direct   89 ff
– indirect   60 ff., 82 ff.
photochemical reaction
– direct   21 f., 35 ff., 50
– in the gas phase   39 ff.
– indirect   21 ff.
photochemical smog formation   21
photochemical transformation
– direct   93, 113
– indirect   93
photolysis   22, 39, 64, 68, 113
photo-transformation, direct   71 ff.
– lifetime   93
Planck constant   90
polluted air   50
polycyclic aromatic hydrocarbons (PAHs)   45, 49
polyfluoroether   5
POP Protocol   7
Pov, see overall persistence
propene   35, 74

q

quadrupole mass spectrometer   69
quality index   108, 111, 114 ff.
quantitative structure activity relationship (QSAR)   1

quantitative structure reactivity relationship (QSRR)   78
quantum efficiency   23, 35 ff., 57, 89, 91, 114 ff.
– determination in the gas phase   71 ff.
– effective quantum efficiency   74
– measurement   72 ff.
quantum yield   108, 114 ff.

r

rate constant   1, 12 f., 22 f., 25 ff., 35
reaction rate constant   12, 57 ff., 79, 83, 89, 108, 114 ff.
– absolute measurement   69
– acetonitrile   27
– acetylene   27
– alternative measurement   77 f.
  benzene   27
– carbon monoxide   23
– direct methods for measuring   60 ff.
– ethane   48
– gas-phase reaction rate constant   107
– hexafluorobenzene   27
– indirect methods for measuring   62
– d-limonene   34
– lindane   14
– methanol   25
– methylchloroform (MCF)   27
– naphthalene   26
– organic substances   11 ff.
– pressure dependence   112
– propene   35
– relative measurement   69
– semi-volatile pesticides   77
– temperature dependence   109, 111
– α-terpinene   26
– toluene   32
– vinyl chloride (VC)   35, 48
reactive trace compound   51
residence time   81
resonance fluorescence method   60 f.

s

semi-volatile organic compounds (SOCs)   3, 22, 39, 43 f., 46 f., 68, 89, 93, 113
semi-volatile pesticides   14
singlet oxygen   23
sink   21 ff., 28, 30, 32 ff.
sinking velocity   46
smog
– formation   33 f.
– photochemical smog   32
smog chamber   14, 27, 35, 39, 62, 70, 74 f.
– advantages   68
– disadvantages   68
– schematic representation   63
sodium chloride   48
solar constant   91
solar irradiation   41 f.
solar radiation   91
solar radiation density   91
spectral irradiance   73
spectral photon irradiance   37, 89, 91 ff.
spectral solar photon irradiance, average   92
Stockholm Convention   7 f.
Stockholm persistence criterion   9
stratosphere   2, 32, 39
sulphur hexafluoride   25, 39, 110
Swedish chemicals policy committee   4

t

terpenes   26, 31, 34, 63
α-terpinene   26
toluene   32
transition metal ions   56
troposphere   2, 21 f., 32, 38, 43 f.

u

Umweltbundesamt   5
UNEP Governing Council   6, 11
United Nations Agenda 21   3
United Nations Conference on the Human Environment   3
urea   58
UV absorption   45
UV radiation   23, 45, 50
UV/VIS absorption   91
UV/VIS absorption spectrum   89

v

Vienna Convention for the Protection of the Ozone Layer   5
vinyl chloride   35, 48
volatile organic substances (VOC)   32 f., 93

w

wavelength   36 f., 41, 89